RECENT RESULTS IN THE THEORY
OF GRAPH SPECTRA

ANNALS OF DISCRETE MATHEMATICS 36

General Editor: Peter L. HAMMER
Rutgers University, New Brunswick, NJ, U.S.A.

Advisory Editors
C. BERGE, Université de Paris, France
M. A. HARRISON, University of California, Berkeley, CA, U.S.A.
V. KLEE, University of Washington, Seattle, WA, U.S.A.
J.-H. VAN LINT, California Institute of Technology, Pasadena, CA, U.S.A.
G.-C. ROTA, Massachusetts Institute of Technology, Cambridge, MA, U.S.A.

NORTH-HOLLAND – AMSTERDAM ● NEW YORK ● OXFORD ●TOKYO

RECENT RESULTS IN
THE THEORY OF GRAPH SPECTRA

Dragos M. CVETKOVIĆ
Department of Electrical Engineering
University of Belgrade
Belgrade, Yugoslavia

Michael DOOB
Department of Mathematics
University of Manitoba
Winnipeg, Canada

Ivan GUTMAN
Department of Chemistry
University of Kragujevac
Kragujevac, Yugoslavia

Aleksandar TORGAŠEV
The Mathematics Institute
Belgrade, Yugoslavia

1988

NORTH-HOLLAND – AMSTERDAM ● NEW YORK ● OXFORD ● TOKYO

ISBN: 0 444 70361 6

Publishers:
ELSEVIER SCIENCE PUBLISHERS B.V.
P.O. BOX 1991
1000 BZ AMSTERDAM
THE NETHERLANDS

Sole distributors for the U.S.A. and Canada:
ELSEVIER SCIENCE PUBLISHING COMPANY, INC.
52 VANDERBILT AVENUE
NEW YORK, N.Y. 10017
U.S.A.

LIBRARY OF CONGRESS
Library of Congress Cataloging-in-Publication Data

Recent results in the theory of graph spectra / Dragoš M. Cvetković
... [et al.].
      p.   cm. -- (Annals of discrete mathematics ; 36)
   Includes index.
   ISBN 0-444-70361-6
   1. Graph theory. 2. Matrices.   I. Cvetković, Dragoš M.
II. Title: Graph spectra. III. Series.
QA166.R43 1988
511'.5--dc19   87-30577
                                                  CIP

PRINTED IN THE NETHERLANDS

Introduction

Almost all of the results related to the theory of graph spectra published before 1978 were summarized in the monograph **Spectra of Graphs** by D. Cvetković, M. Doob, and H. Sachs [CvDSa1]. This book initially appeared in print in 1980, had a second edition in 1982, and was published in a Russian edition in 1984. The purpose of the present book is to review the results in spectral graph theory which have, with only a few exceptions, appeared after 1978. In as much as is possible, we have presented our results in a manner consistent with the terminology and results of [CvDSa1]. In contrast to that reference where the theory of graph spectra was presented in full starting from basic concepts and continuing through to advanced theorems, this book gives new results with no more than a short proof.

The bibliography of [CvDSa1] contains 564 items, most of which were published between 1960 and 1978. In this book there are over 700 new references from the mathematical literature and from the chemical literature; some others come from different areas including physics, mechanical engineering, geography, and the social sciences. We have included all of the references of which we were aware at the end of 1984. The large number of references indicates the rapid growth rate of spectral graph theory. But it must be said that many of these papers have only minor results and hence are presented only briefly. In addition, many results published are in fact rediscoveries of known results. The style of this book has been influenced by this situation.

Since the publication of **Spectra of Graphs**, this topic has been included in several new books dealing with graph theory in general: [Bol1], [CaMo1], [BeCL1], [Lov3]. In addition there have been several expository articles giving either an overview of the subject or a description of some particular topic: [Cam1], [Cve1], [Cve4], [Fin1], [GoHMK], [Hae1], [Mal3], [Moh2], [ScWi1], [Sei1], [SeTe1], [Wils1].

The following directions of investigation were indicated in [Cve4]:

1. Construction of cospectral graphs,

2. Spectral characterizations of graphs,

3. The behaviour of eigenvalues under graph transformations,

4. Relations between spectral and structural properties of graphs,

5. Geometric methods in spectral graph theory,

6. Applications of graph spectra to combinatorial objects,

7. Applications of graphs spectra to other scientific areas,

8. Classifying and ordering graphs,

9. Construction of graphs by spectral techniques,

10. Spectra and the graph reconstruction problem,

11. Special functions and graphs,

12. Solving graph equations by spectral means.

The first seven of these topics, described in [Cve4] as "old" directions, have continued to be investigated in greater depth, as is indicated by the topics in this book. Less progress has been made on the last five topics. New areas of investigation have arisen in the meantime. These include the spectra of random graphs, the matching and other polynomials, the relationship of graph spectra and the Shannon capacity, and extensions of the concept of graph spectra to infinite graphs.

The material of this book has been grouped into five chapters. This means that each chapter is rather broad in scope; while it would have been possible to split several of the chapters, the authors felt that this division reflected most accurately the different directions in which research in spectral graph theory is currently being done.

The problem of characterizing graphs with least eigenvalue -2 was one of the original problems of spectral graph theory. The techniques used in the investigation of this problem have continued to be useful in other contexts including forbidden subgraph techniques as well as geometric methods involving root systems. In the meantime, the particular problem giving rise to these methods has been almost completely solved. This is indicated in Chapter 1.

The study of various combinatorial objects including distance regular and distance transitive graphs, association schemes, and block designs have made use of eigenvalue techniques, usually as a method to show the nonexistence of objects with certain parameters. The basic method is to construct a graph which contains the structure of the combinatorial object and then to use the properties of the eigenvalues of this graph. Methods of this type are given in Chapter 2.

Several topics have been included in Chapter 3. Each of them has experienced some growth, but not enough to devote a separate chapter to any one of them. These include the relationships between the spectrum and automorphism group of a graph, the graph isomorphism and the graph reconstruction problem, spectra of random graphs, and the Shannon capacity problem.

Some graph polynomials related to the characteristic polynomial are described in Chapter 4. These include the matching, distance, and permanental polynomials.

Applications of the theory of graph spectra to Chemistry and other branches of science are described from a mathematical viewpoint in Chapter 5. Although chemical papers form about one third of the references in our bibliography, the mathematical content of these references is somewhat limited because of the differing goals of mathematicians and chemists. Chemical details are generally kept to a minimum in the text.

The last chapter is devoted to the extension of the theory of graph spectra to infinite graphs. Some price has to be paid for this; in order to make a reasonable theory of the spectra of infinite graphs, one must restrict to special classes of graphs, lose the discrete nature of the spectrum, or define a spectrum that is not independent of the labelling of the vertices and keeps only some of the spectral properties as graph invariants.

We have also included a table of graph spectra as an appendix.

The authors would like to express thanks to those who helped this book come into being. The comments of C. Godsil, B. Mohar, P. Rowlinson, and B. McKay have been most useful and improved the presentation in several sections. Financial support has been given by the University of Belgrade and the University of Manitoba, as well as by Sabbatical and Operating Grants from the National Sciences and Engineering Council of Canada. In particular, the Carnegie Research Fellowship at Stirling University allowed one of us (DMC) to have a sabbatical leave. This support not only allowed the research for this book to take place, but also permitted the initiation of T$_E$X, the method by which this book was typeset. It might be of interest to note that because of two sabbatical leaves given to two of the authors, the text for this book was entered on no fewer that five different computers, with virtually no problems of portability.

Contents

Chapter 1

Characterizations of Graphs by their Spectra

The study of spectral graph theory, in essence, is concerned with the relationships between the algebraic properties of the spectra of certain matrices associated with a graph and the topological properties of that graph. By far the most common matrix studied is the 0-1 adjacency matrix, and the spectrum of this matrix is called the *spectrum of the graph*. In some cases this spectrum determines it up to isomorphism, but often this does not happen (perhaps, even, such a characterization occurs only with probability zero). On the other hand, many graphs with a high degree of symmetry do admit characterizations of various types.

Section 1.1: Generalized Line Graphs and Graphs with Least Eigenvalue −2

An early area of study concerned the line graphs $L(G)$ and its least eigenvalue $\lambda(G)$. This was, on the one hand, because there is a rich supply of them. Indeed, since, with only one exception, two connected, nonisomorphic graphs have nonisomorphic line graphs, there are, in a sense, just as many line graphs as there are graphs. On the other hand, it is easy to see that if K is the vertex–edge incidence matrix and $A(L(G))$ is the adjacency matrix of the line graph of G, then $KK^T = A(L(G)) + 2I$, and the least eigenvalue satisfies $\lambda(A(L(G))) \geq -2$. Thus a considerable effort has gone into studying graphs G that satisfy $\lambda(G) \geq -2$.

The most important results concerning graphs with $\lambda(G) \geq -2$ are given in the important paper of Cameron, Goethals, Seidel, and Shult [CAGSS] in [CVDSA1]. The tool used is the theory of real root systems.

Let (u,v) be the usual inner product of two vectors in $\Re^n$. A *real root system* may be defined as a finite set of vectors Φ spanning $\Re^n$ satisfying

(1) $v, kv \in \Phi \Rightarrow k = \pm 1$,

(2) $u, v \in \Phi \Rightarrow 2(u,v)/(u,u) \in \mathbf{Z}$, the integers,

(3) $u, v \in \Phi \Rightarrow f_u(v) \in \Phi$, where $f_u(v) = v - \frac{2(u,v)}{(u,u)}u$ is the reflection of v by the hyperplane through the origin orthogonal to u.

Real root systems have been classified. A representation of each one is given below. The subscript indicates the smallest dimension n of an $\Re^n$ that contains the root system and $\{e_1, \ldots, e_n\}$ is an orthonormal basis.

(1) $A_n : \{e_i - e_j \mid 1 \leq i \neq j \leq n+1\}$.

(2) $D_n : \{e_i \pm e_j \mid 1 \leq i \neq j \leq n\}$.

(3) $E_8 : D_8 \cup \{\pm \frac{1}{2}e_1 \pm \frac{1}{2}e_2 \pm \cdots \pm \frac{1}{2}e_8 \mid$ the number of plus signs is even$\}$.

(4) $E_7 : \{v \in E_8 \mid v \perp u\}$ for some $u \in E_8$.

(5) $E_6 : \{v \in E_8 \mid v \perp u, v \perp w\}$ for $u, w \in E_8$ satisfying $(u,w) = \pm 1$.

It is immediate that any pair of noncollinnear vectors in a root system have an inner product of 0 or ± 1, while two collinear vectors have an inner product of ± 2. Thus if K is a matrix formed by using pairwise noncollinear vectors of the root system as rows, the matrix KK^T is a symmetric matrix taking the value 2 on the diagonal. If the vectors a chosen so that the inner products are not -1, then $A = KK^T - 2I$ is the adjacency matrix of a graph, and since KK^T is positive semidefinite, the graph has its eigenvalues bounded from below by -2. Conversely, if a graph has eigenvalues bounded from below by -2, then $A + 2I$ is a positive semidefinite matrix and hence there exists a matrix K such that $KK^T - 2I = A$. It is then possible to consider the rows of K as a set of vectors and *star-close* them, *i.e.*, for any vector u add the vector $-u$ to the set and for any pair of vectors u, v with inner product -1, add the vector $-(u + v)$ to the set. This will result in a root system, and hence graphs with least eigenvalue $\lambda(G) \geq -2$ are equivalent to sets of vectors from a real root system with pairwise inner products equal to 0 or 1.

As an example, consider the vectors $\{e_0 \pm e_i \mid i = 1, 2, \ldots n\}$. These vectors from D_{n+1} can be used as rows for K, and then $A = KK^T - 2I$ is the adjacency matrix of a *cocktail party graph*, $CP(n)$, the regular graph on $2n$ vertices of degree $2n - 2$. In fact the spectrum consists of $\{2n - 2^1, 0^n, -2^{n-1}\}$ (the exponents are the multiplicities of the respective eigenvalues). The rank of $KK^T - 2I$ is the dimension of the root system from which the row vectors were selected. We say that the root system D_{n+1} *represents the graph* $CP(n)$ since there is a set of vectors in that root system that gives rise to the graph.

Suppose a graph is represented by A_n for some n. Using the representation of A_n given above, in order to have the inner product of two rows nonnegative, each column of the matrix K must have entries with the same sign. Let H be the graph whose vertices correspond to the nonzero columns of K; for each row vector $e_i - e_j$ in K, join the corresponding vertices in H. It is then easy to see that $G = L(H)$, and that the graphs represented by A_n are just the line graphs of bipartite graphs.

A *generalized line graph* $L(G; a_1, \ldots, a_n)$ is defined for graphs G with n vertices $\{1, \ldots, n\}$ and nonnegative integers $(a_1, \ldots, a_n)$ by taking the graphs $L(G)$ and $CP(a_i)$ and adding extra edges. A vertex in $L(G)$ is joined to one in $CP(a_i)$, $i = 1, \ldots, n$ if vertex i is an end point of the vertex in $L(G)$ (viewed as as edge of G). Special cases include an ordinary line graph ($a_1 = a_2 = \cdots = a_n = 0$) and the cocktail party graph $CP(n)$ ($n = 1$ and $a_1 = n$).

Suppose a graph is represented by D_n. The matrix K will still represent the same graph if rows are interchanged, columns are interchanged, or a column is multiplied by -1. Further, each row can be written in the form $e_i \pm e_j$, and so in each case there might exist a row of the form $e_i \mp e_j$. If there is such a row, we shall call the two rows (or corresponding vertices) *paired*. By appropriately permuting rows and columns (and noting that the inner product of two rows of K can never by -1, the nonzero entries of the rows corresponding to paired vertices can be blocked down the diagonal and be a representation of $CP(a_i)$ with the same form as the representation of $CP(n)$ given above. By using appropriate column multiplication, the remaining rows can be of the form $e_i + e_j$. Those with i and j in new columns form the vertex-edge incidence matrix of a line graph while the remaining rows must be the edges added to construct the generalized line graph.

These results, first given in [CAGSS] from [CVDSA1], can be expressed in the following two theorems:

THEOREM 1.1 (P. CAMERON, J. M. GOETHALS, J. J. SEIDEL, E. E. SHULT):
(1) G is represented by A_n if and only if G is the line graph of a bipartite graph, and
(2) G is represented by D_n if and only it G is a generalized line graph.

THEOREM 1.2 (P. CAMERON, J. M. GOETHALS, J. J. SEIDEL, E. E. SHULT):   *If G is a graph with least eigenvalue $\lambda(G) \geq -2$ then*
(i) G is a generalized line graph, or
(ii) G is represented by E_6, E_7, or E_8.

Using Theorem 1.2 as a starting point it is possible to determine all graphs with $\lambda(G) > -2$.

THEOREM 1.3 (M. DOOB, D. CVETKOVIĆ [DOCV1]):   *If G is a connected graph with $\lambda(G) > -2$, then*
(i) $G = L(T; 1, 0, \ldots, 0)$, where T is a tree,
(ii) $G = L(H)$, where H is a tree or H is unicyclic with an odd cycle,
(iii) G is one of 20 graphs that are represented by E_6,
(iv) G is one of 110 graphs that are represented by E_7, or
(v) G is one of 441 graphs that are represented by E_8.

Theorem 1.2 indicates the importance of generalized line graphs in spectral graph theory. Many of the appealing properties of line graphs can be extended to generalized line graphs by spectral means or otherwise. Some examples are given below.

A *generalized cocktail party graph*, denoted GCP, is a connected graph with n vertices in which the degree of each vertex is at least $n - 2$. In such a graph a vertex is called *a-type* if its degree is $n - 1$ and *b-type* if it is of degree $n - 2$.

THEOREM 1.4 (D. CVETKOVIĆ, M. DOOB, S. SIMIĆ [CVDS1]):   *A graph is a generalized line graph if and only if the edges can be partitioned into generalized cocktail party graphs such that*
(1) each vertex is in at most two GCP's,
(2) any two of these GCP's have at most one common vertex, and
(3) a vertex in common with two of these GCP's is a-type.

There is only one pair of connected graphs ($K_{1,3}$ and K_3) that have the same line graph. There are six more pairs for generalized line graphs. These are displayed in [CVDS1].

THEOREM 1.5 (D. CVETKOVIĆ, M. DOOB, S. SIMIĆ [CvDS1]): *There exist seven pairs of root graphs with identical generalized line graphs.*

If G is a connected graph with n vertices and m edges, then the multiplicity of the eigenvalue −2 for $L(G)$ is $m - n$ if G is not bipartite and $m - n + 1$ otherwise.

THEOREM 1.6 (D. CVETKOVIĆ, M. DOOB, S. SIMIĆ [CvDS1]): *If G has n vertices and m edges and not all $a_i = 0$, then the multiplicity of the eigenvalue* −2 *is* $m - n + \sum_{i=1}^{n} a_i$.

There are nine minimal graphs that are not line graphs, *i. e.*, contained as an induced subgraph in any other graph that is not a line graph.

THEOREM 1.7 (D. CVETKOVIĆ, M. DOOB, S. SIMIĆ [CvDS1]): *There exist 31 minimal nongeneralized line graphs.*

The 20 graphs represented by E_6 in Theorem 1.3 are obviously the minimal nongeneralized line graphs with $\lambda(G) = -2$. There are 11 graphs (3 with five vertices and 8 with six vertices) with $\lambda(G) < -2$.

Theorem 1.7 also appears in [RASV1].

A *semiregular bipartite graph* with parameters (n_1, n_2, r_1, r_2) is a bipartite graph with vertex partition of size n_1 and n_2 such that the n_1 vertices are all of degree r_1 and the n_2 vertices are of degree r_2.

Arguments involving root systems and generalized line graphs are used to complete the characterization of line graphs by their spectra. They are summarized by the following theorems:

THEOREM 1.8 (D. CVETKOVIĆ, M. DOOB [CvDo1]): *A regular line graph $L(G)$ is cospectral with a graph which is not a line graph if and only if G is*

(i) $K_{4,4}$, $K_{3,6}$,

(ii) $CP(4)$,

(iii) K_8,

(iv) $\overline{C_8}$,

(v) $\overline{C_3 \cup C_4}$, $\overline{C_4 \cup C_4}$,

(vi) *the semiregular bipartite graph with parameters* $(6, 3, 2, 4)$,

(vii) *the four regular connected graphs on 8 vertices, or*

(viii) *the complements of the five regular graphs on 8 vertices.*

In the previous theorem, $\overline{G}$ denotes the complement of G. The exceptional graphs have a small number of vertices (7, 8, or 9) precisely because of the small dimension of the root systems representing them. It is known ([BuCS] from [CvDSa1]) that there are exactly 68 regular, connected graphs that are cospectral with line graphs but not line graphs themselves (it is perhaps noteworthy that they are constructed in [CvRa1] without the aid of a computer). Some alternatives in constructing these graphs are described in [Cve7].

Theorem 1.8 can be reformulated in the following equivalent manner:

THEOREM 1.8' (D. CVETKOVIĆ, M. DOOB [CvDo1]):   *The spectrum of a graph G determines whether or not it is a regular, connected line graph except for 17 cases. In these cases G has the spectrum of $L(H)$ where H is one of the 3-connected graphs on 8 vertices, or H is a connected, semiregular bipartite graph on $6 + 3$ vertices.*

It is noted in [RaSV2] that the line graph of a regular complete multipartite graph G is characterized by its spectrum unless G is K_8, $K_{4,4}$, or $K_{2,2,2}$.

In [HoJa1] it is shown that $L(K_{n,n,n})$ is characterized by its spectrum.

There is relatively little known about the structure of graphs with least eigenvalue less than -2. It seems that those whose least eigenvalue lies in the interval $(-1 - \sqrt{2}, -2)$ should be interesting.

THEOREM 1.9 (M. DOOB [Doo3]):   *If G is a graph with $\lambda(G) < -2$, then there exists an induced subgraph H contained in G such that $\lambda(H) = -2$.*

The fact that -2 has the property stated in Theorem 1.9 is somewhat surprising. The only other real numbers with that property are 0, -1, and $-\sqrt{2}$.

THEOREM 1.10 (V. KUMAR, S. B. RAO, N. M. SINGHI [KuRS1]):   *If G satisfies $\lambda(G) < -2$ and is minimal with respect to that property, then G has at most 10 vertices.*

There are 11 minimal graphs with 6 vertices (namely those of the 31 graphs mentioned in Theorem 1.7 that are not represented by E_6), and there are 14 minimal graphs with 7 vertices given in [KuRS1]. The remaining graphs have been computed by A. Neumaier and F. Bussemaker (see Section 1.4).

Spectral arguments have been used to tell when a graph G and its complement $\overline{G}$ are generalized line graphs. It is shown in [RSST1] that such a graph G is either a cocktail party graph, the complement of a cocktail party graph, or an induced subgraph of one of 10 graphs (listed in [RSST1]).

Section 1.2:   Other Graph Characterizations

In the last section we looked at graphs G with least eigenvalue $\lambda(G) = -2$. Characterizations by the smallest eigenvalue are very difficult to find in general, and any characterization of that type is certainly a very important contribution to spectral graph theory.

A graph is *strongly regular* with parameters (n, r, λ, μ) if it has n vertices, is regular of degree r, any two adjacent vertices are mutually adjacent to λ other vertices, and any two nonadjacent vertices are mutually adjacent to μ other vertices. The complete graph K_n and its complement $\overline{K_n}$ are usually excluded from the definition; among other things, this implies that a strongly regular graph is connected. The parameters of a strongly regular are not independent since $(n - r - 1)\mu = r(r - \lambda - 1)$. A strongly regular graph has three distinct eigenvalues; they (and their multiplicities) are determined by the parameters, and, conversely, the parameters may be derived from the distinct eigenvalues (see [CvDSA1] or [SEI1]).

There are several types of strongly regular graphs which are described by their parameters. A strongly regular graph is called a *conference graph* if the parameters are of the form $(n, r, \lambda, \mu) = (4t + 1, 2t, t - 1, t)$. The distinct eigenvalues of a conference graph are $2t$ and $-\frac{1}{2}(1 \pm \sqrt{4t + 1})$. If the parameters $(r, \lambda, \mu) = (mn, m(m - 2) + n, m^2)$, the graph has *Steiner parameters*, and if $(r, \lambda, \mu) = (m(n - 1), m(m - 3) + n, m(m - 1))$, the graph has *Latin square parameters*. In these cases the distinct eigenvalues are r, $n - m$ and $-m$. The complete multipartite graph $K_{m,\ldots,m}$ is a strongly regular graph with parameters $(mt, (m - 1)t, (m - 2)t, (m - 1)t)$. The distinct eigenvalues in this case are r, 0, and $-m$.

A. Neumaier has described strongly regular graphs with integral least eigenvalue $\lambda(G) = -m$.

THEOREM 1.11 (A. NEUMAIER [NEU1]):   *A strongly regular graph G with least eigenvalue $\lambda(G) = -m$ is one of the following:*
 (i) *a conference graph with $t = m(m-1)$,*
 (ii) *a complete multipartite graph $K_{m,\dots,m}$,*
(iii) *a graph with Steiner parameters,*
(iv) *a graph with Latin square parameters,*
 (v) *a finite set of exceptional graphs.*

Each of the first four categories in the conclusion of Theorem 1.11 contains an infinite number of graphs. The method used involved *Krein parameters* and the *absolute bound*. These methods are described in Section 2.1.

More can be said about the least eigenvalue of a graph by using *distance matrices*. The matrix $C = (c_{i,j})$ of order n is a distance matrix if the set $X = \{1,\dots,n\}$ and the distance function $d(i,j) = \sqrt{c_{ij}}$ forms a real metric space (X, d). In particular this means that C is real, symmetric matrix with 0 on the diagonal. A matrix is a *spherical distance matrix* if it is a distance matrix such that it is isometrically embeddable into the surface of a sphere in Euclidean space, *i. e.*, there exist points $p_1, \dots, p_n$ such that $c_{i,j} = |p_i - p_j|^2$.

THEOREM 1.12 (A. NEUMAIER [NEU4]):   *If G is a regular, connected graph with least eigenvalue $\lambda(G) = -m$ and G is not a complete graph, then $C = m(J - I) - A$ is a spherical distance matrix.*

The spherical distance matrix of Theorem 1.12 is used for the following result:

THEOREM 1.13 (A. NEUMAIER [NEU4]):   *Let G be a regular graph with $n \geq 6$ vertices and smallest eigenvalue $-m$ with multiplicity g. Then $n \leq 2g + 1$ implies that m is integral or G is a conference graph.*

A conference graph, of course, satisfies $n = 2g + 1$.

Godsil and McKay consider *neighbourhood regular graphs*. For such a graph G, the vertices adjacent to a vertex v induce a regular subgraph $N(v)$ and the vertices not adjacent (and not equal) to v induce a regular subgraph $R(v)$. If a graph is neighbourhood regular, then so is its complement.

THEOREM 1.14 (C. GODSIL, B. MCKAY [GoMK3]): *G is regular and neighbourhood regular if and only if it is strongly regular.*

PROOF: Taking the complement if necessary, we may assume that G is connected. If vertices u and v are adjacent and they are mutually adjacent to λ other vertices, then λ is the degree of $N(v)$. By symmetry it also equals the degree of $N(u)$, and by connectivity the value λ is independent of the choice of u and v. If $N(v)$ has degree r_1, then $R(v)$ has degree $r(n - 2r + r_1)/n - r - 1$. For u and v nonadjacent and mutually adjacent to μ other vertices, the value of μ is the degree of G less the degree of $R(v)$. Hence the value of μ is independent of the choice of u and v.

In Theorem 1.14 the graph formed by taking a disjoint union of complete graphs of the same size is neighbourhood regular (this is sometimes excluded as a strongly regular graph).

Now suppose that G is neighbourhood regular and disconnected. Then if v is a vertex not in a given connected component, then that component is also a component of $R(v)$ and hence is regular and, by Theorem 1.14, strongly regular. Clearly the degree of $R(v)$, which is less than the degree of v, and the degree of any other connected component not containing v are equal. By symmetry, there can be only one component that contains a nontrivial $R(v)$. Thus only one component can not be a complete graph. If there are more than two components, the regularity of $R(v)$ forces G to be regular.

THEOREM 1.15 (C. GODSIL, B. MCKAY [GoMK3]): *If G is neighbourhood regular, disconnected, and not regular, then*
  (i) *G is the union of two complete graphs, or*
  (ii) *G is the union of a strongly regular graph and a complete graph.*

Thus what remains to be determined are neighbourhood regular graphs which are connected with connected complements and not regular. It is shown in [GoMK3] that such a graph must have exactly two vertex degrees r_1 and r_2. The eigenvalues of the subgraphs determined by the vertices of degree r_1 and by the vertices of degree r_2 sum to zero and have, of course, integral multiplicities. This limits the feasibility of these graphs in many cases. A list of possible parameters is given in [GoMK3]; the eigenvalues of the induced subgraphs play a crucial role in characterizing neighbourhood regular graphs.

The second largest eigenvalue of a graph, $\lambda_2(G)$, is of interest. In particular, $\lambda_2(T)$ for a tree T is used by A. Neumaier [NEU6] to classify certain bilinear forms. The critical question is whether or not T satisfies $\lambda_2(T) \leq 2$. Bounds can be given on $\lambda_2(T)$.

THEOREM 1.16 (A. NEUMAIER [NEU6]):   *If T is a tree with n vertices and n is odd, then $\lambda_2(T) \leq \sqrt{(n-3)/2}$. Further, equality is attained for the three classes of graphs which are displayed in Figure 1.1.*

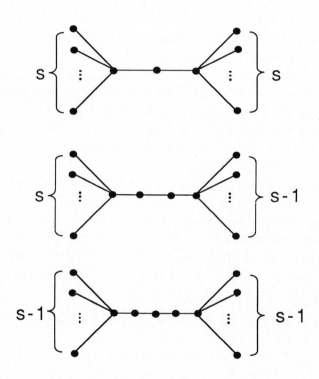

Figure 1.1 Trees for which $\lambda_2(T) = \sqrt{(n-3)/2}$

This result has been further extended by H. Yuan.

THEOREM 1.17 (H. YUAN [YUAN1]):   *If $\lambda_k(T)$ is the kth largest eigenvalue of a tree, with n vertices, then $\lambda_k(T) \leq \sqrt{[(n-2)/k]}$ for $2 \leq k \leq [n/2]$. Further, this bound is the best possible for $n \equiv 1 \bmod k$.*

THEOREM 1.18 (A. NEUMAIER [NEU6]): *Suppose that T is a tree and $\lambda_2(T) \leq \lambda$. Then either*
 (i) *there exists a vertex v of T such that $\lambda_1(T - v) \leq \lambda$, or*
 (ii) *there exists an edge $\{u_1, u_2\}$ of T such that $\lambda_1(T_i - u_i) < \lambda < \lambda_1(T_i)$ where T_1 and T_2 are the two trees obtained by deleting $\{u_1, u_2\}$.*

If we apply this Theorem 1.18 to the case where $\lambda = 2$, we find that for a tree with $\lambda_2(T) \leq 2$, either one vertex or one edge can be deleted to get subtrees with $\lambda_1(T) \leq 2$. These trees are quite well known (see [CVDSA1], pp. 78–79, for example).

A. Neumaier and J. J. Seidel [NESE1] use the second largest eigenvalue of a graph in their description of discrete hyperbolic spaces. *Lorentz space* $\Re^{p,1}$ consists of real vectors $(x_0, x_1, \ldots, x_p)$ with the indefinite inner product $\langle (x_0, x_1, \ldots, x_p), (y_0, y_1, \ldots, y_p) \rangle = -x_0 y_0 + v_1 y_1 + \cdots + x_p y_p$. Graphs with $\lambda_2(G) = 1$ can be represented by taking vectors in the Lorentz space $\Re^{p,1}$ in essentially the same way as is done for the root system graphs in $\Re^n$.

A similar approach is used to find graphs with $\lambda_2(G) \leq 2$. These are called *reflexive graphs* in [NESE1]. The graphs are constructed as graphs extended by a single vertex, much as the trees are in Theorem 1.18.

Graphs with second largest eigenvalue $\lambda_2 \leq 1$ are related to graphs with least eigenvalue $\lambda(G) = -2$. If a graph is regular, the graphs with $\lambda_2 \leq 1$ are just the complements of graphs with least eigenvalue $\lambda(G) \geq -2$. If a graph has $\lambda(G) = -2$ with multiplicity at least 2, then it is still true that $\lambda_2(\overline{G}) = 1$.

THEOREM 1.19 (D. M. CVETKOVIĆ [CVE7]): *Suppose G is a graph with $\lambda_2(G) \leq 1$. Then*
 (i) *$\lambda(\overline{G}) \geq -2$ or*
 (ii) *$\overline{G}$ has exactly one eigenvalue $\lambda_n < -2$.*

A partial converse to Theorem 1.19 exists.

THEOREM 1.20 (D. M. CVETKOVIĆ [CVE7]): *Suppose G is a graph with $\lambda(G) = -2$. Then $\lambda_2(\overline{G}) \leq 1$. Further, equality holds if and only if -2 is a simple eigenvalue with an eigenvector whose coordinates do not sum to zero.*

It has been noted also [CVE7] that (by interlacing arguments) a graph with $\lambda_2(G) \leq 1$ either has girth at most 6 or is a union of trees with diameter at most 4.

There now exists spectral characterizations of both the *Hamming graph* and the *Johnson graph*. The Hamming graph $H(n,q)$ has the q^n n-tuples $(x_1, \ldots, x_n)$, $x_i \in \{1, \ldots, q\}$ as vertices with two vertices adjacent if they differ in exactly one coordinate. The Johnson graph $J(n,q)$ has the $\binom{n}{q}$ subsets of $\{1, \ldots, n\}$ as vertices with two vertices adjacent if as sets they have $q - 1$ elements in common.

THEOREM 1.21 (Y. EGAWA [EGA1]): *The Hamming graph $H(n,q)$ is characterized by its spectrum among all distance regular graphs unless $q = 4$. If $q = 4$ there exist $[n/2]$ cospectral graphs.*

The exceptional graphs were first given by M. Doob ([DOO7] of [CVDSA1]).

THEOREM 1.22 (P. TERWILLIGER [TER2]): *The Johnson graph $J(n,q)$ is characterized by its parameters unless $(n,q) = (2,8)$. If $(n,q) = (2,8)$, there are three other cospectral graphs.*

The three exceptional were first given by Chang ([CHA1] of [CVDSA1]).

The important results of Theorem 1.21 and Theorem 1.22 will be discussed in Chapter 2.

Let $\Lambda(G) = \lambda_1(G)$ denote the largest eigenvalue of a graph. The graphs with $\Lambda(G) \leq 2$ are called the *Smith graphs* and are displayed on page 79 in [CVDSA1]. This result is improved somewhat with the following theorem:

THEOREM 1.23 (D. M. CVETKOVIĆ, M. DOOB, I. GUTMAN [CVDG1]): *Let G be a graph such that $2 < \Lambda(G) \leq \sqrt{2 + \sqrt{5}}$. Then G is a tree with maximum degree 3 and either one or two vertices of degree 3; in particular G is*

(i) $T(i,j,k)$ with $j = 2, k > 5$ or $2 < j < k$,
(ii) $T(2,2,k)$ with $k > 2$,
(iii) $T(2,3,3)$, or
(iv) $S(j,k,l)$, with $k \geq C(j,l)$, a constant depending on j and l.

The graphs $T(i,j,k)$ and $S(j,k,l)$ are displayed in Figure 1.2.

A. Neumaier has also described the largest eigenvalue of a tree.

THEOREM 1.24 (A. NEUMAIER [NEU6]): *If T is a tree with n vertices, then $\lambda_1(T) \leq \sqrt{n-1}$ with equality if and only if $T = K_{1,n-1}$.*

A bipartite graph is characterized by the symmetry of its spectrum with respect to 0, *i.e.*, if λ is an eigenvalue, then so is $-\lambda$; this follows directly from the Perron-Frobenius Theorem:

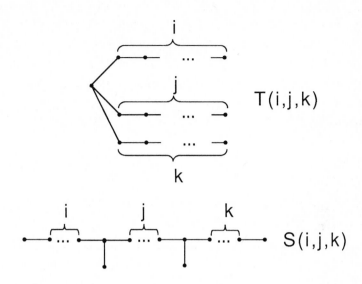

Figure 1.2 Graphs with $2 < \lambda_1(G) \leq \sqrt{2 + \sqrt{5}}$

THEOREM 1.25 (F. ESSER, F. HARARY [ESHA3]): *A digraph D is bipartite if and only if its spectrum is symmetric with respect to 0.*

An equivalent theorem appears in [CVDSA1], page 82. The spectrum of a digraph, of course, need not be real.

Section 1.3: Cospectral Constructions

The construction of cospectral graphs has been prominent in the theory of graph spectra. A variety of constructions exist. There are, roughly speaking, two main methods of constructing sets of cospectral graphs. One method is a kind of "cut and paste" method; a piece of the graph is excised and replaced by a new (possibly isomorphic) piece in such a way that the spectrum is unchanged but the new graph is not isomorphic to the old one. A second method is a "product" method. Products of graphs are taken (often by some variant on the tensor product) where the spectrum of the product is only dependent on the spectra of the factors. One factor is then replaced by cospectral graph to produce a product that is cospectral but nonisomorphic to the original graph.

A. J. Schwenk, W. C. Herndon, and M. L. Ellzey, Jr. use a bipartite graph construction to form cospectral graphs. This involves three graphs: G_1, G_2, and a bipartite graph B with vertices partitioned into R and S with r and s vertices respectively. Subsets $S_1 \subset V(G_1)$ and $S_2 \subset V(G_2)$ are defined arbitrarily. The *generalized composition* $B(G_1, S_1, G_2, S_2)$ is then constructed by taking r copies of G_1 and s copies of G_2 that correspond to the vertices of B; edges are added joining the vertices of S_1 to all of those in S_2 if and only if the corresponding vertices in B are adjacent.

THEOREM 1.26 (A. J. SCHWENK, W. C. HERNDON, AND M. L. ELLZEY, JR. [ScHE1]):   *If either (i) $r = s$, or (ii) G_1 and G_2 are cospectral, then $B(G_1, S_1, G_2, S_2)$ is cospectral with $B(G_2, S_2, G_1, S_1)$.*

PROOF: The k-th spectral moment is $M_k = \sum_{i=1}^{n} \lambda_i^k$; The eigenvalues $\{\lambda_1, \ldots, \lambda_n\}$ determine M_k and the values $\{M_1, \ldots, M_n\}$ determine the eigenvalues. In addition, $M_k = \operatorname{tr} A^k$ which also equals the total number of closed walks of length k. Thus a correspondence between closed walks of a given length is necessary and sufficient for two graphs to be cospectral. For (i) the correspondence is constructed in the following way: a closed walk is entirely within one G_i in the first graph can correspond to the same closed walk in the second graph since $r = s$. A closed walk that traverses several G_i can be viewed as a alternating even sequence of walks within individual G_1's and G_2's separated by edges of B. The corresponding closed walk in the second graph is obtained by leaving the paths within the G_i's unchanged; the edges of B that fit in the gaps between the edges in the G_i's are each advanced to the next gap. For (ii), a closed walk in G_1 corresponds to a closed walk in G_2 since G_1 and G_2 are cospectral. Thus we can have a correspondence between the closed wholly contained with a G_i even if $r \neq s$. For closed walks that traverses several G_i's, the construction is the same.

It is possible, of course that $B(G_1, S_1, G_2, S_2)$ and $B(G_2, S_2, G_1, S_1)$ are isomorphic.

Theorem 1.26 can also be found in [GoM1] of [CvDSa1].

THEOREM 1.27 (A. J. SCHWENK, W. C. HERNDON, AND M. L. ELLZEY, JR. [ScHE1]):   *If G_1 is regular and $|V_1| = 2|S_1|$, then $B(G_1, S_1, G_2, S_2)$ is cospectral with $B(G_1, V_1 - S_1, G_2, S_2)$.*

PROOF: The regularity of G_1 and cardinality of S_1 ensure that the number of walks in G_1 starting and ending in S_1 equals the number of walks starting and ending in $V_1 - S_1$, and so a correspondence may be set up between them. A correspondence between closed walks of length k is then constructed by replacing each part of the walk that starts and ends within S_1 of G_1 by the corresponding walk that starts and ends in $V_1 - S_1$, and by moving the one end point of an edge from S_1 to S_2 so that it is incident to the new walk.

A well known method of producing cospectral graphs is by *Seidel switching*. If D is a subset of the vertices of a graph G and C_1 is the complement of D, then G' is obtained from G by taking the same vertex set and letting two vertices in D or C_1 be adjacent in G' if they are adjacent in G, while a vertex in D and one in C_1 are adjacent in G' if and only if they are not adjacent in G. If G and G' are regular of the same degree, then they are cospectral.

C. Godsil and B. McKay [GoMK1] have given several methods for constructing cospectral graphs. One involves *local switching*, a generalization of Seidel switching. Let $\pi = (C_1, C_2, \ldots, C_k, D)$ be a partition of the vertices of G such that, for every i, j in $\{1, \ldots, k\}$, each $v_i \in C_i$ is adjacent to the same number of vertices $v_j \in C_j$ and each $v \in D$ is adjacent to all, half, or none of the vertices of C_j Now for each $v \in D$ that is adjacent to half of the vertices of C_j, replace the edges from v to C_j by new ones joining v to the half of the vertices that were originally nonadjacent. The new graph so formed is denoted by G^π.

THEOREM 1.28 (C. GODSIL, B. MCKAY [GoMK1]):   *If G^π is constructed from G by local switching, then G and G^π are cospectral with cospectral complements.*

PROOF: The proof, as given in [GoMK1], is clever and almost self-evident. As usual, let J and j be the matrix and vector of appropriate order with every entry equal to one. Define $Q_m = \frac{2}{m} J_m - I_m$. Then it is easy to verify that (i) $Q_m^2 = I_m$, (ii) if X has constant row and column sums, then $Q_m X Q_n = X$, and (iii) if x is a 0-1 vector of size $2m$ with m ones, then $Q_{2m} x = j_m - x$. Now let $Q = \text{diag}(Q_{n_1}, Q_{n_2}, \ldots, Q_{n_k}, I)$.

If A is the adjacency matrix of G (with the rows and columns being ordered consistently with the partition), it then follows easily that QAQ^{-1} is the adjacency matrix of G^{π}. The same matrix works for $\overline{G}$, too.

This method of construction produces a remarkable 72% of the cospectral graphs on nine vertices.

Theorem 1.27 is a special case of Theorem 1.28 by letting $\pi = (C_1, \ldots, C_r, D)$ where C_i a copy of G_1 and D is the union of the G_2's.

Another construction that appears in [GoMK1] involves tensor products.

THEOREM 1.29 (C. GODSIL, B. MCKAY [GoMK1]): *If G_1 and G_2 are cospectral, X and H are square matrices of the same order, then $H \otimes I + X \otimes G_1$ and $H \otimes I + X \otimes G_2$ are cospectral.*

THEOREM 1.30 (C. GODSIL, B. MCKAY [GoMK1]): *If G_1 and G_2 are cospectral with cospectral complements, and C, D, E, and F are square matrices of the same order, then $C \otimes I + D \otimes J + E \otimes G_1 + F \otimes \overline{G_1}$ and $C \otimes I + D \otimes J + E \otimes G_2 + F \otimes \overline{G_2}$ are cospectral.*

Let $A_{1_1}, A_{2_1}, \ldots, A_{k_1}$, $A_{1_2}, A_{2_2}, \ldots, A_{k_2}$ and $B_{1_1}, B_{2_1}, \ldots, B_{k_1}$, $B_{1_2}, B_{2_2}, \ldots, B_{k_2}$ be two sequences of matrices, and suppose that A_{i_1} and B_{i_1} have order n_1 while A_{i_2} and B_{i_2} have order n_2 for $i = 1, \ldots, k$. Let $T(A) = \sum_{i=1}^{k} A_{i_1} \otimes A_{i_2}$ and $T(B) = \sum_{i=1}^{k} B_{i_1} \otimes B_{i_2}$.

THEOREM 1.31 (C. GODSIL, B. MCKAY [GoMK1]): *Suppose that for any f, a monomial in k noncommuting variables,*

$$\mathrm{tr} f(A_{1_1}, A_{2_1}, \ldots, A_{k_1}) = \mathrm{tr} f(B_{1_1}, B_{2_1}, \ldots, B_{k_1})$$

and

$$\mathrm{tr} f(A_{1_2}, A_{2_2}, \ldots, A_{k_2}) = \mathrm{tr} f(B_{1_2}, B_{2_2}, \ldots, B_{k_2}),$$

then $T(A)$ and $T(B)$ are cospectral.

This generalizes many previously known constructions, e. g., the NEPS construction of Cvetković (see [CvDSa1]) and gives an another proof of Theorem 1.26 and Theorem 1.27.

A. Schwenk constructs cospectral graphs by use of *removal-cospectral sets*. Suppose we have graphs G and H with S and T as subsets of the respective vertex sets. Let $\theta: S \rightarrow T$ be a correspondence. If for every $A \subset S$ we have $G - A$ and $H - \theta(A)$ cospectral, then S and T are called removal-cospectral sets.

THEOREM 1.32 (A. SCHWENK [SCH2]): *Let S and T be removal-cospectral sets of size t for G and H. Attach any graph to G and the same graph to H by identifying t of the vertices of the graph with S and with the corresponding vertices in T. The resulting graphs are cospectral.*

C. R. Johnson and M. Newman have investigated graphs that are cospectral using the $(1, x)$ adjacency matrix. This matrix, A_x, takes on values of 1 and x in the positions which would take on the values of 0 and 1 in the usual adjacency matrix.

THEOREM 1.33 (C. R. JOHNSON, M. NEWMAN [JONE1]): *If G and H are cospectral, then the following are equivalent:*
(1) the complements are cospectral,
(2) the $A(1, x)$ characteristic polynomials are identical,
(3) The $(0, 1)$ adjacency matrices are similar via an orthogonal matrix U which has row and column sums equal to 1,
(4) the $A(1, x)$ characteristic polynomials are equal for two values of x unequal to 1.

Section 1.4: Hereditary Characterizations

A property $\mathcal{P}$ is called a *hereditary property* if, whenever G satisfies $\mathcal{P}$ and H is an induced subgraph of G, it follows that H also satisfies $\mathcal{P}$. Unless $\mathcal{P}$ is satisfied by all graphs, the hereditary property implies that there are minimal graphs that do not satisfy $\mathcal{P}$; such graphs are called *forbidden subgraphs*. If we let $\Lambda(G)$ be the largest eigenvalue of G, i.e., the spectral radius, $\lambda(G)$ be the smallest eigenvalue of G, and define the *spectral spread* of G, $S(G)$, to be $\Lambda(G) - \lambda(G)$, then, for any given positive real number r, the following properties are hereditary:
 (i) $\Lambda(G) < r$,
 (ii) $\Lambda(G) \leq r$,
(iii) $\lambda(G) > -r$,
(iv) $\lambda(G) \geq -r$,
 (v) $S(G) < r$,
 (vi) $S(G) \leq r$,
(vii) G has at most r positive eigenvalues,
(viii) G has at most r negative eigenvalues,

(ix) G is a line graph, and

(x) G is a generalized line graph.

For any given hereditary property $\mathcal{P}$ one may consider the (possibly infinite) set of graphs that do not satisfy $\mathcal{P}$. The description of such graphs has often proved to be interesting. L. Beineke [BEIN] of [CvDSA1] has described the 9 minimal graphs that are not line graphs. Similarly, the 31 graphs that are not generalized line graphs and are minimal with respect to that property were first described in [CvDS1] and have also been given in [CvDS2] and [KuRS1]. F. Bussemaker and A. Neumaier have used a computer search to find 1812 minimal graphs satisfying $\lambda(G) < -2$ with the following results:

number of vertices	5	6	7	8	9	10
number of graphs	3	8	14	67	315	1405

J. Smith in [SM,J] of [CvDSA1] shows that a graph has exactly one positive eigenvalue if and only if it is a complete multipartite graph.

THEOREM 1.34 (M. PETROVIĆ [PET1], A. TORGAŠEV [TOR12]): *A connected graph has two positive eigenvalues if and only if it contains the path P_4 or a K_3 with a pendant edge attached. A connected graph has two negative eigenvalues if and only if it contains P_4 or K_3.*

M. Petrović gives a useful construction for determining the number of positive or negative eigenvalues of a graph. For any graph G we construct a *canonical graph* by means of an equivalence relation on the vertices; two vertices are related if they have the same set of neighbours. The canonical graph is then constructed by identifying related vertices.

Let $n_+(G)$ and $n_-(G)$ denote the number of positive and negative eigenvalues of G.

THEOREM 1.35 (M. PETROVIĆ [PET4]): *If G is a graph and G' is its canonical graph, then*

$$n_+(G) = n_+(G') \text{ and } n_-(G) = n_-(G').$$

Thus we see, for example, that a connected graph has one positive eigenvalue if and only if its canonical graph is a complete graph, and it has one negative eigenvalue if and only if its canonical graph is a single edge.

M. Petrović [PET7] has described all graphs with spread $S(G) \leq 4$. They are the induced subgraphs of the Smith graphs and four other graphs with at most six vertices. The minimal graphs that do not have this property have at most 10 vertices; there are 47 of them in all of which 17 are disconnected.

M. Petrović [PET8] has also described graphs with eigenvalues $\lambda_1 \geq \lambda_2 \geq \cdots \geq \lambda_n$ satisfying $\lambda_2 - \lambda_n \leq \frac{3}{2}$. The 22 minimal graphs that do not have this property have at most 7 vertices; one is disconnected.

Further results of this type are given in [TOR9], [TOR13], [PET4], [PET9], and [PET10].

Chapter 2

Distance-Regular and Similar Graphs

Distance-regular graphs have been of interest in the area of spectral graph theory almost from the time of their definition. The spectrum of such graphs has been used not only for their classification but also to derive many specific properties. Much work has been done to identify the different types of distance-regular graphs. In this chapter we shall restrict ourselves to results that are directly relevant to the spectrum of a graph. For a more general description, see [BAIT4]. Much of the information about a distance-regular graph is determined by looking at three bases of the adjacency algebra generated by the adjacency matrix.

Section 2.1:   The Bose–Mesner Algebra

A *distance-regular* (also called *metrically regular* or *perfectly regular*) graph is a graph with a certain regularity in its paths. If the distance between two vertices u and v in a graph satisfies $d(u,v) = k$, then let $p_{ij}^k(u,v)$ be the number of vertices w in the graph such that $d(u,w) = i$ and $d(w,v) = j$ for $0 \leq i,j \leq d$, where d is the diameter of the graph. A graph for which these numbers p_{ij}^k are independent of the choice of vertices u and v is called distance-regular. In particular this implies that the number of vertices at distance k from a given vertex is p_{kk}^0 (which is sometimes denoted n_k) and that the graph is regular. The integers p_{ij}^k are sometimes called the *Bose parameters*. Their integrality is sometimes used to determine the feasiblity of the parameters for certain distance-regular graphs.

For any graph G the *Hoffman polynomial* may be defined. If $r > \lambda_2 > \lambda_3 > \cdots > \lambda_{d+1}$ are the distinct eigenvalues of a graph G with n vertices, then let

$$H(x) = n \prod_{i=2}^{d+1} \frac{x - \lambda_i}{r - \lambda_i}.$$

Hoffman ([HOF3] of [CVDSA1]) has shown that $H(A) = J$ for any regular, connected graph. Since $AJ = rJ$ for a regular graph, it is clear that the algebra (over the complex numbers) of matrices generated by the adjacency matrix of G contains $I, A, A^2, \ldots A^d$, and the Hoffman polynomial ensures that these matrices form a basis and J is in the algebra. Since the basis consists of commuting matrices, they may be simultaneously diagonalized and the eigenspaces are identical for all matrices in the algebra. Also, since the minimal polynomial of a symmetric matrix has degree equal to the number of distinct eigenvalues, the adjacency algebra may also be viewed as $C[x]/(m(x))$ where $m(x)$ is the minimal polynomial over the complex numbers C. This algebra is called the *Bose-Mesner algebra* and has dimension $d + 1$ (which is the number of distinct eigenvalues).

A second basis is defined by *distance matrices*. Let A_k, like the adjacency matrix, have rows and columns corresponding to the vertices of the graph. A given entry in A_k is 1 if the distance between the corresponding vertices is k. Obviously $A_0 = I$ and $A_1 = A$.

It is direct to observe that for any distance-regular graph of diameter d, $A_j A_k = \sum_{k=0}^{d} p_{ij}^k A_k$, and that the A_i commute. Since, for every $0 \le i, j \le d$ there is a unique k such that $(A_k)_{ij} = 1$, it is clear that $A_0, A_1, \ldots, A_d$ is a basis for an algebra of dimension $d + 1$. We have already noted that I and A are in this algebra. Now $A_{ij}^2 = p_{11}^k$ where k is the distance between the corresponding vertices, and hence $A^2 = p_{11}^2 A_2 + p_{11}^1 A_1 + p_{11}^0 A_0$ is in the span of A_0, A_1, A_2. In general, by similar reasoning, A^k is in the span of $A_0, \ldots, A_k$, and the hence the algebra generated by $A_0, \ldots, A_d$ is just the adjacency algebra and $\{A_0, \ldots, A_d\}$ is a basis. In general, if the diameter of a graph is d, then the number of distinct eigenvalues is at least $d + 1$. The two bases of the adjacency algebra show that for distance-regular graphs this inequality is actually equality (A. Neumaier has asked if there exists a characterization of graphs for which this equality holds).

The elementwise product of two matrices is called the *Hadamard product* (also called the *Schur product*). We denote the Hadamard product of A and B by $A \circ B$. It is obvious that for the distance matrices of the adjacency algebra we have $A_k \circ A_l = \delta_{kl} A_k = \delta_{kl} A_l$ (where δ_{kl} is the Kronecker delta).

There is a third basis for the adjacency algebra that is of interest. Suppose that $\lambda_1 = r > \lambda_2 > \cdots \lambda_{d+1}$ are the distinct eigenvalues of a distance-regular graph. Let $S_1, S_2, \ldots S_{d+1}$ be the corresponding eigenspaces, and let p_{i-1} be the projection mapping onto S_i, for $i = 1, \ldots, d+1$. Finally, let E_i be the matrix representing p_i (with respect to the standard basis). Notice that the subscripts of E_i range from 0 to d while those of the eigenvalues range from 1 to $d + 1$; this is unfortunate but is the standard terminology. It is then clear that $E_k E_l = \delta_{kl} E_k = \delta_{kl} E_l$. This certainly implies that each E_i is idempotent, and that they all commute. Also, since a distance-regular graph is connected (except, perhaps, for a disjoint union of complete graphs of the same order, which is sometimes excluded from the definition of distance-regular graphs), an orthonormal basis for S_1 is $\{\frac{1}{\sqrt{n}}(1, \ldots, 1)\}$. Thus it is easy to see that $E_0 = \frac{1}{n} J$.

As an example, consider the complete graph K_n. As was noted in the previous paragraph, $E_0 = \frac{1}{n} J$. The other eigenspace is $S_2 = \{(x_1, \ldots, x_n) \mid \sum_{i=1}^{n} x_i = 0\}$. If $\{e_1, \ldots, e_n\}$ is the standard basis of $\Re^n$, define $y_1, \ldots, y_{n-1}$ by

$$y_k = \frac{1}{\sqrt{k(k+1)}}(e_1 + \cdots + e_k) - \sqrt{\frac{k}{k+1}} e_{k+1}.$$

It is then clear that $\{y_1, \ldots, y_{n-1}\}$ is an orthonormal basis for S_2. The projection mapping $p_1(x) = \sum_{i=1}^{n-1} (x, y_i) y_i$ can be applied to $\{e_1, \ldots, e_n\}$ to conclude that $E_1 = I - \frac{1}{n} J$ (acutally, we shall see shortly that $E_0 + \cdots + E_d = I$).

Suppose we are considering a distance-regular graph with n vertices, and suppose that $\lambda_1 = r > \lambda_2 > \cdots > \lambda_{d+1}$ are the distinct eigenvalues, $S_1, S_2, \ldots S_{d+1}$ are the corresponding eigenspaces, p_{i-1} is the projection mapping onto S_i, for $i = 1, \ldots, d + 1$, and, let E_i is the matrix representing p_i (again, with respect to the standard basis). For any $x \in \Re^n$ we can write $x = \sum_{i=0}^{d} p_i(x)$ where $p_i(x) \in S_{i+1}$. Thus for the adjacency matrix A we have $A^k(x) = A^k(\sum_{i=0}^{d} p_i(x)) = \sum_{i=0}^{d} \lambda_{i+1}^k p_i(x) = \sum_{i=0}^{d} \lambda_{i+1}^k E_i x = (\sum_{i=0}^{d} \lambda_{i+1}^k E_i)x$. Hence $A^k = \sum_{i=0}^{d} \lambda_{i+1}^k E_i$. In particular this implies that A^k is in

the algebra generated by the E_i, and hence the Bose-Mesner algebra
is also. Since $\{E_0, \ldots, E_d\}$ is linearly independent, the algebra generated by $\{E_0, \ldots, E_d\}$ is, in fact, the Bose-Mesner algebra (alterna-
tively, one may note that if $h_i(x) = \prod_{i \neq j} \frac{x - \lambda_j}{\lambda_i - \lambda_j}$, then $E_i = h_{i+1}(A)$ for
$i = 0, \ldots, d$).

If $x \in S_{j-1}$ then $E_i(x) = \delta_{ij}$. Thus the eigenvectors of E_i are
the same as those of A, and the eigenvalues of E_i are 1 and 0. Let q_{ij}^k
be defined by $E_i \circ E_j = \sum_{k=0}^d q_{ij}^k E_k$. The numbers q_{ij}^k are called the
Krein parameters. If $x \in S_{l-1}$ then $(E_i \circ E_j)(x) = \sum_{k=1}^{d+1} q_{ij}^k E_k(x) =
q_{ij}^l(x)$, and so q_{ij}^k is and eigenvalue of $E_i \circ E_j$. Since $E_i \circ E_j$ is a
principal submatrix of the tensor product $E_i \otimes E_j$ which has 0 and 1
as eigenvalues, we see that

$$0 \leq q_{ij}^k \leq 1.$$

These inequalities are called the *Krein conditions*.

It's easy to see, for example, that $q_{0j}^k = \delta_{jk}/n$.

The matrices E_0, E_1, E_2 have been given for strongly regular graphs
by J. J. Seidel [SEI1]. We shall return to these graphs later in this
section. A. Neumaier uses the Krein condition for strongly regular
graphs to characterize graphs with least eigenvalue $\lambda(G) = -m$ (see
Section 1.2 for more details).

It is immediately striking that $A_i A_j = \sum_{k=0}^d p_{ij}^k A_k$ and $A_i \circ A_j =
\delta_{ij} A_1$ while $E_i \circ E_j = \sum_{k=0}^d q_{ij}^k E_k$ and $E_i E_j = \delta_{ij} E_i$. This apparent
duality in fact has deeper mathematical underpinnings as Ph. Delsarte
has shown in his theory of *P-polynomial* and *Q-polynomial* association
schemes (see [DEL1] of [CVDSA1]). *P*-polynomial association schemes
are those that arise from the usual distance relations of a graph.

The Bose parameters p_{ij}^k have a great deal of redundancy built into
them. A shorter representation of them is the *intersection array*. Let
$a_k = p_{1k}^k$, $b_k = p_{1,k+1}^k$, and $c_k = p_{1,k-1}^k$. The intersection array is then

$$B = \begin{pmatrix} * & c_1 & c_2 & \cdots & c_{d-1} & c_d \\ a_0 & a_1 & a_2 & \cdots & a_{d-1} & a_d \\ b_0 & b_1 & b_2 & \cdots & b_{d-1} & * \end{pmatrix}$$

Clearly $a_0 = 0$ and $a_k + b_k + c_k = r = b_0$.

The intersection array in fact contains all the spectral information
for a distance-regular graph.

THEOREM 2.1 (R. C. BOSE, D. MESNER (SEE [CVDSA1])): *Let B be the intersection array of a distance-regular graph G, and let D be the tridiagonal matrix with B as the diagonal entries. Then the eigenvalues of D are distinct and are the eigenvalues of G.*

Not only can the distinct eigenvalue be given, but so can their multiplicities as eigenvalues of the graph, if the eigenvectors of D are computed. If, for example, n_k is the number of vertices at distance k from a given vertex, then it is easy to see that $(1, n_1, n_2, \ldots, n_d)$ is an eigenvector of D with r as corresponding eigenvalue. Thus the number of vertices of G is determined.

THEOREM 2.2 (N. BIGGS (SEE [CVDSA1])): *Let D be the tridiagonal matrix derived from the intersection array of a graph G with n vertices, and let λ be an eigenvalue of D with corresponding eigenvector $(x_0 = 1, x_1, \ldots, x_d)$. Then the multiplicity m_λ of λ as an eigenvalue of G satisfies*

$$m_\lambda = \frac{n}{\sum_{i=0}^{d} x_i{}^2 / n_i}.$$

There are many relations between the parameters of an intersection array. It is easy to see, for example, that the top row is a ascending sequence of integers while the bottom row is an descending sequence, $a_i + b_i + c_i = r$, the degree of the graph, and $n_k b_k = n_{k+1} c_{k+1}$, the number of edges joining a vertices at distance k to those at distance $k + 1$ from a given vertex. Further results may be found in [GAR2].

The determination of the existence and the classification of distance-regular graphs in general by their intersection matrices is very difficult. Hence many special cases have been investigated.

T. Ito has looked at bipartite distance-regular graphs. These are ones, of course, for which $a_k = 0$ for $k = 0, \ldots, d$. Also, the degree of the graph is b_0.

THEOREM 2.3 (T. ITO [ITO1]): *Let G be a bipartite distance-regular graph of degree $r = 3$. Then G is either*
  (i) *$K_{3,3}$, the complete bipartite graph,*
 (ii) *Q_3, the 3-cube,*
(iii) *$PG(2, 2)$, the point-block incidence matrix of the projective plane,*
 (iv) *$3 \cdot K_{3,3}$, a three-fold covering,*
  (v) *the generalized 4-gon,*
 (vi) *$2 \cdot O_3$, a two fold covering of the Petersen graph,*

(vii) the generalized 6-gon, or

(viii) a three-fold covering of the generalized 4-gon.

C. Roos and A. J. van Zanten consider regular graphs of diameter d and girth $2d$ such that any two vertices that attain the diameter are on a unique circuit of length $2d$. It is immediate that such a graph with degree r is distance-regular and has an intersection array of the form

$$B = \begin{pmatrix} * & 1 & 1 & \ldots & 1 & 2 \\ 0 & 0 & 0 & \ldots & 0 & r-2 \\ r & r-1 & r-1 & \ldots & r-1 & * \end{pmatrix}$$

We shall say that a graph has an intersection matrix $L_d(t)$ if it is regular of degree $t+1$, has diameter d and girth $2d$ and any two vertices at maximum distance lie in a unique $2d$-cycle.

THEOREM 2.4 (C. ROOS, A. J. VAN ZANTEN [RoZa1]): *A graph with intersection matrix $L_d(t)$, $t > 1$, exists only if $m \leq 3$. In addition, $L_3(t)$ exists only if $t = \frac{1}{2}a(a+1)$ with $a \in \{1, 2, 3, 5, 10, 17, 27, 38, 82, 115, 577\}$.*

E. Bannai and T. Ito have also studied distance-regular graphs with large girth.

THEOREM 2.5 (E. BANNAI, T. ITO [BaIt1], [BaIt2]): *Let G be a distance-regular graph with intersection array*

$$B = \begin{pmatrix} * & 1 & 1 & \ldots & 1 & c \\ 0 & 0 & 0 & \ldots & 0 & r-c \\ r & r-1 & r-1 & \ldots & r-1 & * \end{pmatrix}, \quad r > 2,$$

and let $m_0, \ldots, m_d$ be the respective multiplicities of $\lambda^0 < \lambda^1 < \cdots < \lambda^d = r$. Then $m_0 < m_{d-1} \leq m_1 < m_{d-2} \leq m_2 < m_{d-3} \leq \cdots m_{\lfloor d/2 \rfloor}$. In addition, equality holds if and only if either $r - c = 0$ or $(d, r, c) = (3, 4, 2)$.

This theorem has several interesting consequences:

COROLLARY 2.5.1 (E. BANNAI, T. ITO [BaIt1], [BaIt2]): *If G is as in Theorem 2.5, then the multiplicities form a unimodal sequence.*

COROLLARY 2.5.2 (E. BANNAI, T. ITO [BaIt1], [BaIt2]): *If G is as in Theorem 2.5 and $r - c \neq 0$, then the eigenvalues are integral.*

This corollary is a consequence of the fact that if $(d, r, c) = (3, 4, 2)$ then $G = O_4$, the odd graph.

D. E. Taylor and R. Levingston characterize distance-regular graphs for which $n_1 = n_2$ and $d > 3$.

THEOREM 2.6 (D. E. TAYLOR, R. LEVINGSTON [TaLe1]): *If G is distance-regular with diameter $d > 3$ and $n_1 = n_2$, then G is a circuit.*

A graph is called *pseudocyclic* if the multiplicities of all the subdominant eigenvalues are equal.

COROLLARY 2.6.1 (M. DOOB [Doo5]): *A pseudocyclic distance-regular graph is either a strongly regular conference graph or a circuit.*

P. Terwilliger has given bounds on the multiplicities of distance-regular graphs.

THEOREM 2.7 (P. TERWILLIGER [Ter1]): *Let G be a distance-regular graph of degree r and girth $g \geq 4$. Then the multiplicity m_λ of the eigenvalue $\lambda \neq \pm r$ satisfies*

$$m_\lambda \geq \begin{cases} r(r-1)^{[g/4]-1} & \text{if } g \equiv 0, 1 \bmod 4 \\ 2(r-1)^{[g/4]} & \text{if } g \equiv 2, 3 \bmod 4. \end{cases}$$

A distance-regular graph of diameter $d = 2$ is just a strongly regular graph. These have been defined in Section 1.2. In that case the distance matrix $A_2 = J - A - I$ is the adjacency matrix of the complement. Further, the eigenspaces of the complement and also of the idempotent matrices E_k are equal. If the eigenvalues of the graph are r, λ_2, λ_3, then the eigenvalues of the complement are $n - r - 1, -\lambda_3 - 1, -\lambda_2 - 1$. Thus we have

$$A_0 = I = E_0 + E_1 + E_2$$
$$A_1 = A = rE_0 + \lambda_2 E_1 + \lambda_3 E_3$$
$$A_2 = J - A - I = (n - r - 1)E_0 + (-\lambda_2 - 1)E_1 + (-\lambda_3 - 1)E_2.$$

Since a strongly regular graph is regular, we know that $E_0 = \frac{1}{n}J$ (this is also implied by the equations of the previous paragraph). Hence we get

$$E_0 = \frac{1}{n}J$$

$$E_1 = \frac{1}{\lambda_2 - \lambda_3}(\frac{r - \lambda_3}{n}J + A - \lambda_3 I)$$

$$E_2 = \frac{1}{\lambda_2 - \lambda_3}(\frac{r - \lambda_2}{n}J - A + \lambda_2 I)$$

and since we obviously have $A_0 + A_1 + A_2 = J$, $A \circ A = A$, $J \circ X = X$, $I \circ I = I$, and $A \circ I = 0$ we can (at least in principle) compute the q_{ij}^k in terms of the eigenvalues of A. Further computations of this type can be found in [SEI2].

A. Neumaier uses the *absolute bound* that comes from the Krein condition in [NEU1].

THEOREM 2.8 (A. NEUMAIER [NEU1]):   *If G is a strongly regular graph with parameters (n, r, λ, μ) and least eigenvalue $\lambda(G) = -m$, then either*

(i) G is a complete multipartite graph, or
(ii) $\mu \leq m^3(2m - 3)$.

Theorem 1.11 characterizes strongly regular graphs with least eigenvalue $-m$.

Section 2.2:   Moore Graphs and their Generalizations

An early and interesting application of spectral graph theory was to *Moore graphs*. A regular graph of degree r and diameter d can have at most $1 + r + r(r - 1) + \cdots + r(r - 1)^{d-1}$ vertices. A Moore graph is one for which the number of vertices attains the bound. It is known that Moore graphs consist of circuits, the Petersen graph, the Hoffman-Singleton graph, and possibly a graph with diameter $d = 2$ and degree $r = 57$ (see [CVDSA1] for details). The girth of a Moore graph is clearly $2r + 1$ There have been a number of interesting results concerning graphs which are "almost" Moore graphs. If G is a graph with n vertices that has degree r and diameter d, the *defect* of G is $1 + r + r(r-1) + \cdots + r(r-1)^d - n$. P. Erdös, S. Fajtowicz, A. J. Hoffman found all graphs with defect 1 and diameter 2, while E. Bannai and T. Ito have generalized this result to graphs of larger diameter.

THEOREM 2.9 (P. ERDÖS, S. FAJTOWICZ, A. J. HOFFMAN [ErFH1]): *If G has maximum degree r, r^2 vertices, and diameter 2, then $G = C_4$.*

THEOREM 2.10 (E. BANNAI, T. ITO [BaIt3]): *If G is a regular graph with defect equal to 1, then $G = C_4$.*

The concept of a Moore graph can also be generalized by fixing the degree and the girth. For a graph with n vertices, degree r, and girth $2r+1$, define the *excess* of G to be $n - (1 + r + r(r-1) + \cdots + r(r-1)^d)$.

THEOREM 2.11 (E. BANNAI, T. ITO [BaIt3]): *There exist no regular graphs of degree r, girth $2r + 1 \geq 5$, and excess equal to 1.*

P. Kovács has looked at the excess of graphs with girth $g = 5$. It is known that there are no regular graphs with girth $g = 5$ and excess $e = 1$.

THEOREM 2.12 (P. KOVÁCS [Kov1]): *Let r be an odd integer, $r \geq 3$, $r \neq l^2 + l + 3$, and $r \neq l^2 + l - 1$. Then there exists no regular graph of degree r, girth $g = 5$, and excess $e = 2$.*

A lower bound for the excess has been given by N. Biggs. For a given k let the polynomials $F_r(x)$ be defined recursively by $F_0(x) = 1$, $F_1(x) = x + 1$, and $F_r(x) = xF_{r-1}(x) - (k-1)F_{r-2}(x)$ for $r \geq 2$.

THEOREM 2.13 (N. BIGGS [Big3]): *Suppose G is a graph that is regular of degree k, has girth $g = 2r + 1$ and excess e. Further, assume that $\lambda \neq k$ is an eigenvalue of G. Then*

$$e \geq |F_r(\lambda)|.$$

PROOF: The recursive definition of $F_r(x)$ implies that $\sum_{i=1}^{r} A_i = F_r(A)$. If we let $E = \sum_{i=r+1}^{d} A_i$, we then have $E + F_r(A) = J$. If x is an eigenvector of A corresponding to an eigenvalue $\lambda \neq k$, then $E(\lambda) + F_r(\lambda) = 0$. In addition, $E(j) + F_r(j) = n$, and $F_r(j) = 1 + k + k(k-1) + \cdots k(k-1)^{d-1}$, so $E(j)$ has constant row sum equal to e, the excess of the graph. Hence $|E(\lambda)| = |F_r(\lambda)| \leq e$.

N. Biggs has used Theorem 2.13 in particular cases to show that excess is unbounded as either the degree or the girth gets large.

The concept of excess can be applied to graphs of even girth, too. In this case, a graph of degree r, diameter d, and girth $2d$ must have at least $n_0 = 2(1 + (r - 1) + (r - 1)^2 + \cdots + (r - 1)^{d-1})$ vertices. N. Biggs and T. Ito define the excess for a graph of even girth $g = 2d$ and degree r as $n - n_0$.

THEOREM 2.14 (N. L. BIGGS, T. ITO [BiIt1]):   *Let G be a regular graph with degree r, girth $2d - 2 \geq 4$, and excess $e \leq r - 1$. Then G is bipartite with diameter d.*

THEOREM 2.15 (N. L. BIGGS, T. ITO [BiIt1]):   *There is no graph with even girth $g \geq 8$ and excess $e = 2$.*

THEOREM 2.16 (N. L. BIGGS, T. ITO [BiIt1]):   *There is no regular graph of degree r with girth $g = 6$, excess $e = 2$ if $r \equiv 5 \bmod 8$ or $r \equiv 7 \bmod 8$.*

There have also been extensions of Moore graphs to the directed case. J. Bosák uses both directed and undirected edges in what he calls a *partially directed graph*. Such a graph is *homogeneous* if, for any vertex v, the number of directed edges going into v equals the number of the directed edges going out of v. A partially directed graph with r edges incident to each vertex, diameter d, and with every pair of vertices joined by a unique directed path with length less than d is called a (r, d) Moore Graph.

THEOREM 2.17 (J. BOSÁK [Bos1]):   *Suppose there exists a partially directed Moore graphs of diameter 2 with n vertices. Then*
  (i) $n = 3$, $n = 5$, or
  (ii) $n = r(r + 1) - s(s + 1) \geq 6$ where r and s are the two largest eigenvalues of G. In this case r is the degree and $s \geq 0$ is an integer.

THEOREM 2.18 (J. BOSÁK [Bos1]):   *A partially directed Moore graph of diameter 2 with n vertices and degree r exists if $n = r(r + 1)$ or $n \in \{3, 5, 10, 18, 50\}$.*

C. Roos and A. J. van Zanten consider *generalized Moore geometries* $GM_m(s, t, c)$. These have $s + 1$ points per line, $t + 1$ lines per point, a (point) graph of diameter m such that two points at distance m are joined by exactly c paths of length m. The graph of the geometry is distance-regular with an intersection array of the form

$$
L_m(s,t) = \begin{pmatrix} * & 1 & 1 & \dots & 1 & s+1 \\ 0 & s-1 & s-1 & \dots & s-1 & st-1 \\ s(t+1) & st & st & \dots & st & * \end{pmatrix}
$$

THEOREM 2.19 (C. ROOS, A. J. VAN ZANTEN [ROZA3]): *A distance-regular graph with intersection array $L_m(s,t)$ does not exist for any odd $m > 7$.*

THEOREM 2.20 (R. M. DAMERELL, M. A. GEORGIACODIS [DAGO1], [DAGO2]): *A distance-regular graph with intersection array $L_m(1,t)$ does not exist if $m > 5$.*

An example of a partially directed Moore graph given in [BOS1] is $B(r)$ formed in the following way: the vertices are the ordered pairs of integers $\{(i,j) \mid 0 \le i \ne j \le r\}$. An undirected edge joins (i,j) and (j,i), while there are directed edges from (i,j) to (k,i) where $j \ne k$. It is then easy to check that $B(r)$ is a partially directed Moore graph of diameter $d = 2$.

W. G. Bridges and R. A. Mena have considered cyclic strongly regular graphs. If $p \equiv 1 \bmod 4$ is a prime, then the *Paley graph* has vertices $\{0, \dots, p-1\}$ with an edge joining i and j if and only if $i - j$ is a quadratic residue modulo p.

THEOREM 2.21 (W. G. BRIDGES, R. A. MENA [BRME1]): *The only cyclic strongly regular graphs are Paley graphs.*

R. P. King [KIN2] has considered graphs where the vector $(n_0, \dots, n_d)$ is equal to the vector of eigenvalue multiplicities $(m_0, \dots, m_d)$. He shows that the only such graphs with diameter $d = 2$ are conference graphs.

Section 2.3: Distance-Transitive Graphs

Let $d(u,v)$ be the distance between vertices u and v. A graph G is *distance-transitive* if, for any vertices u, v, w, x such that $d(u,v) = d(w,x)$ there is an automorphism ϕ of G with $\phi(u) = w$ and $\phi(v) = x$. Now suppose for such a graph we have $d(u,v) = k$ and $\{y_1,\ldots,y_t\}$ are the vertices such that $d(u,y_l) = i$ and $d(y_l,v) = j$ for $l = 1\ldots t$. Then if w and x satisfy $d(w,x) = k$ and ϕ is the automorphism G given by the distance transitivity, the vertices $\{\phi(y_1),\ldots,\phi(y_t)\}$ satisfy $d(w = \phi(u), \phi(y_l)) = i$ and $d(\phi(y_l), x = \phi(v)) = j$ and hence $p_{ij}^k(u,v) = p_{ij}^k(w,x)$. Thus distance-transitive graphs are distance-regular.

The study of distance-transitive and similar graphs is a large and interesting topic in its own right. Here we will restrict our attention to problems that involve their spectra.

The concept of excess given in Section 2.2 can be applied to distance-transitive graphs. N. L. Biggs has shown that among vertex transitive graphs in particular the excess can get very large.

THEOREM 2.22 (N. L. BIGGS [BIG4]): *For any odd integer $r \geq 3$, there exists a sequence $g_1, g_2, \ldots$ of girths such that any vertex transitive graph G_i of girth g_i and degree r has excess $e(G_i) \geq g_i/r$.*

Another variation on the distance-transitive graph is the *distance-transitive digraph*. The definition for digraphs is identical to that of undirected graphs. A directed cycle is distance-transitive; another example of such a digraph is the *Paley digraph*. For any prime $p \equiv 3 \mod 4$, let $\{0,\ldots,p-1\}$ be the vertices with an arc from i to j if $i - j$ is a quadratic residue modulo p. In both cases the diameter d and girth g satisfy $g = d + 1$. Intersection arrays can be defined and have an algebraic structure that is essentially the same as for distance-regular graphs. A general construction is given by C. W. H. Lam.

THEOREM 2.23 (C. W. H. LAM [LAM2]): *Let G be a distance-transitive digraph with diameter d and girth $g = d+1$. For any integer $t \geq 2$ let the vertices of G' be $V(G') = \{(v,i) \mid 0 \leq i \leq t-1\}$ with an arc from (u,i) to (v,j) if and only if (u,v) is an arc in G. Then G' is a distance-transitive digraph with diameter and girth both equal to $d+1$.*

PROOF: The distance from (u,i) to $(v,j) \neq (u,i)$ is the distance from u to v in G if $u \neq v$ and is $d+1$ if $u = v$.

THEOREM 2.24 (C. W. H. LAM [LAM2]): *If G is a distance-transitive digraph with n vertices, degree r and diameter 2, then its adjacency matrix A satisfies*

$$AA^T = (r - \lambda)I + \lambda J$$

where $(n, k, \lambda) = (4t - 1, 2t - 1, t - 1)$ for some integer t.

Thus these digraphs coexist with Hadamard designs.

R. M. Damerell has shown that the girth and diameter of a distance-transitive digraph are directly related. More generally, strongly distance-regular digraphs are ones that satisfy the same conditions on p_{ij}^k as do distance-regular graphs.

THEOREM 2.25 (R. M. DAMERELL [DAM1]): *If G is a strongly distance-regular digraph with diameter d and girth $g \geq 3$, then either*
 (i) $g = d$ and G is called a short digraph, or
(ii) $g = d + 1$ and G is called a long digraph.

The Lam construction of a distance-transitive graph G' from G given in Theorem 2.21 takes a long digraph and produces a short digraph.

THEOREM 2.26 (R. M. DAMERELL [DAM1]): *Every long digraph comes a short digraph by the Lam construction.*

Thus the study of distance-transitive digraph may be restricted to those that are short. Those of odd girth are known.

THEOREM 2.27 (E. BANNAI, P. J. CAMERON, J. KAHN [BaCK1]): *A short distance-transitive digraph of odd girth is a directed cycle or a Paley digraph.*

The conclusion of Theorem 2.27 is also shown to follow under various group theoretic assumptions for the automorphism group.

THEOREM 2.28 (E. BANNAI, P. J. CAMERON, J. KAHN [BaCK1]): *A short distance-transitive digraph with solvable automorphism group girth is a directed cycle or a Paley digraph.*

If Γ is a group and C is a subset of Γ, then the *Cayley digraph* is defined by letting the elements of Γ be the vertices and (g, h) an arc if and only if $gh^{-1} \in C$. C. Godsil [GOD6] has shown that Cayley digraphs have important spectral properties. Godsil gives a construction for embedding a digraph G into a regular digraph H that maintains much of the eigenvalue structure. Let $v_1, \ldots, v_n$ be the vertices of G and Δ be the degree desired in the new graph (obviously at least and usually equal to the maximum degree of G). For each vertex v_i, let d_i be the difference between Δ and the outdegree of v_i, i.e., the number of new arcs going out of v_i that need to be added to make the degree equal to Δ. Take Δ copies of G and add $\sum_{i=1}^{n} d_i$ new isolated vertices. Add an arc from v_1 (in each copy of G) to the first d_1 isolated vertices, an arc from v_2 to the next d_2 isolated vertices and continue until the outdegree of all of the vertices v_i in each copy of G is Δ. Now do the same thing for the indegree and add arcs from the $\sum_{i=1}^{n} d_i$ vertices to the copies of G. It is clear that $\sum_{i=1}^{n} d_i$ is the number of vertices necessary to make all the indegrees equal to Δ. The resulting graph, H, is then regular and contains Δ disjoint copies of G as an induced subgraph. Note that if the digraph G is regular and Δ is the degree of G, then new digraph H and the old digraph G are identical.

THEOREM 2.29 (C. GODSIL [GOD6]): *If G is a digraph with minimum degree greater than 1, then there exists a vertex transitive Cayley digraph whose minimal polynomial divides that of G. Further, if G is a graph, then so is the Cayley digraph.*

The eigenvalues of a graph are algebraic integers.

COROLLARY 2.29.1 (C. GODSIL [GOD6]): *Every algebraic integer is an eigenvalue of a Cayley digraph.*

A. J. Hoffman has asked if it is possible to characterize the algebraic integers that are eigenvalues of graphs.

As we saw in Theorem 2.8, bounds can be given on the multiplicities of the eigenvalues for distance-regular graphs. P. Terwilliger has given bounds for distance-transitive graphs.

THEOREM 2.30 (P. TERWILLIGER [TER1]): *Let G be a regular graph of degree r and girth g that acts transitively on s-arcs for $s = 1, \ldots, [\frac{1}{2}g]$. Then the multiplicity m_λ of the eigenvalue $\lambda \neq \pm r$ satisfies*

$$m_\lambda \geq \begin{cases} r(r-1)^{\frac{s}{2}-1} & \text{if } s \text{ is even} \\ 2(r-1)^{\frac{s-1}{2}} & \text{if } s \text{ is odd} \end{cases}$$

Section 2.4:   Distance-Regular Graphs and other Combinatorial Objects

Distance-regular graphs have a highly symmetric structure; as such it is not surprising that they are highly related to other combinatorial structures. The Hamming graph $H(n,q)$ and the Johnson graph $J(n,q)$ are obvious examples (see Theorem 1.20 and Theorem 1.21). Another example is the *odd graph* $O(k)$. This has as vertices the the $\binom{2k-1}{k-1}$ subsets of size $k-1$ of a set of size $2k-1$. Two vertices are adjacent if, as subsets, they are disjoint. Clearly the distance between two vertices is determined by the size of the intersection of the corresponding sets. In fact, the distance between vertices u and v is $2r$ if the corresponding sets have intersection of size $k-1+r$ and is $2r+1$ if the intersection has size r. N. Biggs [BIG2] has described $O(k)$ in some detail. In [BIG2] the intersection matrix is given:

$$
B = \begin{pmatrix}
* & 1 & 1 & 2 & 2\ldots & \frac{1}{2}k-1 & \frac{1}{2}k-1 & \frac{1}{2}k \\
0 & 0 & 0 & 0 & 0\ldots & 0 & 0 & \frac{1}{2}k \\
k & k-1 & k-1 & k-2 & k-2\ldots & \frac{1}{2}k+1 & \frac{1}{2}k+1 & *
\end{pmatrix}
$$

when k is even and

$$
B = \begin{pmatrix}
* & 1 & 1 & 2 & 2\ldots & \frac{1}{2}(k-1) & \frac{1}{2}(k-1) \\
0 & 0 & 0 & 0 & 0\ldots & 0 & \frac{1}{2}(k+1) \\
k & k-1 & k-1 & k-2 & k-2\ldots & \frac{1}{2}(k+1) & *
\end{pmatrix}
$$

when k is odd.

The eigenvalues of $O(k)$ are $\lambda_i = (-1)^i(k-i)$ with multiplicity $m_i = \binom{2k-1}{i} - \binom{2k-1}{i-1}$.

THEOREM 2.31 (N. BIGGS [BIG2]):   *The automorphism group of $O(k)$ is the symmetric group S_{2k-1} acting in the obvious way on the $(k-1)$ subsets.*

COROLLARY 2.31.1 (N. BIGGS [BIG2]): *The graph $O(k)$ is distance-transitive.*

It is unknown whether or not the graph $O(k)$ is characterized by its spectrum. However A. Moon [MOO1] has shown that it is characterized by its spectrum among all distance-regular graphs.

An *partial geometry* with parameters (R, K, T) is a system of points and line such that
 (i) two points determine at most one line,
(ii) each line contains K points,
(iii) each point is contained in R lines, and
(iv) if a point is not on a line, then there exist T other lines intersecting the that line and containing the point.

A strongly regular graph may be formed from a partial geometry by letting the vertices be the points and joining two vertices if the corresponding points are collinear. A strongly regular graph with parameters consistent with that of a partial geometry is called a *pseudo-geometric graph*. C. Bumiller presents a partial converse.

THEOREM 2.32 (C. BUMILLER [BUM1]): *Let G be a strongly regular graph with parameters (n, r, λ, μ), a rank three permutation group, and $r > s > -m$ as eigenvalues. Further, assume that $\mu \geq 2$, $m \geq 3$, and*

$$s \geq \frac{(4m - 5)(2\mu + 1)}{3} + 3 + \frac{3}{r - 2}.$$

Then G arises from a partial geometry with parameters $(m, \frac{\mu}{m}, \frac{\mu}{m})$.

A. Neumaier extends Theorem 2.30 in [NEU2].

THEOREM 2.33 (A. NEUMAIER [NEU2]): *Suppose that G is an edge transitive strongly regular graph with parameters (n, r, λ, μ) and eigenvalues r, $n - m$, and $-m$, and let s be the smallest integer such that $4m \leq (s + 1)^2$. Further suppose that*

$$\mu \geq 2 + \frac{1}{4}(2m - 1)s,$$

$$n \geq (2m - 1)\mu.$$

Then G is a Latin Square graph or a Steiner graph.

M. S. Shrikhande uses eigenvalue arguments to show that certain graphs are pseudo-geometric. In [SHRI1] a strongly regular graph is defined on a group divisible design.

THEOREM 2.34 (M. S. SHRIKHANDE [SHRI1]): *A graph from a semiregular group divisible design is a strongly regular graph if and only if it is a pseudo-geometric graph with parameters $(q^2 + 1, q + 1, 1)$ or $(2, n + 1, 1)$.*

W. G. Bridges and M. S. Shrikhande [BRSH1] use partially balanced incomplete block designs to construct graphs with six or seven distinct eigenvalues.

H. D. Patterson and E. R. Williams [PAWI1] measure the efficiency of a block design by taking the harmonic mean of the nonzero eigenvalues of $I - \frac{1}{rk} NN^T$ where N is the block incidence matrix of the design with parameters (v, b, r, k).

H. Beker and W. Haemers [BEHA1] have investigated balanced incomplete block designs with parameters (v, b, r, k, λ) that have $k - n$ for as intersection number for some of the blocks ($n = r - \lambda$ is the *design order*). If $k - n = \lambda$, then the design is symmetric; if $k - n = 0$ then the design is quasi-residual, *i.e.*, it has exactly two intersection numbers. For a graph with three intersection numbers $k - n$, ρ_1, and ρ_2, two blocks may be called equivalent if their intersection number is $k - n$. This is an equivalence relation, and the equivalence classes may be used as vertices of a graph; two vertices are adjacent if the blocks are in the different classes have an intersection number of ρ_1. This class graph is strongly regular since it has three distinct eigenvalues. The properties of this graph limit the types of designs that can exist.

D. M. Thompson [THO1] also relates strongly regular graphs to block designs using the spectral properties of the graph. A graph is *design constructible* if the vertices can be partitioned into sets V and B such that V is independent and each vertex of B is adjacent to the same number of vertices of V. This gives a subgraph that is the incidence graph of a block design. The eigenvalues of the graphs limit the existence of design constructible graphs.

A subgraph of a regular graph is called an *eigengraph* in [THO1] if it is regular, say of degree r_1, and every vertex not in the subgraph is adjacent to the same number of vertices in the subgraph, say r_2. Thus design constructible graphs are just the strongly regular ones with independent sets as eigengraphs. A graph G has an eigengraph only if there is an eigenvalue of a special form.

A. Cohen [COH1] uses the maximal cliques in a distance-regular graph to construct regular *near 2d-gons*. The nonexistence of certain regular $2d$-gons is used to prove the nonexistence of certain distance-regular graphs.

K. S. Vijayan [VIJ1] has given a characterization of certain distance-transitive graphs. Define the *projective graph* $G = \mathcal{G}_{d,q}$ by letting the vertices correspond to the points and hyperplanes in the projective space $PG(d, q)$ with two vertices adjacent if they correspond to a hyperplane and a point contained in that hyperplane. The graph $\mathcal{G}_{11}$ has 22 vertices which correspond to the 11 elements in finite field $GF(11)$ and the 11 blocks of the form $B_i = \{1+i, 3+i, 4+i, 5+i, 9+i\}$, for $i = 1, \ldots, 11$. Two vertices are adjacent if one is an element of the field and the other is a block containing that element. Both $G = \mathcal{G}_{d,q}$ and $G = \mathcal{G}_{11}$ are distance-transitive bipartite graphs of diameter 3.

THEOREM 2.35 (K. S. VIJAYAN [VIJ1]):   *Let G be a distance-transitive bipartite graph of diameter 3 with degree r where r or $r - p_{11}^2$ is a prime. Then either $G = \mathcal{G}_{d,q}$ or $G = \mathcal{G}_{11}$.*

A *generalized hexagon* with parameters (s, t) can be viewed as a genrealization of a partial geometry with parameters $(s + 1, t + 1, t)$. It has the additional property that any pair of points of distance 3 are joined by exactly $t + 1$ paths of length 3. W. Haemers and C. Roos use the Krein condition to prove the following theorem:

THEOREM 2.36 (W. HAEMERS, C. ROOS [HARO1]):   *Let H be a generalized hexagon with parameters (s, t). Then $s = 1$ or $t \leq s^3$.*

The technical report by J. J. Seidel, A. Blokhuis, and H. A. Wilbrink [SEBW1] relates distance-regular graphs to several types of combinatorial designs. That reference as well of that of R. Mathon [MAT1] are more general studies of association schemes, and in some cases graphs arise from these schemes (these schemes are called P-polynomial by Ph. Delsarte).

C. D. Godsil and B. D. McKay generalize distance-regular graphs to *walk-regular* graphs. This is a graph where the number of closed walks of length r starting (and ending) at v is the same for any choice of v. This means, of course, that all powers of the adjacency matrix A have a constant diagonal. For a given vertex v the *vertex deleted subgraph* is obtained by deleting v and all edges incident to it.

THEOREM 2.37 (C. D. GODSIL, B. D. MCKAY [GOMK4]): *A graph G is walk-regular if and only if all of the vertex deleted subgraphs G_v are cospectral.*

The crucial concept used is that of an *equitable partition*, also called a *divisor*.

Ph. Delsarte [DEL1] has generalized the concept of an association scheme to K-schemes; many of the main idea of the Bose-Mesner algebra carry over to the more general context. A regular K-scheme generates a commutative algebra which have eigenmatrices and irreducible orthogonal idempotent Hermitian matrices.

For a detailed study of association schemes in general, the reader is referred to [BAIT4].

Section 2.5:   Root Systems and Distance-Regular Graphs

Root systems were defined in Section 1.1; in that section they were shown to be of importance in the spectral characterization of generalized line graphs. Here we shall apply the root systems to other problems that involve graph spectra.

The Johnson schemes $J(n, q)$ was defined in Section 1.2. It was known many years ago that there are three graphs that are cospectral with $J(8, 2)$, and that $J(n, 2)$ is characterized by its spectrum otherwise. The case of $J(n, 3)$ is somewhat more difficult. The graph in the case is called a *tetrahedral graph*.

THEOREM 2.38 (R. A. LIEBLER [LIEB1]): *A tetrahedral graph $J(n, 3)$ is characterized by its spectrum if $n \neq 9, 10$.*

The work of A. Moon has the following special case:

THEOREM 2.39 (A. MOON [MOO2]): *A tetrahedral graph $J(3, 9)$ and $J(3, 10)$ is characterized by its spectrum.*

P. Terwilliger [TER2] uses a very different and interesting approach to characterize $J(n, q)$. Mappings ϕ from the vertex set of the graph to $\Re^n$ are defined such that for two adjacent vertices u and v, the vector $\phi(u) - \phi(v)$ is a member of the root system A_n using the representation given in Section 1.1. Conversely, in order to have such mappings, necessary and sufficient restrictions exist on the adjacency structure. This gives the important result which was also mentioned in Section 1.1.

THEOREM 2.40 (P. TERWILLIGER [TER2]):   *The Johnson graph $J(n,q)$ is characterized among all distance-regular graphs by its parameters unless $(n,q) = (2,8)$.*

A *signed graph* is a graph in which each edge is labelled by one of two values: $+$ or $-$.

T. Zaslavsky relates root systems to signed graphs in his interesting expository article [ZAS2]. The hyperplanes orthogonal to the vectors of the root system divides $\Re^n$ into regions that can be counted using signed graphs.

A. J. Hoffman has extended several ideas used in root systems by considering symmetric matrices A with 0 on the diagonal and 0 or ± 1 in other positions. If λ^1 is the smallest eigenvalue of A, one may ask how closely $A - \lambda^1 I$ may be approximated by the product KK^T where K is a matrix with entries of 0 or ± 1. If $\lambda^1 = -2$, by Theroem 1.2 we have $A - \lambda^1 I = KK^T$ with only a finite number of exceptions.

THEOREM 2.41 (A. J. HOFFMAN [HOF1]):   *There exists a function $g(x)$ such that for any given adjacency matrix A with least eigenvalue λ, there is a matrix K with entries equal to 0 or 1 such that*

$$|A - \lambda I - KK^T| \leq g(\lambda).$$

Further, the number of nonzero entries in each row of K is $-\lambda$.

A. Nuemaier [NEU3] uses root systems to construct n-dimensional designs. These designs give rise to Delsarte matrices which are generalizations of the distance matrices of a distance-regular graph.

It is possible to use combinatorial methods that essentially reflect the graph-theoretic structure of generalized line graphs to prove the classification theorem of real root systems. Results of this type can be found in [CVDO1].

Chapter 3
Miscellaneous Results from the Theory of Graph Spectra

This chapter contains several sections each of which describes a topic for which there has been some notable progress, but not enough to justify a full chapter. It has been known for some time that there are graph theoretic interpretations of the notion of the determinant of a matrix. One of these interpretations implies a theorem expressing the coefficients of the characteristic polynomial of a graph in terms of the graph structure. This theorem, known as the Sachs Theorem, and related results are given in Section 3.1. There are many results that describe the spectra of graphs obtained by various transformations or operations from other graphs. Such results, although quite technical in the majority of cases, are often very useful, and they are described in Section 3.2. Constructions of graphs using the theorems of Section 3.2 are given in Section 3.3. The relations between the automorphism group and the spectrum of a graph and, in particular, bounds for the multiplicities of the eigenvalues are given in Section 3.4. It has been proved that the testing for isomorphism of graphs with eigenvalues with bounded multiplicities can be done in polynomial time. The spectra of graphs play an important role when considering Ulam's reconstruction conjecture. These important problems together with a survey of existing tables of graphs are given in Section 3.5. The problem of Shannon about the capacity of the pentagon has been solved by giving a bound for the capacity of a graph. The bound is related to graph eigenvalues and is given in Section 3.6. A more recent area of study is that of the spectra of random graphs although some related results about random matrices have been known for some time. These are given in

Section 3.7. The number of walks and the number of spanning trees
in a graph can often be determined using eigenvalues of graphs, and
recent results on these two topics are given in Section 3.8 and Section
3.9, respectively. Graph equations are surveyed in Section 3.10, spectra
of tournaments in Section 3.11 and other results are given in Section
3.12.

Section 3.1: The Sachs Theorem

We start with a well known graph theoretic interpretation of the
determinant concept. An easy consequence of this interpretation (which
can also be considered as the definition of the determinant) is the important Sachs theorem giving connections between the characteristic
polynomial of a graph and its circuit structure. Several recent results
exploiting this type of interaction of graph theoretic and matrix theoretic notions are reviewed in this section. These include results for
sigraphs and for skew symmetric adjacency matrices of digraphs.

To a square matrix $A = (a_{ij})$ of order n there corresponds a
weighted digraph G_A. The vertex set is $\{1, 2, \ldots, n\}$ and G_A has all
possible arcs and loops. The *weight* of the arc going from vertex i to
vertex j is a_{ij}. If $a_{ij} = 0$, then the corresponding arc may be omitted.
Conversely, the matrix A can be thought of as the adjacency matrix of
the weighted digraph G_A.

A subgraph of G_A in which the indegree and outdegree of every
vertex is equal to 1 is called a *linear subgraph* of G_A. Obviously a linear
subgraph is a vertex disjoint union of cycles. Let $\mathcal{L}$ be the set of linear
spanning subgraphs of G_A. Then determinant of A may be defined by

$$\det A = (-1)^n \sum_{L \in \mathcal{L}} (-1)^{p(L)} W(L) \qquad (3.1)$$

where $p(L)$ is the number of cycles of L and $W(L)$ is the weight of the
linear subgraph, *i.e.*, the product of the weights of the arcs contained
in that subgraph.

The monograph [CVE3] develops the theory of matrices using
graph theoretic tools. One of the basic features of that book is that the
determinant concept is introduced by the formula (3.1). An outline of

the elementary theory of determinants developed along these lines has been given previously in [CVE15] from [CVDSA1] and [DOO4].

Let $\mathcal{L}_i$ be the set of linear subgraphs of G_A having exactly i vertices.

THEOREM 3.1: *Let $A = (a_{ij})$ be a square matrix and let $P_A(\lambda) = \det(\lambda I - A) = \lambda^n + a_1 \lambda^{n-1} + \ldots + a_n$ be its characteristic polynomial. Then*

$$a_i = \sum_{L \in \mathcal{L}_i} (-1)^{p(L)} W(L), \quad i = 1, 2, \ldots, n.$$

This theorem can be proved easily using (3.1) and the fact that a_i can be expressed (up to sign) as the sum of the principal subdeterminants of A of order i (in [CVE3] even this representation of a_i, which is an elementary result in matrix theory, is derived using graph theoretic methods).

Theorem 3.1 was proven independently and almost simultaneously in 1964 by M. Milić, H. Sachs, and L. Spialter (an electrical engineer, a mathematician, and a chemist respectively). See [CVDSA1], p. 36 for further historical details. The application of Theorem 3.1 to undirected graphs has been called the Sachs theorem in the chemical literature for the last dozen years. This is because Sachs described the extensions of this theorem to the study of the relationships between the cycle structure and characteristic polynomial of a graph (see [SAC3] from [CVDSA1]). We shall follow the example of the chemists and call Theorem 3.1 the Sachs Theorem.

Two important special cases of the Sachs Theorem are the following:

THEOREM 3.2: *Let $\lambda^n + a_1 \lambda^{n-1} + \cdots + a_n$ be the characteristic polynomial of a (multi) digraph. Then*

$$a_i = \sum_{L \in \mathcal{L}_i} (-1)^{p(L)}, \quad i = 1, 2, \ldots, n.$$

A figure is called *elementary* if it is the graph K_2 or a circuit C_n, $n \geq 3$. A basic figure U is a graph where each component is an elementary figure. Let $p(U)$ and $c(U)$ be the number of components and circuits respectively of U, and let $\mathcal{U}_i$ be the set of all basic figures of the graph G with exactly i vertices.

THEOREM 3.3:   Let $\lambda^n + a_1\lambda^{n-1} + \cdots + a_n$ be the characteristic polynomial of a (multi) graph. Then

$$a_i = \sum_{U \in \mathcal{U}_i} (-1)^{p(U)} 2^{c(U)} \quad (i = 1, 2, \ldots, n).$$

The relationship between graphs and matrices expressed by the Sachs theorem has been attractive to many writers. Evidence of this is found in the history of the Sachs Theorem given in [CVDSA1], p. 36. In this section a further elaboration of technical details concerning this theorem is described following the methods of recent papers. With only the exception of [CVE3], the Sachs Theorem is not included in general books on the theory of matrices. For a short exposition on the Sachs Theorem see [CVE1].

Note that B. D. Acharya [ACH3] has rediscovered Theorem 3.1.

The problem of finding coefficients of the characteristic polynomial from the computational point of view has been treated in [RAN2], [RAN5], and [RAN6].

An interpretation of the subdeterminants of $\lambda I - A$ has been given by P. Heinrich [HEIN3]. The rough idea is as follows. Let G be a graph with vertex set $\{1, 2, \ldots, n\}$, A be the adjacency matrix of G, and let S and Z be disjoint subsets of vertices of order k. Let U be the submatrix formed by deleting the rows and columns numbered by S and Z respectively. A k-connection from S to Z is a spanning subgraph of G which consists of k disjoint paths starting at a vertex in S and terminating at a vertex in Z plus, possibly, some cycles. We have

$$\gamma \det U = Y_0 \lambda^{n-2k} + Y_1 \lambda^{n-2k-1} + \cdots + Y_{n-2k},$$

where $Y_i = \sum_{W \in V_{i+k}} \delta(W)$. V_{i+k} is the set of all k-connections from S to Z in G, $\gamma \in \{-1, 1\}$, and $\delta(W) = \pm 1$, the value being determined from the components of W.

The Jordan normal form of the adjacency matrix of an acyclic digraph has been studied by P. Heinrich [HEIN1]. In this context a digraph is called *acyclic* if there are no loops, and the underlying graph (obtained by ignoring the orientations of the arcs) has no circuits. The Jordan normal form of the adjacency matrix of G is then determined by the structure of the sets of disjoint (directed) paths of G.

We now turn to sigraphs. A graph whose edges have weight ± 1 is called a *sigraph*. It is called *circuit balanced* if each circuit has positive weight. Given a matrix M, let M^+ be the matrix whose entries are the absolute values of the corresponding entries of M.

THEOREM 3.4 (B. D. ACHARYA [ACH2]): *The matrices M and M^+ are cospectral if and only if the corresponding sigraph is circuit balanced.*

A sigraph is called *locally circuit balanced* at x if every circuit containing x has positive weight. Now suppose that G is a digraph and x is one of its vertices. Let G_x be the weighted digraph obtained from G by changing the sign of each edge with a negative weight that lies on a circuit with negative weight passing through x.

THEOREM 3.5 (M. K. GILL, B. D. ACHARYA [GIAC1]): *G and G_x are cospectral if and only if G is locally circuit balanced at x.*

Let $S(G)$ be the adjacency matrix of a sigraph G on n vertices and let H be the corresponding graph. In addition, let

$$\det(\lambda I - S(G)) = \sum_{i=0}^{n} s_i(G)\lambda^{n-i}$$

be the corresponding characteristic polynomial of G. A measure β_0 of the *degree of balance* in a sigraph has been defined by

$$\beta_0(G) = 1 - \frac{T(G)}{n-2},$$

where $T(G)$ is the number of indices i for which $s_i(G) \neq s_i(H)$. B. D. Acharia [ACH2] posed the following conjecture.

CONJECTURE 3.1: *Let G_1 and G_2 be sigraphs with the same corresponding graph H and with $\beta_0(G_1) = \beta_0(G_2)$. Then G_1 and G_2 are cospectral.*

This conjecture has been disproved by M. K. Gill [GIL1]. A counterexample is provided by sigraphs whose corresponding graph H satisfies

(1) H has exactly k circuits,
(2) all circuits of H have the same length m, and
(3) all the circuits of H pass through a single vertex.

In [ZEI1] and [ZEI2] the author considers the characteristic polynomial of a skew symmetric adjacency matrix. Define the skew symmetric matrix $S = (s_{ij})$ by assigning an orientation and a positive weight t_{ij} to each edge. Then $s_{ij} = t_{ij}$ if the edge is oriented from vertex i to vertex j and $s_{ij} = -t_{ij}$ otherwise. The remaining entries are equal to zero. Let $\det(\lambda I - S) = \lambda^n + a_1 \lambda^{n-1} + \cdots + a_n$. Since the determinant of a skew symmetric matrix of odd order is zero, we have $a_i = 0$ for odd i. The coefficients a_i where i is even can be expressed in terms of the basic figures of G. Note that the spectrum of G is purely imaginary.

The determinant of the adjacency matrix of a graph and its relations to the graph structure has been studied in a chemical context and is described in Chapter 5. In addition, the relationships between the characteristic and matching polynomial are given in Chapter 4 and are also elaborated in [MOH1].

Section 3.2: Spectra of Graphs Derived by Operations and Transformations

In this section we extend the results given in Chapter 2 of [CVDSA1] where some procedures for determining the spectra and/or characteristic polynomials of graphs derived from some simpler graphs by graph operations or transformations are described.

We start by describing several results which have been united by the concept of a *rooted product* given by C. D. Godsil and B. D. McKay [GOMK2] (see Theorem 3.8). Further results include formulas for the characteristic polynomial of a subdivision graph and for the generalized line graph.

We denote by P_k, as usual, the path with k vertices, and we denote by e_j the edge joining v_i and v_{i+1}. An edge e of a graph G is called a *bridge* if $G - e$ has more connected components than G.

DEFINITION 3.1: A graph Q_k belongs to the class $\mathcal{Q}_K$ if and only if it contains P_k as a subgraph with the edges e_j being bridges for $j = 1, \ldots k - 1$.

Hence every graph with at least one vertex belongs to $\mathcal{Q}_1$, every graph with a bridge belongs to $\mathcal{Q}_2$ and, in general, a graph from $\mathcal{Q}_k$ will have the structure of a graph as indicated in Figure 3.1.

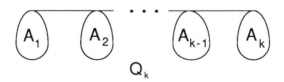

Figure 3.1 A graph in $\mathcal{Q}_k$

The graphs $A_1, A_2, \ldots A_k$ are arbitrary rooted, mutually disjoint graphs. Hence the graph Q_k can be thought of as having fragments $A_1, \ldots, A_k$ with the roots joined by a path. The path P_k itself belongs to $\mathcal{Q}_k$. The subgraph obtained by deleting v_j from A_j will be denoted B_j.

DEFINITION 3.2: A graph R_k belongs to the class $\mathcal{R}_k$ if and only if $R_k \in \mathcal{Q}_k$, $A_2 = A_3 = \cdots = A_{k-1}$, and $B_2 = B_3 = \cdots = B_{k-1}$. A graph R_k^* belongs to the class $\mathcal{R}_k^*$ if and only if $R_k^* \in \mathcal{R}_k$, $A_k = A_{k-1}$, and

$B_k = B_{k-1}$. A graph S_k belongs to the class $\mathcal{S}_k$ if and only if $S_k \in \mathcal{R}_k^*$, $A_1 = A_2$, and $B_1 = B_2$.

It is clear that $\mathcal{S}_k \subseteq \mathcal{R}_k^* \subseteq \mathcal{R}_k \subseteq \mathcal{Q}_k$. In Figure 3.2 the structure of R_k, R_k^*, and S_k are displayed.

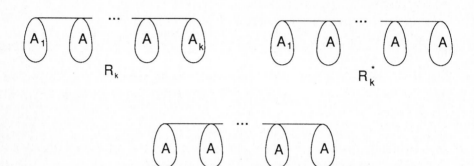

Figure 3.2 Graphs in $\mathcal{R}_k$, $\mathcal{R}_k^*$, and $\mathcal{S}_k$

Since e_k is a bridge, Q_k satisfies the following equality (see, for example, [CVDSA1], p. 59):

$$\Phi(Q_k) = \Phi(A_k)\Phi(Q_{k-1}) - \Phi(B_{k-1})\Phi(B_k)\Phi(Q_{k-2})$$

where $\Phi(G)$ is the characteristic polynomial of G.

This equation can be written in matrix form as

$$\begin{pmatrix} \Phi(Q_k) \\ \Phi(B_k)\Phi(Q_{k-1}) \end{pmatrix} = \begin{pmatrix} \Phi(A_k) & -\Phi(B_k) \\ \Phi(B_k) & 0 \end{pmatrix} \begin{pmatrix} \Phi(Q_{k-1}) \\ \Phi(B_{k-1})\Phi(Q_{k-2}) \end{pmatrix}$$

$$(3.2)$$

A repeated application of (3.2) yields

THEOREM 3.6 (I. GUTMAN [GUT11]): *Using definitions 3.1 and 3.2,*

$$\begin{pmatrix} \Phi(Q_k) \\ \Phi(B_k)\Phi(Q_{k-1}) \end{pmatrix} = T_k T_{k-1} \cdots T_2 T_1 \begin{pmatrix} 1 \\ 0 \end{pmatrix},$$

$$(3.3)$$

where

$$T_j = \begin{pmatrix} \Phi(A_j) & -\Phi(B_j) \\ \Phi(B_j) & 0 \end{pmatrix}, \quad j = 1, 2, \dots k.$$

CONCLUSION 3.6.1 (I. GUTMAN [GUT11]): *The characteristic polynomial of Q_k is completely determined by the characteristic polynomials of A_j and B_j, $j = 1, \ldots, k$.*

This implies the existence of many cospectral pairs of graphs. The two graphs in Figure 3.3, for example, are both in S_2.

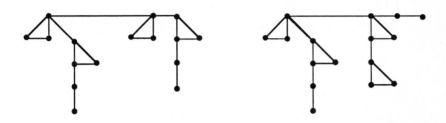

Figure 3.3 Cospectral graphs in S_2

COROLLARY 3.6.2 (I. GUTMAN [GUT11]): *Using the definitions 3.1 and 3.2,*

$$\begin{pmatrix} \Phi(R_k) \\ \Phi(B_k)\Phi(R_{k-1}^*) \end{pmatrix} = T_k T^{k-2} T_1 \begin{pmatrix} 1 \\ 0 \end{pmatrix},$$

and

$$\begin{pmatrix} \Phi(S_k) \\ \Phi(B_k)\Phi(S_{k-1}) \end{pmatrix} = T^k \begin{pmatrix} 1 \\ 0 \end{pmatrix}$$

where

$$T = \begin{pmatrix} \Phi(A) & -\Phi(B) \\ \Phi(B) & 0 \end{pmatrix}.$$

COROLLARY 3.6.3 (I. GUTMAN [GUT11]):

$$\begin{pmatrix} \Phi(P_k) \\ \Phi(P_{k-1}) \end{pmatrix} = \begin{pmatrix} \lambda & -1 \\ 1 & 0 \end{pmatrix}^k \begin{pmatrix} 1 \\ 0 \end{pmatrix}.$$

Results of the form (3.3) can also be found in [KACH1]. Similar results are also contained in [GUT33].

Let G_x denote a graph with rooted vertex x. A composition $G_u \circ H_v$ has been considered in [GOMK1]. The graph $G_u \circ H_v$ consists of disjoint copies of $G - u$ and $H - v$ with additional edges joining each vertex adjacent to u in G to every vertex adjacent to v in H. The the following formula holds:

$$P_{G_u \circ H_v}(\lambda) = P_{G-u}(\lambda)P_{H-v}(\lambda) - (\lambda P_{G-u}(\lambda) - P_G(\lambda))(\lambda P_{H-v}(\lambda) - P_H(\lambda))$$

The spectrum of the corona of two graphs is determined in [CvGS1]. Let G be a graph on n vertices, and let H be a regular graph on m vertices with degree r. The characteristic polynomial of the corona of G and H, $G * H$, is the determinant of the matrix

$$\begin{pmatrix} \lambda I - A & -J_1 & -J_2 & \cdots & -J_n \\ -J_1^T & \lambda I - B & \cdots & & \\ -J_2^T & & \lambda I - B & \cdots & \\ \vdots & & & \ddots & \\ -J_n^T & & & & \lambda I - B \end{pmatrix}$$

where A and B are the adjacency matrices of G and H respectively and J_k is an $n \times m$ matrix with every entry in the k-th row equal to one and all other entries equal to zero. By elementary transformations this matrix can be changed to

$$\begin{pmatrix} (\lambda - \frac{m}{\lambda-r})I - A & 0 & 0 & \cdots & 0 \\ -J_1^T & \lambda I - B & & & \\ -J_2^T & & \lambda I - B & & \\ \vdots & & & \ddots & \\ -J_n^T & & & & \lambda I - B \end{pmatrix}$$

and so

$$P_{G*H} = P_G(\lambda - \frac{m}{\lambda - r})(P_H(\lambda))^n.$$

A special case of this result appears in [CvDSA1], p. 60. These results have been further generalized by C. Godsil and B. D. McKay. Given a graph H with n vertices and a family $\mathcal{G} = \{G_1, G_2, \ldots, G_n\}$ of rooted graphs, they define the *rooted product* $H(\mathcal{G})$ to be the graph obtained by identifying, for $k = 1, 2, \ldots, n$, the root of G_k with the k-th vertex of H.

THEOREM 3.7 (C. GODSIL, B. D. MCKAY [GOMK2]): *If (a_{ij}) is the adjacency matrix of H, and $A_\lambda(H, \mathcal{G}) = (\alpha_{ij})$ is defined by $\alpha_{ii} = P_{G_i}(\lambda)$ and $\alpha_{ij} = -a_{ij}P'_{G_i}(\lambda)$ for $i \neq j$. Then*

$$P_{H(\mathcal{g})}(\lambda) = \det A_\lambda(H, \mathcal{G}).$$

Next we consider the following recursive definition of a class of trees which stems from [GUT28].

Let R be a rooted graph, *i.e.*, a graph with a particular vertex labelled by v_0. Let $d = (d_1, d_2, \ldots, d_m)$ be an m-tuple of positive integers. K_1 will denote the graph consisting of a single vertex. Then we define a graph $G_m = G_m(R, d)$ recursively in the following manner.

DEFINITION 3.3: (i) $G_0 = R$, and (ii) for $k = 0, 1, \ldots, m - 1$, the graph G_{k+1} is obtained by taking d_{k+1} copies of G_k and joining each v_k to a new vertex which is labeled v_{k+1}.

Hence if G_k possesses n_k vertices then G_{k+1} possesses $n_{k+1} = d_{k+1}n_k + 1$ vertices. A recursive formula for the characteristic polynomial of G_m is derived in [GUT28]. Some special cases are treated there in further detail.

The spectrum of *complete k-ary tree* is obtained in [ROU1] by the use of a recursive formula along similar lines.

The following theorem is a graph-theoretic reformulation of a well known matrix-theoretic result (Jacobi's formula). It will be used in Chapter 5.

THEOREM 3.8: *Let u and v be vertices of a graph G. Let $\mathcal{P}_{uv}$ be the set of all paths which connect u and v. Then*

$$P_{G-u}(\lambda)P_{G-v}(\lambda) - P_G(\lambda)P_{G-u-v}(\lambda) = \left(\sum_{T \in \mathcal{P}_{uv}} P_{G-T}(\lambda) \right)^2.$$

See also [TUT2] from [CVDSA1].

We now turn to the subdivision graph and other related graph operations. Let us subdivide each edge of a graph by adding k new vertices. The resulting graph is called the *k-th subdivision graph* of the original graph. Let $S_k(G)$ be the k-th subdivision graph of a graph G and $L_{k+1}(G)$ be $L(S_k(G))$, where $L(H)$ denotes the line graph of a graph H. Further, Let $R_k(G)$ be the graph obtained from G by adding for each edge u of G a copy of K_k and by joining each endpoint of u with each vertex of the corresponding K_k.

THEOREM 3.9 (V. B. MNUHIN [MNU1]):   *We have*

$$P_{S_k(G)}(\lambda) = (U_k(\frac{\lambda}{2}))^{m-n} \det(\lambda U_k(\frac{\lambda}{2})I - U_{k-1}(\frac{\lambda}{2})D - A)$$

where A is the adjacency matrix of G, D is the (diagonal) degree matrix of G, and $U_k(x)$ is the Chebyshev polynomial of the second kind.

Similar formulas are obtained for $L_{k+1}(G)$ and $R_k(G)$, thus generalizing some results of D. Cvetković (cf. [CVDSA1], pp. 63–64). Results of D. Cvetković are special cases of $k = 1$. Those for $S_1(G)$ also appear in [SHI1], [SHI2]. All these results are extended to digraphs in [MNU2].

The definition of a generalized line graph $L(G; a_1, a_2, \ldots, a_n)$ of a graph G on n vertices (where $a_1, a_2, \ldots, a_n$ are nonnegative integers) is given in Chapter 1. *Generalized line graphs* were introduced by A. J. Hoffman, and they play an important role in spectral graph theory (see Chapter 1). It has been proved that the least eigenvalue of a generalized line graph is bounded from below by -2 just as it is for line graphs. Here we present a result giving, in some cases, the whole spectrum of a generalized line graph.

THEOREM 3.10:   *Let G be a graph having vertex degrees $d_1, d_2, \ldots, d_n$. If $a_1, a_2, \ldots, a_n$ are nonnegative integers such that $d_i + 2a_i = d$, $i = 1, 2, \ldots, n$, then*

$$P_{L(G;a_1,a_2,\ldots,a_n)}(\lambda) = \lambda^{\sum_{i=1}^n a_i}(\lambda + 2)^{m-n+\sum_{i=1}^n a_i} P_G(\lambda - d + 2).$$

PROOF: Consider the matrix

$$\begin{pmatrix} R & L_1 & L_2 & \cdots & L_n \\ 0 & M_1 & 0 & \cdots & 0 \\ 0 & 0 & M_2 & \cdots & 0 \\ \vdots & \vdots & \vdots & \ddots & \vdots \\ 0 & 0 & 0 & \cdots & M_n \end{pmatrix}$$

where L_i, $i = 1, \ldots, n$ is an n by $2a_i$ matrix in which i-th row has all entries equal to 1, all other entries being equal to 0, and M_i, ($i = 1, \ldots, n$), is an a_i by $2a_i$ matrix of the form $(\,I_{a_i} \quad -I_{a_i}\,)$, I_m being a unit matrix of order m. The theorem follows from *Lemma 2.1* from [CVDSA1] and the fact that $S^T S = B + 2I$ where B is the adjacency matrix of $L(G; a_n, \ldots, a_n)$.

We continue with miscellaneous results.

THEOREM 3.11 (M. BOROWIECKI, T. JÓŹWIAK [BoJÓ1]): *Let G be a multidigraph, $V(G)$ its vertex set, and let x be one of its vertices.*

Let $\mathcal{C}(x)$ be the set of all cycles of G containing x. Then

$$P_G(\lambda) = \lambda P_{G-x}(\lambda) - \sum_{C \in \mathcal{C}(x)} P_{G-V(C)}(\lambda).$$

This formula extends a previous result of A. J. Schwenk (see [CvDSA1], p. 78) to multidigraphs. The authors proceed in [BoJÓ1] in the same spirit and give generalized versions of the known reduction formula when one deletes an edge (arc) from a multidigraph as well as a formula for the coalescence of rooted multidigraphs (cf. [CvDSA1], p. 159). The proof is carried out easily by the use of the Sachs theorem.

These results have also been presented in [JÓz1].

Similar technical generalizations of the Schwenk formula are given in [GiAc2] for the characteristic polynomial of sigraphs, defined in Section 3.1, and in [GiL2] for the characteristic polynomial of an arbitrary matrix. See also [WAT1], where a reformulation of the Sachs theorem has been used to generalize to multigraphs two results of A. J. Schwenk ([SCHW3] from [CvDSA1]).

A disadvantage of formulas like the one in Theorem 3.11 is that one needs to construct the set of cycles $\mathcal{C}(x)$. This is avoided in the recent formula by P. Rowlinson [ROW5]. Suppose that G is a multigraph with m edges connecting vertices u and v, $G - uv$ is obtained from G by deleting the edges between u and v, and G^* is constructed from $G - uv$ by identifying the vertices u and v (with vertices adjacent to both u and v producing multiple edges). We then have

$$P_G(\lambda) = P_{G-uv}(\lambda) + P_{G^*}(\lambda) + m(\lambda - m)P_{G-u-v}(\lambda)$$
$$- mP_{G-u}(\lambda) - mP_{G-v}(\lambda).$$

Spectra or characteristic polynomials for graphs obtained by different compositions of graphs are given in papers [GoMK1], [SCHW2], and [ScHE1]. Since the main goal of these papers is the construction of cospectral graphs, they are described in Section 1.4.

A *path polynomial* of a graph with respect to an initial and a final vertex has been introduced in [SiSR1]. The path polynomial is used to compute characteristic polynomials of some particular graphs. Results are more or less in terms of previously known facts.

Let $G_{k,m}$ be a graph obtained by attaching the end points of the paths P_k and P_m at a vertex of a nontrivial connected graph G. It is proved in [LIFE1] that for $1 \leq m \leq k$, the largest eigenvalue of $G_{k,m}$ is greater than the largest eigenvalue of $G_{k+1,m-1}$. As a corollary, the largest eigenvalue of P_n is less than the largest eigenvalue of any other connected graph on n vertices. It also follows that among unicyclic graphs obtained by attaching a tree on n vertices to a vertex of a circuit C_m, the index is smallest when the tree attached is a path. However, the authors' conjecture that the last graph has the smallest index among all unicyclic graphs with the same number of vertices has been disproved in [CVE6] using an counter example constructed by a computer.

Graph transformations arising from the application of formal grammars to graphs have been studied in [MIC1]. The characteristic polynomials of the resulting graphs have been expressed in terms of algebraic operations on the polynomials of the initial graphs and their subgraphs.

The definition of a very general n-ary graph operation, called NEPS (noncomplete extended p-sum of graphs), including one of its special cases, the product of graphs, is reproduced in the next section. Here we note that the same operation can be defined for digraphs (cf. [CVPE2], where eigenvalues and the strong connectedness of the digraph obtained as a NEPS of other digraphs have been studied, [ESHA2], and a review of that paper in **Mathematical Reviews** (MR 81m: 05096) for some bibliographical data).

The product $G \times K_2$ of a graph G and the graph K_2 is called the *bipartite square* $G \circ G$ of G in [CVDSA1]. It is noted in [POR1] that $P_{G \circ G}(\lambda) = (-1)^n P_G(\lambda) P_G(-\lambda)$. If G_1 and G_2 are nonisomorphic cospectral graphs then $G_1 \circ G_1$ and $G_2 \circ G_2$ are also nonisomorphic and cospectral, $G \circ G$ is always a bipartite graph, and it is disconnected if G is bipartite. These results appeared in [CVDSA1], p. 70.

Section 3.3: Constructions of Graphs Using Spectra

The idea of using graph spectra for the construction of graphs with given properties has been outlined in [CVDSA1], p. 190, with some examples being given there. Suppose that properties of the desired graph determine the spectrum or a spectral property of the graph. Now the problem is to construct a graph with the given spectrum or spectral property. One could start with graphs with known spectra and perform graph operations on them in order to obtain a new graph with the desired spectrum. Examples of such constructions of strongly regular graphs have been given in [CVE4] and some of them are generalized in [CVE2].

We shall reproduce these constructions here. First we follow [CVE4].

There is an n-ary composition on graphs which is called NEPS (noncomplete extended p-sum of graphs).

DEFINITION 3.4: Let B be a set of n-tuples $(\beta_1, \ldots, \beta_n)$ of symbols 0 and 1 which does not contain the n-tuple $(0, \ldots, 0)$. The *NEPS* with basis B of the graphs $G_1, \ldots, G_n$ is the graph, whose vertex set is equal to the Cartesian product of the vertex sets of graphs $G_1, \ldots, G_n$ in which two vertices $(x_1, \ldots, x_n)$ and $(y_1, \ldots, y_n)$ are adjacent if and only if there is an n-tuple $(\beta_1, \ldots, \beta_n)$ in B such that $x_i = y_i$ holds when $\beta_i = 0$ and x_i is adjacent to y_i in G_i when $\beta_i = 1$.

If B contains all n-tuples having exactly p coordinates equal to 1, then the NEPS is called the *p-sum*.

The spectrum of a NEPS can be determined by the spectra of the original factor graphs (see [CVDSA1], p. 69).

Let $\lambda_{i1}, \ldots, \lambda_{in_i}$ be the spectrum of the graph G_i, $i = 1, \ldots, n$. Let $\beta = (\beta_1, \ldots, \beta_n)$. Then the spectrum of the NEPS with basis B of the graphs $G_1, \ldots, G_n$ consists of all possible values of $\Lambda_{i_1, \ldots, i_n}$, where

$$\Lambda_{i_1, \ldots, i_n} = \sum_{\beta \in B} \lambda_{1i_1}^{\beta_1} \lambda_{2i_2}^{\beta_2} \ldots \lambda_{ni_n}^{\beta_n}, \quad i_k = 1, 2, \ldots n_k, \ k = 1, 2, \ldots n.$$

In particular, the spectrum of the p-sum of graphs $G_1, \ldots, G_n$ consists of all the values of the elementary symmetric function of order p in variables $x_1, \ldots, x_n$, where the variable x_i runs through the eigenvalues of G_i, $i = 1, \ldots, n$.

We shall now construct some regular connected graphs with 3 distinct eigenvalues. Such graphs are known to be strongly regular (see, for example, [CVDSA1], p. 103). The difficulties in our constructions arise from the fact that the NEPS generally contains many more distinct eigenvalues than do the starting graphs, and that only in exceptional cases do some eigenvalues become equal. We start with some examples.

EXAMPLE 3.1 Since the spectrum of K_n consists of the simple eigenvalue $n-1$ as well as $n-1$ eigenvalues equal to -1 we can readily check the following statements.

- The 1-sum of two complete graphs K_n yields the graph $L(K_{n,n})$ which is strongly regular with distinct eigenvalues $2n-2$, $n-2$, and -2.
- The 2-sum of three copies of the graph K_4 results in a strongly regular graph on 64 vertices with the distinct eigenvalues 27, 3, -5.
- The 2-sum of four copies of K_3 gives a graph on 81 vertices with distinct eigenvalues 24, 6, -3.
- The 3-sum of four copies of K_3 gives a strongly regular graph on 81 vertices having distinct eigenvalues 32, 5, -4.
- The 4-sum of five copies of K_2 yields a disconnected graph with two components, each being the complement of the Clebsch graph.

We proceed to more general constructions. The next few constructions have been given according to [CVE2].

DEFINITION 3.5: The *odd (even) sum* of graphs is the NEPS with the basis containing all the n-tuples with an odd (even) number of 1's.

DEFINITION 3.6: The *mixed sum* of graphs is the NEPS with the basis containing all the n-tuples in which the number of 1's is congruent to 1 or 2 modulo 4.

Note that the odd, the even, and the mixed sum of two graphs is called the sum, the product, and the strong product, respectively.

In Theorems 3.12 and 3.14 we construct two more infinite series of strongly regular graphs by means of the NEPS.

THEOREM 3.12 (D. CVETKOVIĆ [CVE2]):   *For all $n \geq 2$ the odd sum F_n of n copies of the graph K_4 is a strongly regular graph with the eigenvalues $2^{2n-1} + (-1)^{n-1}2^{n-1}, 2^{n-1}, -2^{n-1}$.*

PROOF: The distinct eigenvalues of K_4 are 3 and -1. Let S_p be the elementary symmetric function of order p on the variables $x_1, \ldots, x_n$, and let these variables take the values 3 or -1. If k variables take the value 3 and if the remaining $n-k$ ones take value -1, the value of S_i is equal to the coefficient of x^{n-i} in the polynomial $P_k(x) = (x-3)^k(x+1)^{n-k}$. The eigenvalues of the odd sum are given by $\sum_i S_i$

where the summation goes over all odd numbers i not greater than n. If $P_k(x) = a_0 x^n + a_1 x^{n-1} + \cdots + a_{n-1}x + a_n$, then we have

$$\Lambda_k = -(a_1 + a_3 + \cdots) = \frac{1}{2}((-1)^n P_k(-1) - P_k(1)), \quad k = 0, 1, \ldots n,$$

Λ_k are the eigenvalues of F_n, and we immediately get $\Lambda_k = (-1)^{k+1}2^{n-1}$, $k = 0, 1, \ldots, n-1$, and $\Lambda_n = 2^{2n-1} + (-1)^{n-1}2^{n-1}$, which proves the theorem.

F_n is a regular graph of degree $2^{2n-1} + (-1)^{n-1}2^{n-1}$ on 4^n vertices. Any two distinct vertices have $2^{2n-2} + (-1)^{n-1}2^{n-1}$ common neighbours. F_n can be visualized in the following way. The 4^n n-tuples of 4 distinct symbols are the vertices and two n-tuples are adjacent if they differ in an odd number of coordinates.

A result on spectral characterizations of block designs (cf. [CvDSa1], p. 167) can be reformulated in the following way.

THEOREM 3.13: *A symmetric BIBD with the parameters (v, k, λ) exists if and only if there exists a graph with the spectrum consisting of eigenvalues k, $(k - \lambda)^{1/2}$, $-(k - \lambda)^{1/2}$, $-k$ with the multiplicities $1, v-1, v-1, 1$, respectively.*

The even sum of graphs F_n and K_2 is a regular bipartite graph with four distinct eigenvalues $\pm(2^{2n-1}+(-1)^{n-1}2^{n-1})$, $\pm 2^{n-1}$. Hence, by Theorem 3.13 we have constructed a family of symmetric BIBDs with the parameters $v = b = 4^n$, $r = k = 2^{2n-1} + (-1)^{n-1}2^{n-1}$, $\lambda = 2^{2n-2} + (-1)^{n-1}2^{n-1}$.

The graphs F_n and the corresponding block designs have been constructed in the literature in many different ways; the construction given above is a spectral one.

THEOREM 3.14 (D. CVETKOVIĆ [Cve2]): *For $s \geq 1$, the mixed sum H_s of $4s$ copies of the graph K_2 is a strongly regular graph with eigenvalues $2^{4s-1} - (-1)^s 2^{2s-1}$, 2^{2s-1}, and -2^{2s-1}.*

PROOF: The eigenvalues of K_2 are 1 and -1. In order to obtain the eigenvalues of the mixed sum of $n = 4s$ copies of K_2 consider the polynomial

$$Q(x) = (x - 1)^k (x + 1)^{n-k} = b_0 x^n + b_1 x^{n-1} + \cdots + b_{n-1}x + b_n. \quad (3.4)$$

Define α_0, α_1, α_2, α_3 by the equations

$$\alpha_0 = b_0 + b_4 + b_8 + \cdots + b_n,$$
$$\alpha_1 = b_1 + b_5 + b_9 + \cdots + b_{n-1},$$
$$\alpha_2 = b_2 + b_6 + b_{10} + \cdots + b_{n-2},$$
$$\alpha_3 = b_3 + b_7 + b_{11} + \cdots + b_{n-3}.$$

The values of $-\alpha_1 + \alpha_2$ for $k = 0, 1, \ldots, n$ represent the eigenvalues of H_s. First we have

$$Q_k(1) = \alpha_0 + \alpha_1 + \alpha_2 + \alpha_3,$$
$$Q_k(i) = \alpha_0 - i\alpha_1 - \alpha_2 + i\alpha_3,$$
$$Q_k(-1) = \alpha_0 - \alpha_1 + \alpha_2 - \alpha_3,$$
$$Q_k(-i) = \alpha_0 + i\alpha_1 - \alpha_2 - i\alpha_3.$$

Further we obtain

$$-\alpha_1 + \alpha_2 = \frac{1}{4}(2Q_k(-1) - (i+1)Q_k(i) - (1-i)Q_k(-i)) = \Lambda_k \quad (3.5)$$

Using (3.4) and (3.5) we get $\Lambda_k = (-1)^{s-1+\binom{k}{2}}2^{2s-1}$ for $k = 0, 1, \ldots 4s - 1$ and $\lambda_{4s} = 2^{4s-1} - (-1)^s 2^{2s-1}$, which proves the theorem.

The *Seidel spectrum* of H_s contains the eigenvalues $2^{2s} - 1$ and $-2^{2s} - 1$, and again we have a regular two-graph with the same eigenvalues as with F_{2s} of Theorem 3.12.

We now mention constructions of another type following [CVE2]. Distance-regular graphs are defined in Chapter 2.

If a graph G is distance-regular with adjacency matrix A, we can find a linear combination of matrices A^2, A and I which represents the adjacency matrix of graph G^2 defined on the vertex set of G, with two vertices being adjacent in G^2 if they are at distance 2 in G.

If G is the cubic lattice graph (see [CVDSA1], p. 178), with characteristic 4, the graph G^2 coincides with the graph of the second type in Example 3.1. If we take for G the exceptional graph in the characterization of the cubic lattice graph of characteristic 4, then G^2 represents a graph with the same spectrum as above.

Starting with G equal to the tetrahedral graph (see [CvDSA1],p. 180) with characteristic 10, we get for G^2 a strongly regular graph on 120 vertices with eigenvalues 56, 8, -4 which is related to the exceptional root system E_8 (see Chapter 1).

$L(K_n)$ is a strongly regular graph. In particular, the complement of $L(K_5)$ is the Petersen graph. The 2-sum of the Petersen graph and the graph K_2 yields Desargues graph. Tutte's 8-cage can be constructed as a certain square root of a graph. Namely, if G is the 8-cage, then G^2 is isomorphic to the complement of $2L(K_6)$.

If G is the Hoffman-Singleton graph, then $(L(G))^2$ is a strongly regular graph on 175 vertices having the eigenvalues 72, 2, and -18.

Spectral constructions of graphs have been used by H. Sachs and M. Stiebitz [SAST2] to obtain transitive graphs which satisfy an upper bound for the number of simple eigenvalues (see Section 3.4). It is easy to see that the NEPS of transitive graphs is again a transitive graph. The construction starts again with complete graphs which are transitive and whose spectrum is known.

Let $\tau(G)$ be the number of simple eigenvalues of a graph G. A graph is called an I-graph if the following holds:

1. The eigenvalues λ_i of G are integers and $\lambda_i \neq -2, 0$, $i = 1, \ldots, n$, and

2. $\lambda_i - \lambda_j \neq 2$, $i, j = 1, \ldots, n$.

Let $G_1 + G_2$ and $G_1 \cdot G_2$ be the sum and the strong product of graphs G_1 and G_2. The sum and the strong product are special cases of the NEPS. If λ_i are eigenvalues of G_1 and if μ_j are eigenvalues of G_2 then $\lambda_i + \mu_j$ are eigenvalues of $G_1 + G_2$ and $\lambda_i\mu_j + \lambda_i + \mu_j$ are eigenvalues of $G_1 \cdot G_2$ (see Section 3.6 and [CvDSA1], p. 70).

It is proved in [SAST2] that if G is an I-graph, then for any $m \geq 3$ the graph $(G + K_2) \cdot K_m$ is also an I-graph.

Let us define $\mathcal{G}_0 = \{K_{2m+1} \mid m \geq 1\}$, and $\mathcal{G}_{k+1} = \{(G + K_2) \times K_{2m+1} \mid G \in \mathcal{G}_k, m \geq 1\}$.

THEOREM 3.15 (H. SACHS, M. STIEBITZ [SAST2]): *For any $G \in \mathcal{G}_k$ we have $\tau(G) = 2^k$ and $\tau(G + K_2) = 2^{k+1}$.*

With these graphs, the upper bound for $\tau(G)$ described in Section 3.4 is attained.

See [ROW2] for an extension of these constructions.

Spectral constructions of graphs are used to construct integral graphs and Gaussian digraphs (see Section 3.12). A suitable operation for such purposes is the NEPS. Since the eigenvalues of a NEPS are polynomials in eigenvalues of graphs on which the operation is performed, the NEPS of integral graphs is also integral. If the NEPS is defined for digraphs (as is done in [CvPE2]), then the same applies to the Gaussian digraphs.

Spectral constructions of graphs are applied in constructing cospectral graphs and digraphs having particular properties (see Section 1.4). We shall describe here a result of this type which is related to the fact that eigenvalues of the product of two graphs are products of eigenvalues of these graphs (product of two graphs is just the NEPS of these graphs in which the basis consists only of the pair $(1,1)$).

The components of the product of P_3 with itself (see Figure 7.1 of [CvDSA1]) are just the graph C_4 and $K_{1,4}$ which give rise to the smallest pair of cospectral nonisomorphic graphs: $C_4 \cup K_1$ and $K_{1,4}$. It was noted in [Cve6] that this fact inspired a more general conjecture which, after computer experiments with special cases, has been proved. In order to formulate the result we give the following definition.

DEFINITION 3.7: Two graphs are said to be *almost cospectral* if their nonzero eigenvalues (including multiplicities) coincide.

THEOREM 3.16: *If G and H are connected bipartite graphs, then the graph $G \times H$ has exactly two components, and these are almost cospectral.*

It turns out that Theorem 3.16 has already been discovered by C. Godsil and B. McKay (see [GoM1] of [CvDSA1]). The following corollary and conjecture has been formulated in [Cve6].

COROLLARY 3.16.1: *The components of $G \times H$ from Theorem 3.16 are cospectral if and only if one of the bipartite graphs G, H has the same number of vertices in both parts of its vertex bipartition.*

CONJECTURE 3.2: *If the NEPS of bipartite graphs (see [CvDSA1], p. 204) is disconnected, then its components are almost cospectral.*

A spectral construction of certain bipartite graphs called expanders is given in [ALMi1]. An (n, k, c)-expander is a bipartite graph with at most kn edges, with vertex bipartition X, Y satisfying $|X| = |Y| = n$, and such that for every $A \subset X$ we have $|N(A)| \geq (1 + c(1 - |A|/n))|A|$, $N(A)$ being the set of neighbours of vertices of A.

THEOREM 3.17 (N. ALON, V. D. MILMAN [ALMI1]): *Let G be a regular graph of degree k on n vertices with the second largest eigenvalue $\lambda_2 \leq k - \epsilon$, and let $B = \{(1,0),(1,1)\}$. Then the NEPS with the basis B of graphs K_2 and G is an $(n, k+1, c)$-expander for $c = 4\epsilon/(k+4\epsilon)$.*

Expanders appear in a branch of computer science. See also [TAN1] where concentrators are considered. Concentrators are, in fact, graphs with a small second largest eigenvalue. The authors point out that generalized polygons (cf. [CvDSA1], p. 193) have good concentration properties.

See Section 3.5 for information on graphs with a small λ_2.

Section 3.4: The Automorphism Group and the Spectrum of a Graph

Let G be a graph with vertex set X. A bijection $\gamma : X \rightarrow X$ is called an *automorphism* of G if it preserves the adjacency relation of G. The set Γ of all automorphisms γ is a group with respect to the composition of mappings. There are many relations between the spectrum of a graph and the group of automorphisms as described in Chapter 5 of [CvDSA1].

The results described in this section are, generally speaking, of the following two types: first, procedures for factoring the characteristic polynomial by means of the automorphism group, and, second, bounds for the number of simple eigenvalues (or eigenvalues of a small multiplicity).

The work of [YAP1] was not mentioned in [CvDSA1]. The notion of a (front) divisor of a multigraph (see Chapter 4 of [CvDSA1]) has been rediscovered there using different terminology. The fact that the characteristic polynomial of a divisor divides the characteristic polynomial of the digraphs is proved.

A factorization theorem for characteristic polynomial of weighted pseudograph has been proved in [WAN3]. A *weighted pseudograph G* may be viewed as a pair (V, w) where V is the vertex set of G and w is the weight function $w : V^2 \rightarrow C$ (C is the set of complex numbers) such that the corresponding weighted adjacency matrix of G is symmetric. If Γ is a finite abelian group, then we say that G admits a Γ-*action* if Γ is a subgroup of the automorphism group of G. Let V^* be the set of orbits under Γ and let F be the set of fixed points under all elements of Γ. We say that Γ *acts freely* on G if $\gamma(v) = v$ for any $v \in V$ implies that γ is the identity, and that Γ *acts semifreely* on G if it acts freely on $V - F$. In the second case the *orbit (pseudo)graph* G^* of G is defined by $G^* = (V^*, w^*)$, where for $u^*, v^* \in V^*$

$$w^*(u^*, v^*) = \begin{cases} w(u,v) & \text{if both } u \text{ and } v \text{ are in } F \\ |\Gamma|^{1/2} w(u,v) & \text{if one of } u \text{ and } v \text{ are in } F \\ \sum_{\gamma \in \Gamma} w(u, \gamma(v)) & \text{if neither } u \text{ nor } v \text{ is in } F. \end{cases}$$

THEOREM 3.18 (K. WANG [WAN3]): *Under the conditions described above, we have*

$$P_G(\lambda) = P_{G^*}(\lambda) \frac{P_{G-F}(\lambda)}{P_{(G-F)^*}(\lambda)}.$$

A special case of this result related to pseudographs whose automorphism group contains an *involution* has been previously studied by the same author [WAN4]. However, the statement of Theorem 3.18 holds under more general conditions (see review by B. D. McKay of [WAN3] in **Zentralblatt für Mathematik**, 516, 05038).

Let $P = P_1, P_2, \ldots$ be a partition of the vertex set of a graph and let $Q = Q_1, Q_2, \ldots$ be a partition of the edge set E. A pair (P, Q) is called a *colouring* of a graph (V, E) [SIE1]. A colouring is a *regular colouring* if any two vertices in P_i are end vertices of the same number of edges in Q_j for all i, j, and any two edges in Q_j have the same number of end vertices in P_i for all i and j. The importance of regular colourings for studying graph spectra is pointed out in [SIE1].

The *walk partitions* of the vertex set of a graph are introduced in [POSU1] and studied in relation with spectral properties of graphs. Two vertices belong to the same set of a walk partition if the numbers of walks of any given length starting at these vertices are the same. Relations between walk partitions and divisors of a graph are discussed.

For a review of several investigations concerning the factorization of characteristic polynomials in a chemical context see section 5.4.

All simple eigenvalues of a graph whose automorphism group has t orbits, with $r_1, r_2, \ldots, r_t$ being the vertex degrees of the different orbits, are elements of a finite set $S_{r_1, r_2, \ldots, r_t}$ (cf.[CVDSA1] p. 140 and p. 270).

If $t = 2$ and $(r_1, r_2) \in \{(1,2), (1,3), (2,2), (2,3), (3,3)\}$, then for each number $\alpha \in S_{r_1, r_2}$ there exists a loopless graph with orbit degrees

r_1 and r_2 having α as a simple eigenvalue. For the case $\alpha = \sqrt{3} \in \mathcal{S}_{3,3}$ only graphs with loops where known until M. Schulz [SCHU1], [SCHU2] constructed a cubic graph on 30 vertices having two orbits and $\sqrt{3}$ as a simple eigenvalue.

Considerable attention has been paid to upper bounds for the number of simple eigenvalues (and, more generally, to eigenvalues of small multiplicity) in transitive (but also in more general) graphs.

The paper [SAST1] is an abbreviated version of [SAST] from [CVDSA1]. The main results are already mentioned in [CVDSA1], Section 5.1, pp. 135–138.

THEOREM 3.19 (H SACHS, M. STIEBITZ [SAST1], [STI2]): *Let G be a transitive multigraph with $n = 2^p k$ vertices, where p is a nonnegative integer and k is odd. Then the number of simple eigenvalues of G is at most 2^p.*

It has been already noted in [CVDSA1] (see footnote on p. 137) that the bound from Theorem 3.19 is attained by multigraphs $G_0, G_1, G_2, \ldots$ defined as follows: $G_0 = K_1$, and if A_m is the adjacency matrix of G_m then

$$A_{m+1} = \begin{pmatrix} 2A_m & I \\ I & 2A_m \end{pmatrix}, \quad m = 0, 1, \ldots.$$

The graph G_m becomes an m-dimensional cube if all multiple edges are replaced by single ones.

A natural question arises: for which n is the bound from Theorem 3.19 attained by a graph without multiple edges or loops [SAST2]? For $n = 2^p$, $p \geq 2$, such graphs do not exist; this follows from a theorem of M. Petersdorf and H. Sachs ([CVDSA1], p. 136) which states that for even n the simple eigenvalues of a transitive graph of degree r are of the form $2q - r$ with $q \in \{0, 1, \ldots, r\}$. Trivial examples of graphs for which the bound is attained are graphs C_n for $n = 4k + 2$, and K_n for n odd and $n = 2$. The paper [SAST2] deals with the problem of constructions of such graphs.

THEOREM 3.20 (H. SACHS, M. STIEBITZ [SAST2]): *Let s be a positive integer and let $m_i \geq 3$, $i = 1, 2, \ldots, s$, be odd positive integers. If $n = 2^s m_1 m_2 \ldots m_s$, then there exist transitive graphs on n vertices with 2^s simple eigenvalues.*

The construction makes use of operations on graphs starting with complete graphs (see also Section 3.3 for further details on spectral constructions of graphs). This problem is not completely solved and the following conjecture is posed in [SAST2].

CONJECTURE 3.3: *The bound from Theorem 3.20 is attained by graphs without multiple edges or loops if and only if $k \geq 2^{p-1}$.*

H. Sachs in [SAC3] and M. Stiebitz [STI1] describe briefly some of the results from the paper [SAST] from [CVDSA1]. See also [STI2]. Theorem 3.19 has been generalized to graphs that are not transitive.

THEOREM 3.21 (P. ROWLINSON [ROW2]): *Let G be a graph whose automorphism group has orbits $\delta_1, \delta_2, \ldots, \delta_s$. If $|\delta_i| = 2^{m_i} k_i$, where k_i is odd, $i = 1, 2, \ldots, s$, then the number of simple eigenvalues of G is at most $2^{m_1} + 2^{m_2} + \ldots + 2^{m_s}$.*

It should be noted that the proof is short and includes the transitive case.

It is also proved that for any nonnegative integers $m_1, m_2, \ldots, m_s$ there exists a connected graph G (without loops or multiple edges) for which the bound of Theorem 3.21 is attained.

J. H. Smith [SMI1] gives in the following three theorems some upper bounds for the number of eigenvalues of small multiplicity.

THEOREM 3.22 (J. H. SMITH [SMI1]): *Let Γ be the automorphism group of a graph G. Let Γ_1 be the subgroup generated by squares of elements of Γ. Then the number of simple eigenvalues does not exceed the number of Γ_1-orbits.*

This theorem is a variation of an old result of M. Petersdorf and H. Sachs (see [PES2] from [CVDSA1]). It is also contained in the proof of Theorem 3.21 in [ROW2].

THEOREM 3.23(J. H. SMITH [SMI1]): *Let Γ_1 be defined as in Theorem 3.22 and let $\Gamma_2 = \Gamma_1'$ (the commutator subgroup). Then the number of simple eigenvalues plus twice the number of double eigenvalues does not exceed the number of Γ_2-orbits.*

THEOREM 3.24 (J. H. SMITH [SMI1]): *If Γ contains a (nonabelian) simple subgroup Γ_3 (different from A_5), then the sum of the multiplicities of eigenvalues of multiplicity at most 3 does not exceed the number of Γ_3-orbits.*

An *s*-arc in a graph is a sequence of $s + 1$ vertices $x_1, x_2, \ldots, x_{s+1}$ with each adjacent to the next and with no "retracing", *i.e.*, $x_i \neq x_{i+2}$.

THEOREM 3.25 (J. H. SMITH [SMI1]): *If G is a regular graph of degree r, $r \geq 3$, and if the automorphism group of G is transitive on s-arcs, then the only eigenvalues of multiplicity less then $s + 1$ are r and, if G is bipartite, $-r$.*

The special case of this theorem when $s = 1$ due to N. L. Biggs and J. H. Smith has been reported in [CVDSA1] p.138.

A theorem concerning a lower bound for eigenvalues different from $\pm r$ in regular graphs of degree r with the automorphism group acting transitively on "unordered" 2-arcs, is also proved in [SMI1].

Let Γ be the automorphism group of a graph G. An orbit of Γ on which Γ acts as an elementary abelian 2-group is called an *exceptional orbit*.

THEOREM 3.26 (P. ROWLINSON, [ROW2]): *If G is a graph with n vertices, then the number of simple eigenvalues of G is at most $\frac{n+t}{2}$ where t is the number of vertices lying in exceptional orbits of the automorphism group of G.*

This theorem improves an earlier bound in [ROW1].

Exceptional orbits include any fixed points of Γ and any pairs of points fixed by Γ.

For further results in this direction see [ROW2].

Let Γ be a finite abelian group. Graphs admitting Γ as a sharply 1-transitive automorphism group and all of whose eigenvalues are rational have been investigated in [BRME4]. The study is made via the rational algebra $\mathcal{A}(\Gamma)$ of rational matrices with rational eigenvalues commuting with the regular matrix representation of Γ. Techniques for the study of $\mathcal{A}(\Gamma)$ are developed in [BRME1] and [BRME5].

If a regular graph acts transitively on *s*-arcs, then a lower bound for the multiplicities can be given (see Theorem 2.29).

Relations between the minimal polynomial of a graph and the automorphism group have been studied in [CKMT1]. An upper bound for the order of the automorphism group is derived for graphs in which the minimal and the characterisitc polynomial coincide. It is shown that in transitive graphs the degree of the minimal polynomial is bounded from above by the number of orbits of the stabilizer of any given vertex.

Section 3.5: Identification and Reconstruction of Graphs

Problems involving graph identification belong to the class of most important problems in graph theory. In this section we treat the graph isomorphism problem, the graph reconstruction problem, the problem of ordering and classification of graphs, all of them from the viewpoint of spectral graph theory. Finally we give a survey of existing tables of graphs containing some spectral information.

We start with the graph isomorphism problem.

As is well known, graphs are not characterized by their spectra in general. There exist many examples and many constructions of (infinite sequences of) sets of nonisomorphic graphs with the same spectrum (see Section 1.4). However, there are several attempts to improve the spectral approach to the graph isomorphism problem by introducing into the consideration other graph invariants besides the spectrum.

As is mentioned in [CvDSA1], p. 44, a multigraph is completely determined by its eigenvalues and the corresponding eigenvectors. However, eigenspaces of eigenvalues are usually described in terms of a basis which, of course, is not unique, and it is difficult to determine whether two cospectral graphs have the same (or, more precisely, permutationally equivalent) eigenspaces by looking at the bases by which they are defined.

In the excellent article *The graph isomorphism disease* [RECo1] the authors, R. C. Read and D. G. Corneil, describe the spectral approach to the graph isomorphism problem as "an apparently attractive approach ... but which seems to be only a snare and delusion." "Variations of this approach are to use the eigenvectors or to consider matrix functions other than determinant ... Progress in that direction seems to be out of the question, even apart from the fact that except for the determinant, these generalized matrix functions do not seem to be computable in polynomial time."

The discovery of a sufficient graph invariant (a complete graph invariant) would solve the graph isomorphism problem if it were computable in polynomial time. Efforts in this direction can be found in the literature (cf. titles of [RiMW1] and [BaPa1]. See also [Kri1] and [Pöt1], [Pöt2], [Pöt3]. The following attempt has been done in [Ran3]. Let $a_{ij}^{(k)}$ be the (i,j) entry of A^k, where A is adjacency matrix of a graph. As is well known, $a_{ij}^{(k)}$ is the number of walks of length k starting at vertex i and terminating at vertex j. It was conjectured that for a graph with n vertices, the set of n-tuples

$$\mathcal{A} = \{(a_{ij}^{(1)}, a_{ij}^{(2)}, \ldots, a_{ij}^{(n)}) \mid 1 \le i,j \le n\}$$

is a complete graph invariant. Counterexamples have been found by A. J. Schwenk (private communication). Any pair of distance-regular cospectral graphs will give identical sets $\mathcal{A}$. See also [Ran6].

A method to represent each graph by a single number has been developed in [ChLe1]. This number is the determinant of a modified adjacency matrix, *i. e.*, the value of a modified characteristic polynomial. The argument of the polynomial is usually taken to be a nonintegral real number so that the graphs are coded by real numbers. Although it has not been proved that different graphs can always be distinguished by this code, the authors did succeed in solving some practical problems in the computer manipulation of graphs.

A polynomial time algorithm (namely one of the complexity $O(n^4)$ based on eigenvalues and eigenvectors is described in [SeFK1], but the reader might be not quite convinced of the validity of the arguments.

In a series of papers [Bab1], [Bab2], [Bab3], [BaGr1], L. Babai made some progress. Using group theoretic methods he proved the existence of polynomial time algorithm for testing the isomorphism of graphs in the case of simple eigenvalues [Bab1], [Bab2], and in the case of eigenvalues with bounded multiplicities [BaGr1].

The algorithm for testing graph isomorphism, described in [JoLe1] uses eigenvalues and eigenvectors of a modified adjacency matrix of a graph. At the beginning labels and degrees of vertices are used to detect a possible nonisomorphism. Further, eigenvalues and eigenvectors are considered. If this is not sufficient the eigenvectors of vertex deleted subgraphs are computed.

The *complete product* $G_1 \nabla G_2$ of graphs G_1 and G_2 is the graph obtained from G_1 and G_2 by joining each vertex of G_1 with every vertex of G_2. It was noticed in [PrDe1] that for cospectral graphs G_1 and G_2, it may happen that graphs $G_1 \nabla K_1$ and $G_2 \nabla K_1$ have different characteristic polynomials and, of course, this fact can be used to detect nonisomorphism of graphs G_1 and G_2. It was also noted that if $G_1 \nabla K_1$ and $G_2 \nabla K_1$ are cospectral, then $G_1 \nabla K_n$ and $G_2 \nabla K_n$ are cospectral too, and hence there is no point to introducing more than one vertex to be joined to all vertices of the original graphs. We can give the right explanation of this phenomenon by referring to Theorem 2.7 of [CvDSa1]. This theorem expresses the characteristic polynomial of $G \nabla H$ in terms of characteristic polynomials of G, H and of their complements $\overline{G}$, and $\overline{H}$. Hence, for cospectral graphs G_1 and G_2 and for any graph G, the graphs $G_1 \nabla G$ and $G_2 \nabla G$ are cospectral if G_1 and G_2 have cospectral complements.

The *Ádám conjecture* concerning isomorphisms of digraphs with circulant adjacency matrix has been disproved in general (cf. [CvDSa1], pp. 223–224), but there has been discussion in the literature concerning the classes of digraphs for which the conjecture is valid. Using spectral techinques, S. Toida [Toi1] has shown that Ádám conjecture is true for a class of cubic graphs with circulant adjacency matrices as well as for some regular graphs of degree four with a prime power number of vertices.

We turn now to the graph reconstruction problem.

The reconstruction problem is perhaps the most famous open problem in graph theory. This is the problem of establishing whether or not every graph is uniquely determined by the collection of its vertex deleted subgraphs. There are several variations of this problem.

A short summary of the appearance of graph spectra in the graph reconstruction problem is given in [CvDSa1], p. 267. The most important result in this field is due to W. T. Tutte who proved that the characteristic polynomial of a graph is reconstructable from the collection of vertex deleted subgraphs. For some comments see also [CvE4] and [BoHe1].

It is known that the reconstruction conjecture is not true for digraphs. According to [Sto1] let us call a graph reconstruction result *significant* if the analogous statement for digraphs is false. In this terminology Tutte's result is significant. Also determinants of the adjacency matrices (*i. e.*, constant terms in characteristic polynomials) of nonreconstructable digraphs are discussed in examples in [Sto1].

A very interesting expository article [Schw1] considers in parallel the following four reconstruction problems: we may be given the collection of vertex deleted subgraphs or we may be given the collection of characteristic polynomials of vertex deleted subgraphs. We may be asked to reconstruct the graph or we may settle for the less ambitious goal of reconstructing the characteristic polynomial of the graph. The problems are depicted in the following table.

	Given vertex Seleted subgraphs	Given characteristic polynomials of vertex deleted subgraphs
Reconstruct graph	Problem A	Problem B
Reconstruct Characteristic polynomial	Problem C	Problem D

Problem A is the original reconstruction conjecture. Problem C is solved in the affirmative by W. T. Tutte. Problem B has a negative solution. $L(K_{4,4})$ and the Clebsch graph (which are switching equivalent) form a nonreconstructable pair of graphs for this problem. Both graphs are transitive and since they are cospectral all vertex deleted subgraphs in both graphs have the same characteristic polynomial. Problem D, posed by D. Cvetković at the XVIII International Scientific Colloquium in Ilmenau in 1973 and first considered by I. Gutman and D. Cvetković (see, for example, [CvDSA1], p. 267), remains unsolved. A. J. Schwenk suspects in [SCHW1] that the type D conjectures are false but that counterexamples will be difficult to find.

Some enumeration techniques, used by W. T. Tutte in his proof that the characteristic polynomial is reconstructable, are simplified in [POU1]. It is proved that if A and A' are square, symmetric matrices of order n, $n \geq 3$, with entries from a commutative, unitary algebra over Q (the set of rational numbers) such that for any i, $i = 1, 2, \ldots, n$, the principal submatrices A_{ii} and A'_{ii} (obtained by deleting the i-th row and column) are permutationally equivalent, then A and A' have the same determinant.

Using an algebra of graph invariants V. B. Mnuhin [MNU2] has further simplified the proofs of W. T. Tutte and of M. Pouzet.

The characteristic polynomial of a line graph is reconstructable [DED1]. It is proved in the same paper that the R-polynomial [CvDSA1], p. 29, of a graph is reconstructable.

Various sufficient conditions on the eigenvalues and eigenvectors of a graph for the reconstructability of the graph are found in [GOMK5].

In particular, a graph is reconstructable if all but at most one of its eigenvalues are main (see Section 3.8 for a definition) and simple. It is further observed in [HON1] that the graph is also reconstructable when a vertex deleted subgraph exists whose eigenvalues are simple and main.

It is proved in [FAGR1] that the circuit polynomial (a generalization of the characteristic polynomial) of graph is reconstructable from circuit polynomials of circuit deleted subgraphs.

We shall describe how eigenvalues can be used to classify and to order graphs.

For example, strongly regular graphs are classified by their parameters or by their (distinct) eigenvalues since the parameters uniquely determine the eigenvalues and vice versa. Exactly the same thing holds for the parameters of a symmetric block design and the eigenvalues of the corresponding bipartite graphs on the varieties and the blocks.

The graphs can be ordered lexicographically according to their eigenvalues in nonincreasing order. This ordering was first applied in tables of cubic graphs by F. C. Bussemaker, S. Čobeljić, D. Cvetković, and J. J. Siedel (see, for example, [CvDSA1], pp. 268–269). Since the largest eigenvalue λ_1 is equal to 3 in cubic graphs, the second largest eigenvalue λ_2 determines roughly the ordering of graphs. Decreasing λ_2 shows graphs of more "round" shape (smaller diameter, higher connectivity and girth). In the set of cubic graph on n vertices with $n \leq 14$ the connectivity varies monotonically with λ_2. However, as is noted in [CvE4], in cubic graphs with $n \geq 16$ the connectivity is not a monotonic function of λ_2 (F. C. Bussemaker, private communication).

Taking a graph (with sufficiently many vertices) with a maximal girth and adding a little piece (with 4 or 6 vertices) which reduces the connectivity to 2 or 1, F. C. Bussemaker has constructed graphs in which λ_2 is still in the range of values of 3-connected graphs. By the same examples we see that Fielder's algebraic connectivity (*i. e.*, the second smallest eigenvalue of the admittance matrix; cf. [CvDSA1], p. 265) compared to the usual connectivity does not vary monotonically.

The following theorem has been offered as a partial explanation for the behaviour of λ_2.

THEOREM 3.27 (D. CVETKOVIĆ [CVE4]): *Let G be a regular graph of degree r with n vertices. Let x be any vertex of G and let μ be the average vertex degree of the subgraph induced by the vertices not adjacent to x. Then, if λ_2 is the second largest eigenvalue of G, we have*

$$\mu \leq r \frac{\lambda_2{}^2 + \lambda_2(n - r)}{\lambda_2(n - 1) + r}. \tag{3.6}$$

PROOF: Let the vertex set of G be partitioned into three parts: the vertex x, the vertices adjacent to x and the vertices not adjacent to x and let the adjacency matrix of G be partitioned into the corresponding blocks. The average row sums in the blocks form the following matrix

$$B = \begin{pmatrix} 0 & r & 0 \\ 1 & r - \nu - 1 & \nu \\ 0 & r - \mu & \mu \end{pmatrix}$$

where ν is the average number of edges going from a vertex adjacent to x to vertices nonadjacent to x. Obviously we have $r\nu = (n-1-r)(r-\mu)$. According to [HAEM] from [CVDSA1] the spectrum of B interlaces the spectrum of G. The largest eigenvalue of B is r and the remaining two eigenvalues satisfy the equation $\lambda^2 - (\mu - \nu - 1)\lambda - \mu = 0$. Because of the interlacing of eigenvalues we get $\lambda_2{}^2 - (\mu - \nu - 1)\lambda_2 - \mu \leq 0$ which after the elimination of ν yields (3.6).

The function on the right hand side of (3.6) is increasing for each n and r. Hence, decreasing λ_2 decreases μ, which means that the edges are coming closer to x! On the other hand, decreasing μ causes the the quantity $\tau = r - \nu - 1$ to decrease, and this quantity is the average vertex degree in the subgraph induced by the vertices adjacent to x. Hence the edges are getting away from both subgraphs induced by adjacent and nonadjacent vertices. They tend to connect adjacent and nonadjacent vertices and this phenomenon intuitively corresponds to the idea of becoming more "round". In the case of cubic graphs up to 14 vertices, the bound given in (3.6) is sufficiently sharp, to force the edges to the desired place.

Some inequalities for the algebraic connectivity of a graph (which is equivalent to the second largest eigenvalue in regular graphs) have been recently obtained in [ALMI1]. These inequalities are related to the girth, the connectivity, and the diameter and further support the idea of ordering graphs by eigenvalues.

It was noticed in [CvPE1] that graphs are even "better" ordered if we use the lexicographic ordering by spectral moments in nonincreasing order instead of eigenvalues (the tables in the appendix are ordered this way). Let $\lambda_1, \ldots, \lambda_n$ be the eigenvalues of a graph G. Then for $k \geq 0$, the quantity

$$S_k = \sum_{i=1}^{n} \lambda_i{}^k$$

is called the *k-th spectral moment* of G. We have $S_0 = n$, $S_1 = l$, $S_2 = 2m$, $S_3 = 6t$, where n, l, m, t denote respectively the number of vertices, the number of loops, the number of edges, and the number of triangles of G. Of course, two graphs are cospectral if and only if they have the same spectral moments. The two cospectral graphs on 6 vertices are ordered by the sum of coordinates of the (positive) normalized eigenvector belonging to the largest eigenvalue. In sets of graphs with a large number of vertices, cospectral graphs occur very often, and some other graph invariants based on the eigenvectors must be used to classify graphs.

Experiences with the interactive programming system **Graph** (an expert system for graph theory developed at the University of Belgrade, Faculty of Electrical Engineering, during the period 1980-1984; see [CVE8], [CvKs1], [CVE6], [CVEP1], [CVE9]) in creating graph catalogues shows that such an ordering of graphs is very useful. For example, regular graphs of degree 4 on 10 vertices cannot be easily recognized by their pictures. However, if we calculate the spectrum of our graph it can be easily found in a catalogue where graphs are spectrally ordered.

We conclude this section by surveying tables of graphs which are relevant to spectral graph theory.

[CvPE1] contains a table of the 112 connected graphs on 6 vertices together with their eigenvalues, characteristic polynomials, spectral moments and many other types of data. In [MERW1] permanental polynomials (see section 4.6) of graphs on 6 vertices are given.

[CvRA1] contains a table which explains the construction of the 68 connected regular graphs nonisomorphic but cospectral to line graphs (cf. Chapter 1).

The paper [BuMS1] contains several tables related to switching classes of graphs (*i. e.*, to two-graphs, cf. [CvDSA1], pp. 199–203): a table of all nonisomorphic two-graphs on $n \leq 9$ vertices, a table of all nonisomorphic two-graphs on $n \leq 10$ vertices whose automorphism group does not fix any graph in its switching class, and data on known regular two-graphs on $n \leq 50$ vertices. The two-graph is represented by its $(-1, 1, 0)$ adjacency matrix and some of its invariants (e. g.the eigenvalues and the order of automorphism group) are also given.

The appendix of [GHMK1] contains a small but very interesting table of cospectral graphs.

Tables of transitive graphs with fewer than 20 vertices are given in [McKA1]. Along with many other types of data, the eigenvalues of these graphs are given.

A table of spectra of hexagonal systems (animals) is published in [GuPM1], [GuPM2]. Matching polynomials and their zeros are given as well.

Tables of characteristic polynomials of some infinite periodic plane graphs (see Chapter 6) are given in [Col1]. These graphs include the 11 known regular plane coverings of Laves.

Section 3.6: The Shannon Capacity and Spectral Bounds for Graph Invariants

Several graph invariants can be bounded from above and/or below by some functions of the eigenvalues of the graph. The central result of this section is Theorem 3.28, which gives an upper bound for the capacity of a graph. This bound enables the determination of the Shannon capacity of a pentagon, thus providing a solution to a problem open for many years. The concept of the capacity of a graph is also relevant in the theory of communication channels and networks.

The strong product $G_1 \cdot G_2$ of graphs G_1 and G_2 is the NEPS of these graphs with the basis $\{(1, 1), (1, 0), (0, 1)\}$ (cf. [CvDSA1], p. 66, and Section 3.3). As is known, the vertex set of NEPS of several graphs is the Cartesian product of vertex sets of these graphs. In the case of strong product of G_1 and G_2, vertices (x_1, y_1) and (x_2, y_2) are adjacent if and only if either x_1 and y_1 are adjacent in G_1 and y_1 and y_2 are adjacent in G_2, or x_1 and x_2 are adjacent in G_1 and $y_1 = y_2$, or $x_1 = x_2$ and y_1 and y_2 are adjacent in G_2.

Let $\alpha(G)$ be the *internal stability number* of the graph G, *i.e.*, the maximum number of mutually nonadjacent vertices in G.

Let us consider vertices of a graph G as letters in an alphabet. The letters are used to code some information which is sent through a telecommunication channel. Due to the noise in the channel, certain letters may be mixed up before reaching the output of the channel, *i.e.*, they are "cofoundable". Two vertices of G are adjacent if an only if x and y are cofoundable letters.

Let G^k be the k-th power of G using the strong product. Then $\alpha(G^k)$ is the maximum number of k-letter words such that any two of them are not cofoundable in at least one position.

Having this view in mind, the quantity $\Theta(G)$ is defined by

$$\Theta(G) = \sup_k \sqrt[k]{\alpha(G)^k} \tag{3.7}$$

and is called the *Shannon capacity* of a graph after C. E. Shannon [SHA1].

Let $G = K_5$. We then have $\alpha(G) = 2$, while $\alpha(G^2) = 5 > 2^2$. This can be seen in Figure 3.4 by a chess interpretation. The graph $G = K_5$ can be interpreted as the graph of king's motion on the left chessboard in Figure 3.4. This is a cylindric chessboard, *i.e.*, the edges 1—1' are to be glued. There corresponds to the graph G^2 a toroidal chessboard, the right one in Figure 3.4, where first the top and the bottom edges are to be glued to form a cylinder, and then the opposite edges of this cylindric board are to be glued to form a torus. The positions of kings on the board correspond to an internally stable set. Pairs of signals that can be reconstructed at the output of the channel are (a, a), (b, d), (c, b), (d, e), (e, c).

Hence, in general we have $\alpha(G) \leq \Theta(G)$.

The problem of finding $\Theta(K_5)$ was open for over 20 years until it was solved by L. Lovász [LOV2] who used linear algebraic techniques. Now we know that $\Theta(K_5) = \sqrt{5}$. The technique used by L. Lovász enables us to determine the capacity for many graphs.

There is an excellent expository article on the Shannon capacity problem and related topics by W. Haemers [HAE1]. However, we shall follow the thread of the original articles.

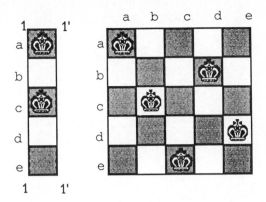

Figure 3.4 Nonattacking kings on a toroidal chessboard

An *orthonormal representation* of a graph G on n vertices is a system $v_1, v_2, \ldots, v_n$ of unit vectors in an Euclidean space such that v_i and v_j are orthogonal whenever vertices i and j are nonadjacent. For example, if $\lambda_n < 0$ is the least eigenvalue of the adjacency matrix of a graph then $I + \frac{1}{\lambda_n} A$ is the Gram matrix of an orthogonal representation.

The *value* of an orthonormal representation $v_1, v_2, \ldots, v_n$ of a graph G is defined by

$$\min_c \max_{1 \leq i \leq n} \frac{1}{(cv_i)^2}$$

where c runs over all unit vectors. Let $\vartheta(G)$ be the minimum value of representations of G.

Let $u_1, u_2, \ldots, u_n$ and $v_1, v_2, \ldots, v_m$ be orthonormal representations of graphs G and H, respectively. It is easy to prove that $u_i \otimes v_j$, $i = 1, 2, \ldots, n$, $j = 1, 2, \ldots, m$, is an orthonormal representation of the strong product of G and H. One can also prove that

$$\vartheta(G \cdot H) = \vartheta(G)\vartheta(H),$$

$$\alpha(G^k) \leq \vartheta(G^k) = (\vartheta(G))^k,$$

and we get the following theorem.

THEOREM 3.28 (L. LOVÁSZ [LOV2]):   *For any graph G*

$$\Theta(G) \le \vartheta(G).$$

Now it is easy to derive that $\vartheta(K_5) = \sqrt{5}$ and hence $\Theta(K_5) = \sqrt{5}$.

The Lovász theta-bound $\vartheta(G)$ can be evaluated in many ways via several optimizations on matrices associated to a graph (see [LOV2] and [HAE1]).

For a regular graph the quantity $\vartheta(G)$ can be bounded in terms of the eigenvalues of G.

THEOREM 3.29 (L. LOVÁSZ [LOV2]):   *Let G be a regular graph on n vertices with eigenvalues* $\lambda_1 \ge \lambda_2 \ge \ldots \ge \lambda_n$. *Then*

$$\vartheta(G) \le \frac{-n\lambda_n}{\lambda_1 - \lambda_n}.$$

If G is edge-transitive then equality holds.

COROLLARY 3.5:   *For odd n we have*

$$\vartheta(C_n) = n\frac{\cos\frac{\pi}{n}}{1 + \cos\frac{\pi}{n}}.$$

The upper bound for $\vartheta(G)$ in Theorem 3.29 was known to be an upper bound for the internal stability number $\alpha(G)$ of a graph G (see a note on p. 115 of [CVDSA1]).

Another upper bound for the Shannon capacity $\Theta(G)$ of a graph G has been found in [HAE4], [HAE3].

Let $r(A)$ denote the rank of the matrix A.

THEOREM 3.30 (W. HAEMERS [HAE4]):   *Let* $A = (a_{ij})$ *be a symmetric matrix of order n with* $a_{ii} = 1$. *Let G(A) be the graph with vertex set* $\{1, 2, \ldots, n\}$ *where for any* $i \ne j$, *i and j are adjacent if* $a_{ij} \ne 0$. *Then*

$$\Theta(G(A)) \le r(A).$$

The proof is based on the fact that for symmetric matrices $r(A \otimes B) = r(A)r(B)$.

Now it is useful to introduce the quantity $R(G)$ for a graph G:

$$R(G) = \min_A \{r(A) \mid G(A) \cong G\}.$$

Of course, we have $\Theta(G) \leq R(G)$.

It is shown in [HAE4] that the Shannon capacity of the Schläfli graph is strictly less than the Lovász bound.

The function $\vartheta(G)$ has been compared in [SCHR2] with the previously developed Delsarte linear programming bound for the cardinality of cliques in association schemes (see [DEL1] of [CvDSA1]). It appears that the two upper bounds can be treated in a unified fashion. It is also shown that there exist graphs G such that $\vartheta(G)\vartheta(\overline{G}) = n$, the number of of vertices of G.

Using these considerations A. Schrijver has proved in [SCHR1] that if $K_{n,m,r}$ is the graph whose vertices are all m-subsets of a fixed n-set, with two vertices being adjacent if their intersection contains less than r elements, then $\vartheta(K_{n,m,r}) = \binom{n-r}{m-r}$ if n is large compared to m.

Bounds mainly involving the rank of the adjacency matrix for several graph invariants have been elaborated in [NUF1]–[NUF6] (seealso [NUF2] from [CvDSA1]).

A set of vertices is called a *dominating set* of a graph G if every vertex is either in the set or adjacent to a vertex in the set. The *domination number* $\beta(G)$ of a graph G is the size of the smallest dominating set of G. Determination of $\beta(G)$ is NP-hard. Let $r(G)$ be the rank of the adjacency matrix of G. It is proved in [NUF5] that $\beta(G) \leq r(G)$ and $\beta(G) \leq r(\overline{G})$.

The *clique number* $K(G)$ of a graph G is defined as the maximum number of vertices in a complete subgraph of G. It is proved in [NUF6] that $K(G) \leq r(G)$. The equality holds if and only if G is complete k-partite, where $r(G) = k$. Similar relationships can be shown for the chromatic number, the internal stability number, the diameter, etc. Most of the inequalities are not very sharp, but note that the determination of most of these numbers is NP-hard.

Several inequalities for the internal stability number , the clique number, the chromatic number, etc., have also been obtained in [HAE5]. Using these inequalities and other spectral techniques, W. Haemers has determined all strongly regular graphs with chromatic number at most 4. A new proof of a known inequality for the internal stability number of a strongly regular graph (cf. [CvDSA1] p. 115) has been given in

[HALI1]. The same inequality is mentioned in [PETR1]. The author constructs some strongly regular graphs and related block designs in which the inequality is actually an equality.

Some lower bounds for the size $K(G)$ of a maximal clique and the chromatic number $\chi(G)$ of a graph G in terms of eigenvalues are given in [EDEL1]. For example, it is proved that

$$K(G) > \frac{n}{n - \lambda_1} - \frac{1}{3},$$

where n is the number of vertices and λ_1 the largest eigenvalue of G. The first term in the lower bound for $K(G)$ is known to be a lower bound for $\chi(G)$ (cf. [CVDSA1], p. 92). Several upper bounds for $\chi(G)$ have been compared in [DRFR1]. The spectral bound, $\chi(G) \leq 1 + \lambda_1$, ([CVDSA1], p. 92) usually came out in the middle of the list.

Graph partitions are treated in [YUMO1] from the viewpoint of eigenvalues. Let G be a graph with the vertex set X. For any $X_1, X_2 \subseteq X$ let us define

$$d(X_1, X_2) = \frac{1}{|X_1||X_2|} \sum_{\substack{i \in X_1 \\ k \in X_2}} a_{ik},$$

where $a_{ii} = 1$ and a_{ik} is equal to the number of edges connecting i and k for each $i \neq k$.

If X_1 and X_2 partition X, then the quantity

$$\frac{|X_1||X_2|}{|X|^2} (d(X_1, X_1) - 2d(X_1, X_2) + d(X_2, X_2))$$

is called the *quality of the partition*.

THEOREM 3.31 (F. JUHÁSZ, K. MALYUSZ [JUMA1]): *Let $S = \{X_1, X_2\}$ be a partition of the n vertices of G. Let μ be the largest eigenvalue of the matrix PAP where $A = (a_{ij})$ and $P = (p_{ij})$ with $p_{ij} = \delta_{ij} - \frac{1}{n}$. Then the quality of S is at most μ/n.*

Another related theorem has been proved in [JUMA1].

Section 3.7: Spectra of Random Graphs

Random graphs have been studied for many years from several points of view. However, only very recently has the behaviour of eigenvalues of random graphs attracted researchers in graph theory and its applications (F. Juhász, B. D. McKay and others in mathematics, B. E. Eichinger and others in physics). However, eigenvalues of random matrices have been studied earlier (see [ARN1]–[ARN3], [GIR1]). One of the significant early results is the Wigner's semicircle law for the distribution of eigenvalues (see e. g.[JUH2]).

One of the major topics is the investigation of the asymptotic behaviour of eigenvalues (or of some functions of eigenvalues) of random graphs as the number of vertices gets large. (However, for spectra of infinite graphs see Chapter 6).

Let $A_n = (a_{ij})$ be an n by n symmetric random 0-1 matrix, where $a_{ii} = 0$, and a_{ij}, $i > j$ are independent random variables which equal 1 with probability p:

$$P(a_{ij} = 1) = p \quad (0 < p < 1).$$

Of course, A_n can be interpreted as the adjacency matrix of a graph.

If λ_1 is the largest eigenvalue of the matrix A_n, and λ_2 is the second largest eigenvalue, then [JUH2]

$$\lim_{n \to \infty} \frac{\lambda_1}{n} = p, \quad \text{with probability 1, and}$$

$$\lim_{n \to \infty} P(\lambda_2 > n^{1/2+\epsilon}) = 0$$

for each $\epsilon > 0$. Hence, $\lambda_2 = o(n^{1/2+\epsilon})$ where $\epsilon > 0$.

If the entries a_{ij} of A_n are independent random variables with the same distribution $H(x)$, if the mean value $M(a_{ij}) = 0$, and if moments $M(a_{ij}^k)$ are finite for all $k \geq 2$, then

$$\lim_{n \to \infty} P(|\lambda_1| > n^{1/2+\epsilon}) = 0$$

for each $\epsilon > 0$.

The results of [JUH2] are extended in [JUH1] and [JUH3] to non-symmetric random 0-1 matrices, i. e., to digraphs. ([JUH3] seems to be a version of [JUH1]).

Let $A_n = (a_{ij})$ be a (not necessarily symmetric) random 0-1 matrix, where $a_{ii} = 0$, $P(a_{ij} = 1) = p$, $P(a_{ij} = 0) = 1 - p$, for $i \neq j$. Let $\lambda_1 \geq \lambda_2 \geq \ldots \geq \lambda_n$ be the spectrum of A_n.

In [JUH3] it is proved that, with probability approaching 1, $\lambda_1 = np + o(n^{1/2+\epsilon})$, i.e., $P(|\lambda_1 - np| > \delta n^{1/2+\epsilon}) \to 0$ for each $\delta > 0$. In addition, we have

$$\lambda_i = o(n^{-1/2+\epsilon}) \tag{3.8}$$

for $i = 2, 3, \ldots, n$.

It is proved that (3.8) holds for each $i = 1, 2, \ldots, n$ if the entries a_{ij}, $i \neq j$, of A_n are independent random variables with the same distribution function, if $a_{ii} = 0$, $M a_{ij} = 0$, and all moments $M(a_{ij}^k)$, $k \geq 2$, are finite for $i \neq j$.

F. Juhász [JUH4] also investigates the asymptotic behaviour of the Lovász theta function $\vartheta = \vartheta(G)$ of a graph G. The function $\vartheta(G)$ is a function related to eigenvalues and it has been defined in [LOV2] (see section 3.6).

Let G be a random graph on $n = n(G)$ vertices, $p(a_{ij} = 1) = p$, and $q = 1 - p$. With probability $1 - o(1)$ the following estimations hold [JUH4]

$$\frac{1}{2}\sqrt{2n/p} + O(n^{1/3}\log n) \leq \vartheta(G) \leq 2\sqrt{2n/p} + O(n^{1/3}\log n).$$

Other problems concerning spectra of random regular graphs are treated in papers [MCKA2], [MCKA3] and [MCKA5]. Note that [MCKA3] is an extended version of [MCKA2].

Let G be a regular graph of degree d on $n = n(G)$ vertices having $c_k(G)$ circuits of length k, $k \geq 3$. For any $x \in \Re$, $F(G, x)$ denotes $1/n$ times the number of eigenvalues of G (including the multiplicities) which do not exceed x.

Obviously we have $F(G, x) = 0$ if $x < -d$, and $F(G, x) = 1$ for $x \geq d$, and $F(G, x)$ is a monotonically nondecreasing and right continuous function for all $x \in \Re$.

Let G_i, $i = 1, 2, \ldots$, be an arbitrary sequence of regular graphs of degree $d \geq 2$ such that

$$n(G_i) \to \infty \text{ as } i \to \infty, \text{ and} \tag{3.9}$$

$$\frac{c_k(G_i)}{n(G_i)} \to 0 \text{ as } i \to \infty \text{ for all } k \geq 3 \qquad (3.10).$$

Then it is shown in [McKa3] that, for any $x \in \Re$, $F(G_i, x) \to F(x)$ as $i \to \infty$, where $F(x) = 0$ for $x \leq -2\sqrt{d-1}$, $F(x) = 1$ for $x \geq 2\sqrt{d-1}$ and for all other x we get

$$F(x) = \int_{-2\sqrt{d-1}}^{x} \frac{d\sqrt{4(d-1) - t^2}}{2\pi(d^2 - t^2)} \, dt.$$

Conversely, if condition (3.9) holds and if $F(G_i, x)$ does not converge to $F(x)$ for some $x \in \Re$, then condition (3.10) fails for some $k \geq 3$.

Spanning trees in random graphs are investigated in [McKa5]. Let $n_1 < n_2 < \cdots$ be the numbers of vertices for a sequence of connected graphs of a fixed degree d, $d \geq 3$. For each i, choose a graph G at random from the set of all connected regular graphs of degree d on n_i vertices. If $t(G)$ denotes the number of spanning trees of G, then with probability 1,

$$\lim_{i \to \infty} t(G_i)^{1/n_i} = \frac{(d-1)^{d-1}}{(d^2 - 2d)^{\frac{d-2}{2}}}.$$

The result is based on results on the number of spanning trees from [McKa4], mentioned in Section 3.9. The result is further extended to graph having prescribed degree sequences [McKa6].

A statistical model of polymer molecules has been considered in papers [Eic1], [Eic2], [Eic3], [EiMa1], [MaEi1]. Random networks are represented by graphs and the admittance matrix (the Kirchhoff matrix) of these graphs is studied. Many physical properties of molecules considered depend on C-eigenvalue (eigenvalues of the Kirchhoff matrix) distribution if the graph is chosen at random from a certain set of graphs. It is hard to extract mathematical statements from the context of physics. See also Section 5.16.

Section 3.8: The Number of Walks in a Graph

The enumeration of walks in a graph can be performed by using the eigenvalues and the eigenvectors of the graph. This is useful since many interesting mathematical problems can be reduced to the problem of walk enumeration.

By a *walk of length k* in a digraph we mean any sequence of (not necessarily different) arcs $u_1, u_2, \ldots, u_k$, such that for $i = 1, 2, \ldots, k-1$ the terminating vertex of u_i is the starting vertex of u_{i+1}. Similarly, a walk of length k in an (undirected) graph is an alternating sequence $x_1, u_1, x_2, u_2, \ldots, x_k, u_k, x_{k+1}$ of vertices $x_1, x_2, \ldots, x_k, x_{k+1}$ and edges $u_1, u_2, \ldots, u_k$ such that for any $i = 1, 2, \ldots, k$ the x_i and x_{i+1} are the endpoints of the edge u_i. The walk is closed if $x_{i+1} = x_1$.

Counting the number of walks with specified properties in a graph (or digraph) is related to graph spectra by the well known result mentioned previously; if $A = (a_{ij})$ is the adjacency matrix of the graph, then the entry $a_{ij}^{(k)}$ of the matrix A^k is equal to the number of walks of length k that start at vertex i and terminate at vertex j. Thus, for example, the number of closed walks of length k is equal to the k-th spectral moment (see Section 3.5).

Many combinatorial enumeration problems can be reduced to the enumeration of walks in a suitably chosen graph or digraph. Also, formulas giving the number of walks in terms of eigenvalues and eigenvectors represent a link between spectral and structural properties of a graph which are very useful auxiliary tools in treating many problems on graphs. An important notion related to the number of walks is the main part of the spectrum (see below).

Previous results on the number of walks are described in [CvDSa1], in Section 1.8 and Section 7.5. We shall mention shortly the recent results from this part of spectral graph theory.

A sequence of k symbols is called a *k-sequence*. Let C_k be a property of k-sequences of zeros and ones. A sequence of zeros and ones in which each subsequence of k consecutive symbols has property C_k is called a C_k-*sequence*. Consider the digraph G whose vertices are all k-sequences having property C_k. An arc goes from x to y if the k-sequence x can be extended by a zero or a one so that last k symbols form the k-sequence y. A C_k-sequence is a walk in G and known formulas for the numbers of walks can be applied to this case [Ein1].

Corresponding generating functions including the C_k-sequences with a prescribed number of zeros are derived too.

Given a graph G, let $S(G)$, $L(G)$ and G^* be, respectively, the (first) subdivision graph of G (see Section 3.2), the line graph of G and the graph obtained from G by adding to each vertex, as many loops as is the degree of the vertex. Let $H_G(t) = \sum_{k=0}^{\infty} N_k t^k$ be the generating function for numbers N_k of walks of length k in the graph G. The following formulas were proved in [PE1], [PE2]

$$H_{S(G)}(t) = \frac{(2t+1)^2}{4t^2} H_{G^*}(t^2) - \frac{n+4t}{4t^2},$$

$$H_{L(G)}(t) = \frac{1}{4t} H_{G^*}\left(\frac{t}{2t+1}\right) - \frac{n}{4t}.$$

In a chemical context, M. Randić [RAN1] introduces a coding of the n vertices of a graph by the numbers of walks. To each vertex there corresponds a sequence of integers $N_1, N_2, \ldots, N_n$ where N_k represents the number of closed walks of length k starting and terminating at the vertex being considered. Properties of such vertex codes are discussed in the examples.

Let $N_{k,i}$ be the number of closed walks of length k that start and end at vertex i of a graph G and let $H_{G,i}(t) = \sum_{k=0}^{\infty} N_{k,i} t^k$ be the corresponding generating function. Further, let $H'_{G,i}(t)$ be the generating function for the numbers of walks of length k which start at i and terminate elsewhere.

THEOREM 3.32 (C. D. GODSIL, B. D. MCKAY [GOMK5]):

$$H_{G,i}(t) = \frac{1}{t} \frac{P_{G-i}\left(\frac{1}{t}\right)}{P_G\left(\frac{1}{t}\right)},$$

$$H'_{G,i}(t) = \frac{1}{t} \frac{(-1)^n P_{G-i}\left(\frac{1}{t}\right)}{P_G\left(\frac{1}{t}\right)} \left(\frac{P_{\overline{G}}\left(-\frac{1}{t}-1\right)}{P_G\left(\frac{1}{t}\right)} + \frac{P_{\overline{G}-i}\left(-\frac{1}{t}-1\right)}{P_{G-i}\left(\frac{1}{t}\right)} \right)^{1/2}.$$

The first formula follows also from an early result by P. W. Kasteleyn (see [CVDSA1], p. 47).

A graph is called *walk-regular* if the generating function $H_{G,i}(t)$ is the same for any vertex i. By Theorem 3.32 all vertex deleted subgraphs of walk-regular graphs have the same characteristic polynomial. Walk-regular graphs are, of course, regular graphs. Transitive graphs, strongly regular graphs and distance-regular graphs are examples of walk-regular graphs. Walk-regular graphs are introduced in [GoMк4] and their existence is studied (see also Chapter 2).

A formula for the number of walks in a graph has been derived in [ABR1]. In that context *co-circuit graphs* (*i. e.*, graphs with the same number of circuits of any specified length) are studied.

A formula for counting the number of circuits of an arbitrary length is derived in [HoSe2]. Traces of the adjacency matrix are used and the properties of the characteristic polynomial are exploited.

An eigenvalue is called a *main eigenvalue* if its eigenspace contains an eigenvector the sum of whose coordinates is different from zero (cf., for example, [CvDSa1], p. 46).

Let $M(\lambda) = (\lambda - \mu_1) \ldots (\lambda - \mu_k)$, where $\mu_1, \ldots, \mu_k$ are the main eigenvalues of a graph G. Then coefficients of $M(\lambda)$ are rational [CvPe1].

It is known that eigenvalues $\Lambda_{i_1,\ldots,i_n}$ of the NEPS (non-complete extended p-sum of graphs) of graphs $G_1, G_2, \ldots, G_n$ can be expressed in terms of eigenvalues λ_{ij_i} of G_i, $i = 1, 2, \ldots, n$ as a certain function $\Lambda_{i_1,i_2,\ldots,i_n} = F(\lambda_{1i_1}, \lambda_{2i_2}, \ldots, \lambda_{ni_n})$ (see Section 3.3). It was observed in [Pe1],[Pe2] that $\Lambda_{i_1,i_2,\ldots,i_n}$ is a main eigenvalue if and only if all eigenvalues $\lambda_{1i_1}, \lambda_{2i_2}, \ldots, \lambda_{ni_n}$ are main.

There was once a conjecture (see [HAS1] from [CvDSa1]) that the spectrum of a divisor with the smallest number of vertices was just the main part of the spectrum of the graph considered (for the definition of a divisor of a graph, see Chapter 4 of [CvDSa1]). This conjecture has been disproved in [PoSu1] by a counterexample with 7 vertices. However, the conjecture had already been disproved a few years earlier (see [CvE18] from [CvDSa1]) although the counterexample was a graph with 21 vertices.

The transfer matrix method has been used in [AHR1] to enumerate the number of monomer-dimer configurations on m by n rectangular boards. As explained in [CvDSa1], pp. 247–251, this number can be interpreted as the number of walks in a digraph called the *transfer graph*. The largest eigenvalue of the transfer graph is used in the study of the asymptotics of the number of walks.

The asympotic behaviour of the number of walks of length k between two fixed vertices in a graph as k gets large has been described in terms of the largest eigenvalue of the graph and in terms of the coordinates of the corresponding eigenvector in [LIFE1] (see **Mathematical Reviews** 80:k 05079).

The existence of eigenvectors and eigenvalues of graphs when a generalized adjacency matrix is constructed with entries from a semiring is discussed in [GOMI1], along with the connections of this problem with walks in a graph. See [GOMI2] also.

Section 3.9: The Number of Spanning Trees in a Graph

The number of spanning trees in a graph can be calculated using several spectral techniques (cf. Sections 1.5 and 7.6 of [CVDSA1]).

Let G be a regular graph of degree r on n vertices, let $P_G(\lambda)$ be its characteristic polynomial and $\lambda_1 = r, \lambda_2, \ldots, \lambda_n$ its eigenvalues. As is well known, the number of spanning trees $t(G)$ in a graph G, can be express by the following formula

$$t(G) = \frac{1}{n} \prod_{i=2}^{n} (r - \lambda_i) = \frac{1}{n} P_G{}'(r). \tag{3.11}$$

The fact that the number of spanning trees of a regular graph can be determined from the graph spectrum, is used in [SED1] to compute the number of spanning trees for cubic graphs on 12 vertices. The same techinque has been used in [WAYA1] to develop several formulas for the number of spanning trees in graphs with a circulant adjacency matrix. Note that the eigenvalues of such graphs can be calculated easily (cf. [CVDSA1], p. 53). Using eigenvalues, B. D. McKay [MCKA4] has found a tight upper bound for the number of spanning trees in regular graphs in terms of the number of small circuits and other subgraphs. For spanning trees in random graphs see Section 3.7. We shall now describe two extensions of formula (3.11) to nonregular graphs.

Graphs in which all vertices but one have a fixed degree r are called *nearly regular* graphs of degree r. The vertex not having degree r is called the *exceptional vertex*.

THEOREM 3.33 (D. CVETKOVIĆ, I. GUTMAN [CVGU1]): *Let G be a nearly regular graph of degree r and let H be the subgraph obtained by removing the exceptional vertex. Then*

$$t(G) = P_H(r). \qquad (3.12)$$

PROOF: By the matrix-tree theorem (see, for example, [CVDSA1], p.38) $t(G)$ is equal to any cofactor in the admittance matrix $D - A$ (A is the adjacency matrix and D is the matrix of vertex degrees). Formula (3.12) is obtained if we take the cofactor of the diagonal element corresponding to the exceptional vertex of G.

EXAMPLE 3.2. The characteristic polynomial of a circuit C_n is known to be equal to $2T_n(\frac{\lambda}{2}) - 2$, where $T_n(\lambda)$ is the Chebyshev polynomial of the first kind. Introducing a new vertex connected by an edge to all vertices of C_n we get a wheel W_n, which is a nearly regular graph. By Theorem 3.33 we have $t(W_n) = 2T_n(\frac{3}{2}) - 2$.

The *inner dual* G' of a plane graph G is the subgraph of the usual dual G^*, obtained by deleting the vertex corresponding to the infinite region of the original plane graph.

Let G be a plane graph in which any finite region is bounded by a circuit of a fixed length r. (The so called animals belong to this class of graphs). Then G^* is a nearly regular graph. On the other hand, it is known that the graphs G and G^* have the same number of spanning trees. Combining these observations with Theorem 3.33 we immediately reach the following theorem.

THEOREM 3.34 (D. CVETKOVIĆ, I. GUTMAN [CVGU1]): *If G is a plane graph in which any finite region is bounded by a circuit of length r, then*

$$t(G) = P_{G'}(r).$$

EXAMPLE 3.3. For the graph G in Figure 3.5, the inner dual G' is the path P_n on n vertices. Since $P_{P_n}(\lambda) = U_n(\frac{\lambda}{2})$, where $U_n(\lambda)$ is the

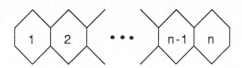

Figure 3.5 A graph with inner dual equal to a path

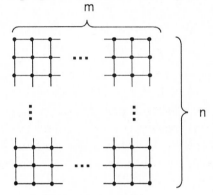

Figure 3.6 $G'_{m,n} = G_{m-1,n-1}$

Chebyshev polynomial of the second kind, we have $t(G) = U_n(3) = \sum_{k=0}^{[n/2]} (-1)^k \binom{n-k}{k} 6^{n-2k}$.

If in Figure 3.5 we replace hexagons with k-gons, then we get $t(G) = U_n(\frac{k}{2})$.

EXAMPLE 3.4. For the graph $G_{m,n}$ in Figure 3.6 we have $G'_{m,n} = G_{m-1,n-1}$.

The spectrum of $G_{m-1,n-1}$ consists of eigenvalues (cf. [CVDSA1], p. 74) $2\cos\frac{i\pi}{m} + 2\cos\frac{j\pi}{m}$, $i = 1, 2, \ldots, m-1$, $j = 1, 2, \ldots, n-1$, and therefore,

$$t(G_{m,n}) = 4^{(m-1)(n-1)} \prod_{i=1}^{m-1} \prod_{j=1}^{n-1} \sin^2\frac{i\pi}{2m} + \sin^2\frac{j\pi}{2n}.$$

The result was obtained by other spectral methods in [KREW] from [CVDSA1].

Nonisomorphic plane graphs can have isomorphic inner duals. Consequently, such graphs have the same number of spanning trees.

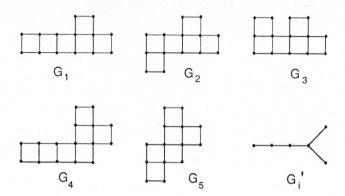

Figure 3.7 Graphs with equal numbers of spanning trees

For example, the five graphs (animals) $G_1, \ldots, G_5$ in Figure 3.7 have an equal number of spanning trees, namely, $4^6 - 5 \cdot 4^4 + 5 \cdot 4^2 = 2896$, since $r = 4$ and $P_{G_i'}(\lambda) = \lambda^6 - 5\lambda^4 + 5\lambda^2$.

Another extension of the formula (3.11) to the nonregular case is given by the following theorem.

THEOREM 3.35 (I. GUTMAN [GUT32]): *Let G be a semiregular bipartite graph with parameters n_1, n_2, d_1, d_2, $d_1 \geq d_2$, $n_1 + n_2 = n$. Then the number of spanning trees $t(G)$ is given by*

$$t(G) = d_2{}^{n_2 - n_1} \frac{d_1 + d_2}{n_1 + n_2} \prod_{i=2}^{n_1} (d_1 d_2 - \lambda_i{}^2).$$

The characteristic polynomial and the eigenvalues of the admittance matrix $C = D - A$ are called *C-polynomial* and *C-eigenvalues* of G. It is known (cf. Proposition 1.3 from [CVDSA1]) that the number of spanning trees in a graph is equal to the product of nonzero *C*-eigenvalues divided by the number of vertices.

A graph is called *equiarboreal* if the number of spanning trees containing a specified edge is independent of the choice of the edge. It is proved by the use of the *C*-polynomial in [GOD4] that the graph of any relation of an association scheme (and thus any distance regular graph) is equiarboreal.

The *C*-spectrum is used in [CHE1] to find a graph G with a maximum number of spanning trees among the multigraphs with n vertices and m edges. The unique such graph H is either a regular complete bipartite graph or graph obtained by adding a constant number of edges to each pair of vertices in a regular complete bipartite graph. In fact the following more general theorem has been proved.

THEOREM 3.36 (C. S. CHENG [CHE1]): *Let* $\lambda_1, \lambda_2, \ldots, \lambda_n = 0$ *be the C-eigenvalues of a multigraph G. Let f be any real-valued function on* $[0, 2m]$ *which is strictly convex and continuously differentiable on* $(0, 2m)$ *with* f' *strictly concave on* $(0, 2m)$. *Then H is the unique multigraph which minimizes* $\sum_{i=1}^{n-1} f(\lambda_i)$ *over all multigraphs G with n vertices and m edges.*

A similar theorem for graphs is proved as well.

Let $I = [0, \infty)$ and let x and y be vectors belonging to I^n. It is said that y majorises x, $x < y$ if $\sum_{i=1}^{k} x_i \leq \sum_{i=1}^{k} y_i$ for $k = 1, \ldots, n - 1$, and $\sum_{i=1}^{n} x_i = \sum_{i=1}^{n} y_i$ where $x_1, x_2, \ldots, x_n$ and $y_1, y_2, \ldots, y_n$ are coordinates of x and y in nonincreasing order respectively.

A function $\Phi : I^n \to \Re$ is called *Schur-convex* if $x < y$ implies $\Phi(x) \leq \Phi(y)$.

The C-spectrum and Schur-convex functions are used in [CON1] to solve extremal problems with spanning trees.

The graph obtained by removal of k disjoint edges from the complete graph on n vertices is shown to minimize all Schur-convex functions defined on C-spectra of graphs, over all simple graphs on n vertices and $\binom{n}{2} - k$ edges. When n is odd, $k = \frac{n-1}{2}$ and one edge is removed from the vertex of degree $n - 1$ in the graph mentioned above, the resulting graph enjoys the same property among all simple graphs on n vertices and $\binom{n}{2} - \frac{n+1}{2}$ edges.

The reciprocal of the number of spanning trees of a graph is an example of Schur-convex function on the spectrum.

The results described on extremal problems with spanning trees, obtained by the use of convex functions are in accordance with previous results by Kel'mans and Chelnokov (see [CvDSA1], p.224) who used different methods.

Section 3.10: The Use of Spectra to Solve Graph Equations

Graph equations are the equations in which the unknowns are graphs [CvSi1], [CvSi2]. More precisely, let $f(G_1, \ldots, G_n)$ and $g(G_1, \ldots, G_n)$ be graphs formed by graph operations starting with the graphs $G_1, \ldots, G_n$. Then the equation $f(G_1, \ldots, G_n) = g(G_1, \ldots, G_n)$ is called a *graph equation*, where the equality sign indicates graph isomorphism.

Many details on graph equations can be found in the papers mentioned above. We shall primarily restrict ourselves here to relations between graph equations and graph spectra.

An obvious method for solving graph equations is to compare graph invariants of $f(G_1, \ldots, G_n)$ and $g(G_1, \ldots, G_n)$. Sometimes, the graph spectrum is a very suitable invariant for such purposes. The use of graph spectra is mentioned in [CVSI1], [CISI2] and [CVE4] as one of the possible methods to solve graph equations, and early references related to these topics are quoted in [CVSI2]. See also [CVDSA1], p. 223, for several examples.

We start with a graph equation involving a unary graph operation related to 1-factors.

The *pseudo-inverse graph* $PI(G)$ of a graph G is a graph, defined on the same vertex set as G in which the vertices x and y are adjacent if and only if $G - x - y$ has a 1-factor. The graph equation $PI(G) = G$ has been studied in [CVGS1], both by nonspectral and spectral means. The solutions G of this equation are called *self pseudo-inverse graphs*. Two different characterizations of self pseudo-inverse graphs having a 1-factor are given [CVGS1]. In particular, all self pseudo-inverse graphs having a 1-factor and belonging to the set of trees or unicyclic graphs have been found. Some applications of these results to chemistry are indicated.

The motive for introducing such a peculiar unary graph operation as $PI(G)$ stems from chemistry and will be explained here since graph spectra are involved.

Many relations between the largest eigenvalue of a graph and the graph structure are known. It has been proposed, for example, to take just the largest eigenvalue as a measure of branching a graph. In addition to the largest eigenvalue, the smallest positive eigenvalue plays an important role in quantum chemistry (*i.e.*, Hückel's theory, see Chapter 5), too. Unfortunately, not much is known about the dependence of this smallest positive eigenvalue on the graph structure.

One idea to put some light on this dependence is to look at the inverse matrix A^{-1} of the adjacency matrix A of a graph G, provided that A is nonsingular. Then the smallest positive eigenvalue of A is then mapped onto the largest eigenvalue of A^{-1}. Unfortunately, A^{-1}

need not be the adjacency matrix of a graph and it need not be even nonnegative. Therefore we again cannot say too much about our problem.

However, an obvious question arises at this point: for which graphs is the inverse of the adjacency matrix again the adjacency matrix of a graph. We shall call such a graph an *inverse graph*. For which graphs this inverse graph is isomorphic to the starting graph? Surprisingly, this question can help in the investigation of the original problem.

In order to find the matrix A^{-1} it is useful to introduce the concept of a *1-connection* of a graph. In the case of undirected graphs a 1-connection W_{ij} between the vertices i and j in a graph G is a spanning subgraph of G whose components are circuits, graphs K_2 and a path between i and j. The number of circuits and connected components of W_{ij} is denoted by $c(W_{ij})$ and $p(W_{ij})$ respectively. Then the cofactor of the (i,j) entry of A is equal to the sum

$$(-1)^{n-1} \sum_{W_{ij}} (-1)^{p(W_{ij})} 2^{c(W_{ij})} \qquad (3.13)$$

where the summation is taken over all 1-connections W_{ij} and n is the number of vertices of G.

If G is a tree, the number of vertices must be even, say $n = 2m$, since otherwise A would be singular, and for each i and j there exists at most one 1-connection W_{ij} and it does not contain any circuits. If the distance between i and j is equal to d, then the sum (3.13) is equal to 0 for even d, and, if $d = 2s - 1$, is equal to $(-1)^{m+s}$.

The value of det A for trees is known to be 1, 0 or -1. Hence the entries of A^{-1} in trees are equal to 0, 1 or -1. Obviously, the only tree in which A^{-1} does not have entries with different signs is K_2 and in that case we have $A^{-1} = A$. Thus among trees only K_2 has an inverse graph. One could ask now if there are other graphs which have an inverse graph (or multigraph).

Consider now the matrix A^* which is obtained from A^{-1} by replacing all -1's by 1's. Then A^* is the adjacency matrix of a graph if G is a tree. Let us call this graph the *pseudo-inverse graph* of G and denote it by $PI(G)$. It is obvious that in trees this definition of $PI(G)$ is equivalent to the our previous definition. For other graphs the definition involving 1-factors is more suitable.

If both G and $PI(G)$ are trees then the characteristic polynomials of A^{-1} and A^* are the same which can easily be seen from the coefficient theorem for characteristic polynomial of a graph (see Section 3.1). It can be proved (see [CvGS1]) that if both G and $PI(G)$ are trees then $PI(G) = G$ and G is a self pseudo-inverse graph. Hence, in a self pseudo-inverse tree the smallest positive eigenvalue is just the inverse of the largest eigenvalue, and, more generally, if λ is an eigenvalue, then $\frac{1}{\lambda}$ is an eigenvalue, too.

We give next one of the results concerning the graph equation $PI(G) = G$.

THEOREM 3.37 (D. CVETKOVIĆ, I. GUTMAN, S. SIMIĆ [CvGS1]): *The only self pseudo-inverse trees having a 1-factor are the trees of the form $T * K_1$ where T is any tree and $*$ denotes the corona of the graphs.*

Note that C. D. Godsil has derived an explicit expression for the adjoint matrix of $\lambda I - A$ when A is the adjacency matrix of a forest [GOD9]. It is proved in [GOD8] that the matrices A^* and A^{-1} mentioned above are similar through a diagonal matrix with diagonal entries equal to ± 1.

Next we shall consider graph equations involving matrices. There is a class of graph equations closely related to matrix equations. Consider a matrix equation of the form

$$f(X) = g(X). \tag{3.14}$$

Let G^X be the weighted digraph associated to the matrix X in the usual way. Let $\mathcal{F}(G^X) = G^{f(X)}$ and $\mathcal{G}(G^X) = G^{g(X)}$. Then the graph equation

$$\mathcal{F}(X) = \mathcal{G}(X) \tag{3.15}$$

corresponds to the matrix equation (3.14). The equality sign in (3.15) should be understood as the equality of weighted labelled digraphs.

An interesting case appears in matrix equation $f(X) = C$, where C is a fixed matrix for which the relation $P^{-1}CP = C$ holds for every permutation matrix P. In this equation the existence of a solution $X = R$ implies the existence of the solution $X = P^{-1}RP$ since

$$f(P^{-1}RP) = P^{-1}f(R)P = P^{-1}CP = C. \tag{3.16}$$

In the corresponding graph equation $\mathcal{F}(G) = G^C$ the equality sign can be interpreted as the isomorphism of weighted digraphs. These are real graph equations and the corresponding matrix equations can be solved by graph theoretical methods. If we limit ourselves to solutions in (0,1) (symmetric) matrices, then we have graph equations for (graphs) digraphs.

A similar situation appears in equation (3.15) if

$$f(P^{-1}XP) = P^{-1}f(X)P, \quad g(P^{-1}XP) = P^{-1}g(X)P$$

for any permutation matrix P. Then the equality sign in (3.15) again stands for a (di)graph isomorphism.

The equations $X^T = aX$ and $X^T = a(J - X)$, $a \neq 0$, J the matrix all of whose entries are equal to 1, have been solved in (0,1) matrices in [CVE3], pp. 56–57, by graph theoretic means.

Let j be a vector all of whose entries are equal to 1. The graph equation corresponding to the matrix equation

$$A^k j = rj, \quad r, k \text{ positive integers} \tag{3.17}$$

has been considered in [SYS1]. Let $\mathcal{G}_k$ be the class of all graphs which satisfy (3.17). Obviously $\mathcal{G}_1$ is the class of all regular graphs of degree r and $\mathcal{G}_2$ is the class of graphs in which every connected component is either regular of degree $\sqrt{r}$ or semi-regular bipartite of degrees p, q where $pq = r$.

THEOREM 3.38 (M. M. SYSLO [SYS1]): *For $k \geq 1$ we have $\mathcal{G}_k = \mathcal{G}_1$ if k is odd and $\mathcal{G}_k = \mathcal{G}_2$ is k is even.*

The proof uses the fact that (3.17) implies that r is an eigenvalue of A^k. Equation (3.17) is also solved for digraphs in [SYS1].

Let $Q(x)$ be a polynomial in x. The polynomial $Q(G)$ of a multigraph G with the adjacency matrix A is the multigraph whose adjacency matrix is equal to $Q(A)$ (cf. [CVDSA1], p. 52). Let P_n be a path on n vertices. All polynomials $Q(x)$ of degree less than n such that $Q(P_n)$ is a graph are found in [BEE1].

The equation $A^k = dI + \lambda J$ has been considered in [LAM1] and some similar equations in [LALI1], [BRME3], [WAN1], [WAN5].

Section 3.11: Spectra of Tournaments

A *tournament* is a directed graph in which every pair of vertices is joined by exactly one arc. In other words, a tournament is an oriented complete graph.

The question of what can be said about the spectrum of a tournament was raised in [CVDSA1], p. 266, quoting [WIRJ1] in [CVDSA1]. In fact, A. Brauer and I. C. Gentry had already considered spectra of tournaments in 1968 [BRGE1], as is pointed out in [BRU1]. They proved that the real parts of eigenvalues of a tournament on n vertices lie between $-\frac{1}{2}$ and $\frac{n-1}{2}$. The eigenvalue with the largest absolute value is real and positive, and it is called the *dominant eigenvalue*. Several inequalities concerning the eigenvalues of a tournament are derived in [BRGE2], while the original result from [BRGE1] has been improved in the following way.

THEOREM 3.39 (G. N. DE OLIVEIRA [OLI1]): *Let* $\omega_1, \omega_2, \ldots, \omega_n$ *be eigenvalues of a tournament and let* $R(\omega_i)$ *denote the real part of* ω_i. *Then*

$$R(\omega_i) \leq \frac{1}{2}n - m - 1,$$

where m is any nonnegative number which satisfies $m \leq \omega_1$, ω_1 being the dominant eigenvalue.

It may be, for example, that m is the least outdegree of the tournament.

R. Brualdi [BRU1] locates the dominant eigenvalue of a strongly connected tournament between the geometric means of the three largest and of the three smallest outdegrees of the tournament, thereby improving some of the estimates in [BRGE2].

If we consider tournaments of special types, then more can be said about the spectrum.

A tournament with adjacency matrix A is said to be *associated with the permutation matrix P* if AP is also the adjacency matrix of a tournament. The existence and construction of such tournaments is considered in [ZAG1]. When P corresponds to an n-cycle, a tournament with n vertices is associated with P if and only if $AP = A^T$, the transpose of A. The spectrum of the tournament is completely determined in this special case.

A tournament is called a *regular tournament* if all its vertices have the same outdegree. It is easily seen that if each vertex has outdegree k then the indegree of each vertex is also k. Such a tournament has $2k + 1$ vertices and is said to be regular of degree k. The adjacency matrix A of any tournament satisfies $A + A^T = J - I$, where each entry of J is 1 and I is a unit matrix. For a regular tournament of degree k we have $AJ = kJ = JA$. Consequently $AA^T = A^T A$ and A can be diagonalized by a unitary matrix. It follows that if $\lambda_1, \ldots, \lambda_n$ are the eigenvalues of the tournament, then $\lambda_1 + \overline{\lambda_1}, \ldots, \lambda_n + \overline{\lambda_n}$ are the eigenvalues of a complete graph. Thus the dominant eigenvalue of the tournament is $k = \frac{n-1}{2}$ and the nondominant eigenvalues have real part $-\frac{1}{2}$ (cf. [BRGE1] and [ROW4]). It follows that the characteristic polynomial of a regular tournament of degree k has the form

$$p(\lambda) = (\lambda - k) \prod_{i=1}^{k} (\lambda^2 + \lambda + d_i)$$

where $d_1, \ldots, d_k$ are real numbers greater than $\frac{1}{4}$. Note the similarity to the characteristic polynomial of a self-complementary graph ([CVDSA1] p. 98) whose roots are, of course, real. In each case the nondominant eigenvalues are symmetric about the point $-\frac{1}{2}$ of the complex plane.

We shall say that, in a tournament, the vertex x *dominates* the vertex y if the arc connecting x and y is oriented from x to y. A tournament is called *doubly regular* with subdegree t if it has more than one vertex and any two vertices jointly dominate precisely t vertices. One can prove that in such a tournament, the number of vertices which dominate any two given vertices is also equal to t; moreover each arc lies in precisely $t + 1$ 3-cycles. Doubly regular tournaments with subdegree t are regular of degree $2t + 1$ and hence have $4t + 3$ vertices. They are very interesting but also very rare objects. It has been known for some time that the existence of a doubly regular tournament is equivalent to the existence of a skew-Hadamard matrix (see e.g.[REBR1]). In the following theorem, doubly regular tournaments are characterized by an interesting spectral property.

THEOREM 3.40 (N. ZAGAGLIA-SALVI [ZAG2]): *A tournament is doubly regular of subdegree t, if and only if it has eigenvalues $2t + 1$, $\frac{-1 \pm i\sqrt{4t+3}}{2}$, the first one being simple and the other two having multiplicities $2t + 1$.*

This nice theorem provides an analogy with strongly regular graphs in the undirected case: these too are characterized by the property of having exactly three distinct eigenvalues. Equivalently, the minimal polynomial has degree 3, a property of doubly regular tournaments implicitly noted in an earlier paper by J. Plesnik [PLE1]. He also proved that in a doubly regular tournament, each arc lies in a fixed number of 4-cycles and in a fixed number of 5-cycles. The converse was established by P. Rowlinson [ROW4], exploiting further properties of the adjacency matrix.

Section 3.12: Other Results

We conclude this chapter by reporting on some other recent results from the theory of graph spectra which did not fit into previous sections. These include results on integral graphs, relations between graphs and special functions, some data on expository articles, etc.

• A graph is called *integral* if its spectrum consists entirely of integers (see, for example, [CVDSA1], p. 266).

It is known that there are exactly 13 connected cubic integral graphs, found independently by F. C. Bussemaker and D. Cvetković, on one hand, and A. J. Schwenk on the other. See also [CAMO1], p. 204.

The study of integral graphs with bounded vertex degrees has been continued by Z. Radosavljević and S. Simić [RASI1]. They found all connected, nonregular, nonbipartite, integral graphs with maximal vertex degree equal to 4. There are again exactly 13 such graphs. Eight of them are generalized line graphs and the other are exceptional graphs arising from the root system E_8 (see Chapter 1). These graphs were found by a mixture of mathematical reasoning and computer search. Once an upper bound for the number of vertices of such graphs has been found theoretically (that bound is equal to 13), the graphs have been constructed by the use of the interactive programming system **Graph**, an expert system for graph theory [CVE8], [CVKS1], [CVE6], [CVEP1], [CVE9].

Integral and Gaussian digraphs have been considered in [EsHA2]. A digraph is called *Gaussian* if, for all its eigenvalues $\lambda = \alpha + i\beta$, α and β are integers, *i. e.*, all the eigenvalues are Gaussian integers. Of course, integral digraphs are Gaussian. Several examples of Gaussian digraphs are given and procedures which use graph operations for constructing Gaussian digraphs from smaller ones are described.

Integral trees have been studied in [WAT2] and [WASc1]. It was proved in [WAT2] that the only integral tree with a complete matching is K_2. Also a family of integral trees of diameter 4 has been constructed.

All integral trees homeomorphic to a star have been found in [WASc1]. They are, except for K_1, stars K_{1,k^2} and subdivisions of stars K_{1,k^2-1}, where k is an integer. Also integral double stars, and some integral trees of large diameter have been found.

An infinite family of integral complete tripartite graphs is constructed in [RoI1]. Let $n_1 = 4u^2(u^2 + v^2)^3$, $n_2 = 4v^2(u^2 + v^2)^3$, and $n_3 = 3u^2v^2(34u^2v^2 - u^4 - v^4)$ where u, v are positive integers such that $(3 - \sqrt{8})v < u < v$. Nonzero eigenvalues of K_{n_1,n_2,n_3} are roots of the equation (cf. [CvDSA1], p. 74)

$$\frac{n_1}{\lambda + n_1} + \frac{n_2}{\lambda + n_2} + \frac{n_3}{\lambda + n_3} = 1.$$

It is easy to verify that in the above case the roots of this equation are given by $\lambda_1 = 24u^2v^2(u^2+v^2)^2$, $\lambda_2 = -2uv(u^2+v^2)^2(u^2+6uv+v^2)$, and $\lambda_3 = -2uv(u^2 + v^2)^2(-u^2 + 6uv - v^2)$.

Hence, we really have a family of integral graphs.

Tables of two-graphs (for the definition see, for example [CvDSA1], p. 200) given in [BuMS1] contain data about two-graphs (and also about graphs) with integral spectra.

• I. Gutman (see [CvDSA1], p. 267) posed the problem of finding a graph whose eigenvalues can not be expressed in terms of radicals. Using an old result by I. Schur, C. Godsil [GoD5] proved that Hermite polynomials $H_n(x)$ are not solvable by radicals for $n > 25$. Further it is proved that for each positive integer $n > 3$ there is a tree T_n such that $H_m(x)$ divides the characteristic polynomial of T_n for all m such that $2 \leq m \leq n$. Hence, Gutman's problem is solved. Moreover, it is proved that $\lim_{n\to\infty} p_n = 1$ where p_n is the proportion of trees on n vertices which have characteristic polynomials not solvable by radicals.

The first tree that is constructed by Godsil's method is one with about 1.1×10^{27} vertices. However, it is proved in [KEq1] that trees from Figure 3.8 have characteristic polynomials that are not solvable by radicals. These trees can be found under numbers 2.191 and 2.198 in a table of trees in [CvDSa1] on p. 291 and their characteristic polynomials are

$$\lambda^{10} - 9\lambda^8 + 26\lambda^6 - 29\lambda^4 + 11\lambda^2 - 1, \quad \text{and}$$

$$\lambda^{10} - 9\lambda^8 + 27\lambda^6 - 32\lambda^4 + 13\lambda^2 - 1.$$

It is shown in [Keq1] that the Galois groups of these polynomials are isomorphic to S_5, the symmetric group of order 5. It is well known that S_5 is an insolvable group. Hence, the polynomials cannot be solved by radicals. Of course, any tree with less than 10 vertices has a solvable characteristic polynomial.

Figure 3.8 Trees with characteristic polynomial not solvable by radicals

• Relations between special functions and the theory of graph spectra were mentioned for the first time in [Cve4]. Characteristic polynomials of a circuit and of a path are in fact the Chebyshev polynomials of the first and of the second kind respectively (see [CvDSa1], pp. 72–73). Hermite and generalized Laguerre polynomials have an interpretation as matching polynomials of a complete graph and of a complete bipartite graph (see Chapter 4). Many formulas concerning these polynomials can be interpreted and even derived using the corresponding graphs [GuCv1], [God3]. These examples support the idea of the usefulness of graph theoretical methods in the theory of special functions.

As an example, let us find a relation between the Legendre polynomials $P_n(x)$ and graphs [GuCv1].

The Legendre polynomials can be defined by the Rodrigues formula

$$P_n(x) = \frac{1}{n!\, 2^n} \frac{d^n}{dx^n} (x^2 - 1)^n.$$

Since $\Phi(K_2) = x^2 - 1$, we immediately obtain

$$P_n(x) = \frac{1}{n!\,2^n} \frac{d^n}{dx^n} \Phi(nK_2), \tag{3.18}$$

where $\Phi(G)$ is the characteristic polynomial of G and where nK_2 denotes the graph composed of n copies of K_2.

We consider all n-vertex induced subgraphs of nK_2. Let $n = 2p+q$ with $p = 0, 1, \ldots, [n/2]$. Then the n-vertex subgraphs of nK_2 are of the form $pK_2 \dotplus qK_1 = pK_2 \dotplus (n - 2p)K_1$.

There are $\binom{n}{p}\binom{n-p}{n-2p} 2^{n-2p} = \binom{n}{p}\binom{n-p}{p} 2^{n-2p}$ such subgraphs. Using a formula for the n-th derivative of the characteristic polynomial of a graph (3.18) reduces to

$$P_n(x) = \frac{n!}{n!\,2^n} \sum_{p=0}^{[n/2]} \binom{n}{p} \binom{n-p}{p} 2^{n-2p} \Phi(pK_2) \Phi((n-2p)K_1)$$

$$= \sum_{p=0}^{[n/2]} \binom{n}{p} \binom{n-p}{p} 4^{-p} (\Phi(K_2))^p (\Phi(K_1))^{n-2p}$$

$$= \sum_{p=0}^{[n/2]} \binom{n}{p} \binom{n-p}{p} \left(\frac{x^2-1}{4}\right)^p x^{n-2p}.$$

• Let A be the adjacency matrix of a graph G and let J be a square matrix where all entries are equal to 1. Let $P_X(\lambda)$ be the characteristic polynomial of the matrix X. The function

$$\Psi_G(\lambda) = P_{A-J}(\lambda) - P_A(\lambda).$$

has been introduced and studied in [WAN2].

The following formulas are proved :

$$P_{G\nabla H}(\lambda) = P_G(\lambda) P_H(\lambda) - \Psi_G(\lambda) \Psi_H(\lambda),$$

$$P_{\overline{G}}(\lambda) = (-1)^n (P_G(-\lambda - 1) + \Psi_G(-\lambda - 1)),$$

$$P_{G[H]}(\lambda) = (\Psi_H(\lambda))^n P_G\left(\frac{P_H(\lambda)}{\Psi_H(\lambda)}\right),$$

where $G\nabla H$ and $G[H]$ denote the complete (cf. [CvDSA1], p. 54) and the lexicographic product of graphs, respectively.

• Let $P_G(\lambda)$ be the characteristic polynomial of a graph G on vertices $x_1, x_2, \ldots, x_n$. The polynomial $P_G(\lambda) - \lambda P_{G-x_i}(\lambda)$ has been introduced in [BER1]. It is associated to the vertex x_i. It is proved that graphs $G - x_i$ and $G - x_j$ are cospectral if and only if in each eigenvector of G, the absolute values of coordinates corresponding to x_i and x_j are equal (see also Theorem 7.3 of [ScW1]).

• Let G be a simply connected (without holes) induced subgraph of a planar square lattice graph. Let A be the adjacency matrix of G. Then the determinant of A is equal to 0, 1, or -1 [DoTo1]. Since G is bipartite, it can be properly coloured by two colours. Remove from G any number of vertices of the same colour. Then the statement about the determinant of A continues to hold.

• Let G be a digraph on n vertices having real eigenvalues. Let G have h loops and z cycles of length 2. Then eigenvalues of G are then bounded from above and from below by the numbers

$$\frac{1}{n}\left(h \pm \sqrt{(n-1)(2zn + nh - h^2)}\right)$$

[HaTa1]. For undirected graphs with m edges and without loops these bounds can be reduced to $\pm\sqrt{2m(1 - \frac{1}{n})}$. which is a known result (cf. [CvDSa1], p. 221).

Several bounds for eigenvalues of graphs have been derived in [BrDu1]. These bounds depend on the number of vertices, the number of edges, numbers of positive and negative eigenvalues, vertex covering number, the internal stability number, traces of the adjacency matrices, the index of the eigenvalue, etc. For example, it is proved that

$$-\sqrt{\frac{2mn_+}{(n-r+1)(n-r+1+n_+)}} \leq \lambda_r \leq \sqrt{\frac{2mn_-}{r(r+n_-)}},$$

where n is the number of vertices, m is the number of edges, n_+ is the number of positive eigenvalues, n_- is the number of negative eigenvalues and r is the index of λ_r.

• According to an abstract published in the Graph Theory Newsletter the paper [ACH1] outlines "a combinatorial algorithm which may be used to write a computer program for producing the list of all sigraphs on p or less points, their characteristic polynomials and their spectra." Such a list is given in the paper for $p \leq 4$.

• A heuristic algorithm for a proper colouring vertices of a graph which is based on eigenvectors of the adjacency matrix is proposed in [ASGI1]. Let $\lambda_1 \geq \lambda_2 \geq \cdots \geq \lambda_n$ be eigenvalues and $u_1, u_2, \ldots, u_n$ the corresponding orthogonal eigenvectors. At the beginning vertices are coloured by two colours. Vertices corresponding to nonnegative coordinates of u_n get one colour and those corresponding to negative coordinates get the other one. This colouring is defined using u_{n-1} and other eigenvectors. Whenever there are vertices with the same colour but opposite signs in the coordinates of the next eigenvector, a new colour is introduced. If all eigenvectors are included we get a proper colouring (with n colours) since sign patterns of coordinates of all eigenvectors associated to a vertex must be different (due to orthogonality condition). Hence, starting from u_n and including new eigenvectors we do get a proper colouring at some moment. It is conjectured that eigenvectors of negative eigenvalues will suffice. It is proved that in some classes of graphs the algorithm gives proper colourings with a reasonably small number of colours.

• Expository articles on graph spectra have already mentioned in the introduction. General expository articles on the theory of graph spectra are [CVE4], [GOHMK], [MOH2], and [SCWI1]. Strongly regular graphs are surveyed in [COM1], [SEI1], and [SETE1]. Applications to Chemistry are described in [CVE1], [FIN1], [MAL3], [WILS1]. The paper [MERW1] contains a review on the permanental polynomial of a matrix and of a graph (see also Section 4.6).

The expository article [CVDO2] has been written to complement this book. The article surveys selected topics covered by this book with a somewhat different perspective. These include root systems and spectral characterizations of graphs as well as some new developments in the theory of graph spectra: graph invariants based on the eigenvectors of the adjacency matrix, the Shannon capacity problem, spectra of random graphs, the permanental and the matching polynomial of a graph, and spectra of infinite graphs.

Chapter 4

The Matching Polynomial and Other Graph Polynomials

In this chapter we shall be concerned with some graph polynomials which are related to the characteristic polynomial and thus to the spectrum of a graph. The most interesting among these polynomials seems to be the matching polynomial.

Section 4.1: The Matching Polynomial

Let G be a graph with n vertices and m edges. Let $p(G,k)$ be the number of k-matchings of G, $k = 1,\ldots,m$. Thus $p(G,k)$ is equal to the number of ways in which k independent edges can be selected in G. In addition, let $p(G,0) = 1$. More about the numbers $p(G,k)$ will be given later. We now define the *matching polynomial* of a graph G with n vertices and m edges to be

$$M_G(x) = \sum_{k=0}^{m} (-1)^k p(G,k) x^{n-2k}. \qquad (4.1)$$

For brevity we shall sometimes write M_G instead of $M_G(x)$. As an example, consider the graph G_0 displayed in Figure 4.1. In this case $p(G_0,0) = 1$, $p(G_0,1) = 8$, $p(G_0,2) = 14$, $p(G_0,3) = 2$, and $p(G_0,k) = 0$ for $k \geq 4$. Hence $M_{G_0}(x) = x^6 - 8x^4 + 14x^2 - 2$.

THEOREM 4.1:   *Let $G_1 + G_2$ be the direct sum of the graphs G_1 and G_2. Then*

$$M_{G_1+G_2}(x) = M_{G_1}(x)M_{G_2}(x). \qquad (4.2)$$

PROOF: The number of selections of k independent edges from $G_1 + G_2$ such that j of them are edges of G_1 and $k - j$ of them are from G_2 is obviously $p(G_1, j) \cdot p(G_2, k - j)$. Consequently we have

$$p(G_1 + G_2, k) = \sum_{j=0}^{k} p(G_1, j) \cdot p(G_2, k - j).$$

We obtain (4.2) by substituting this identity into (4.1).

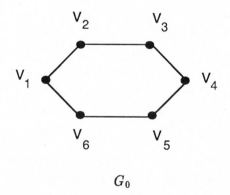

G_0

Figure 4.1

THEOREM 4.2:   *Let G and H be isomorphic. Then*

$$M_G(x) = M_H(x). \qquad (4.3)$$

The converse of Theorem 4.2 is, however, not true. There are numerous pairs of graphs with equal matching polynomials [WAH1]. The smallest such pair is the graphs G_a and G_b in Figure 4.2 ($M_{G_a} = M_{G_b} = x^4 - 3x^2$). The smallest pair of connected graphs having identical matching polynomials is G_c and G_d of Figure 4.2 ($M_{G_c} = M_{G_d} = x^5 - 5x^3 + 4x$).

Algorithms and computer programs for the calculation of the matching polynomial are described in [MOTR1] and [RABA1].

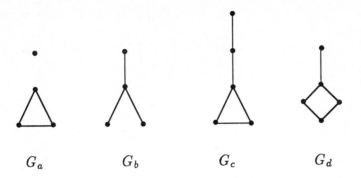

G_a G_b G_c G_d

Figure 4.2 Graphs with Identical Matching Polynomials

The History of the Matching Polynomial

Because of its numerous applications to physics and chemistry, the matching polynomial (or polynomials closely related to it) have been independently introduced several times. M_G can be interpreted as a generating function for the number of matchings of the graph G. Keeping in mind that the concept of a matching is a classical one in graph theory, it would not be unreasonable to expect that mathematical objects like the matching polynomial would occur in the mathematical literature frequently.

In statistical physics matchings are a natural way to represent coverings of crystal lattices with diatomic molecules, or dimers. In 1970, in the first reference to the matching polynomial itself, it was independently established ([HELI1] and [KUN1]) that all roots of M_G are real. The physical applications of M_G (along with several new mathematical results) are given in some detail in [HELI2]. For some more recent results see [HOMO1].

The first chemical application of M_G was proposed in 1971 ([HOS1] from [CVDSA1]) where the *Z-counting polynomial*

$$\sum_k p(G,k)x^k$$

and the quantity $Z_G = \sum_k p(G,k)$ were used to describe the thermodynamic properties of saturated hydrocarbons. The quantity Z_G is now called *Hosoya's topological index*.

Another chemical application of M_G in the theory of aromaticity was independently developed in [AIH2] and [GUMT2]. In [AIH2] M_G

was called the *reference polynomial* while in [GuMT2] it was the *acyclic polynomial*.

Finally, another independent introduction to M_G was given in [FAR3], this time in a mathematical context. In fact [FAR3] uses the two variable polynomial $\sum_k p(G,k)x^{n-2k}y^k$ and relates it to $M_G(x)$ in [FAR6]. The term matching polynomial was first used in [FAR3].

Matchings in Graphs

It is not our purpose in this chapter to present all results on matchings in graphs. Although each such result can be (re)interpreted as a statement involving M_G, we shall immediately concern ourselves with those properties that are "essentially polynomial", e.g., the fact that all the roots are real, or those which relate M_G to the characteristic polynomial. Nevertheless, we shall list some of the basic properties of the numbers $p(G,k)$.

If G is a graph with n vertices and m edges, then $p(G,0) = 1$ (by definition), $p(G,1) = m$, and $p(G,k) = 0$ for all $k > m$. If n is even, then $p(G,n/2)$ is just the number of 1-factors of G. If $p(G,k) \leq k$, then $p(G,k+1) = 0$. In particular, if $p(G,k) = 0$ then $p(G,k+1) = 0$. The sequence of numbers $p(G,k)$, $k = 1 \ldots m$ is unimodal [SCHW3], i.e., for every graph G there exists a number $K = K(G)$ such that $p(G,k) \leq p(G,k+1)$ for all $k < K$ and $p(G,k) \geq p(G,k+1)$ for all $k \geq K$. If the number of vertices of G is large compared to the maximum degree of G, then the numbers $p(G,k)$ are approximately normally distributed [GOD2]. Further results of this type are in [GUT32].

Let H be a subgraph of G. The graph $G - H$ is obtained from G by deleting from G the vertices of H (and all edges incident to these vertices). If G and H have the same vertex set, then it is both consistent and convenient to treat $G - H$ as a graph without vertices having $p(G - H, 0) = 1$ and $M_{G-H}(x) = 1$, the constant polynomial. This convention will be followed for the rest of the present chapter.

It is shown in [CvDSA1], p.87 that if F is a forest, then $(-1)^k a_{2k}$ is equal to the number of linear subgraphs of F with exactly k edges. Here a_i is the i-th coefficient of the characteristic polynomial of F. In other words, we have $(-1)^k a_{2k} = p(F,k)$, which is the basis of the following immediate but important result:

THEOREM 4.3 (I. GUTMAN [GUT1], E. J. FARRELL [FAR3]): *Let $M_G(x)$ and $P_G(x)$ be the matching and characteristic polynomials of a graph G. Then*

$$P_G(x) = M_G(x) \iff G \text{ is a forest.} \qquad (4.4)$$

THEOREM 4.4: *Let $\mathcal{C}$ be the set of all regular graphs of degree two which are contained in the graph G. Let $p(C)$ denote the number of components of $C \in \mathcal{C}$. Then*

$$P_G(x) = M_G(x) + \sum_{C \in \mathcal{C}} (-2)^{p(C)} M_{G-C}(x) \qquad (4.5)$$

and

$$M_G(x) = P_G(x) + \sum_{C \in \mathcal{C}} (+2)^{p(C)} P_{G-C}(x). \qquad (4.6)$$

If G is a forest, then $\mathcal{C} = \emptyset$ and both (4.5) and (4.6) reduce to (4.4). Equation (4.5) was first formulated in [HOS2] from [CVDSA1] whereas equation (4.6) was given in [GUT32].

Because of Theorem 4.3, M_G can be thought of as a means of extending the spectral properties of forests to all graphs. A number of results which hold for the characteristic polynomial of a forest have analogous extensions to the matching polynomial of all graphs. The following result is typical:

Let G be a graph and u and v be two of its vertices. Let $G_1, G_2, \ldots, G_n$ be graphs isomorphic to G, and let u_i and v_i be the vertices in G_i corresponding to the vertices u and v in G, $i = 1, 2, \ldots n$. Construct the graph R_n by taking the union of G_1 to G_n and joining u_i with v_{i+1}, $i = 1, \ldots, n-1$ and also u_n to v_1. Now let $A = P_G - P_{G-u-v}$ and $B = 2(P_{G-u}P_{G-v} - P_G P_{G-u-v})^{1/2}$. Then by group theoretic means one deduces that (see [CVDSA1], p. 141–149 and [GPTY1])

$$P_{R_n} = \prod_{j=0}^{n-1} (A - B \cos \theta_j)$$

where $\theta_j = 2j\pi/n$. An analogous statement for the matching polynomial is obtained [GRBA1] when P is replaced in the above formula by M and $\theta_j = (j + 1/2)\pi/n$.

THEOREM 4.5 ([CVDSA1], P.78):   *Let uv be the edge whose end points are u and v. Then*

$$M_G(x) = M_{G-uv}(x) - M_{G-u-v}(x), \text{ and} \qquad (4.7)$$

$$M_G(x) = x M_{G-v}(x) - \sum_{u'v} M_{G-u'-v}(x). \qquad (4.8)$$

The summation on the right hand side is over all vertices u' adjacent to v.

PROOF: Among the $p(G, k)$ selections of k independent edges in G, there are $p(G - uv, k)$ selections which do not contain uv and an additional $p(G - u - v, k - 1)$ selections which do contain uv. Hence $p(G, k) = p(G - uv, k) + p(G - u - v, k - 1)$. Substituting this identity back into (4.1) gives (4.7). A repeated application of (4.7) to all of the edges adjacent to the vertex v gives

$$M_G(x) = M_H(x) - \sum_{u'v} M_{G-u'-v}(x),$$

where H is the graph obtained from G by deleting all edges adjacent to v. Obviously $H = G + K_1$, and since $M_{K_1}(x) = x$, we obtain (4.8) using Theorem 4.1.

THEOREM 4.6 (I. GUTMAN[GUT32]):   *Let N_n be the graph consisting of n isolated points. Then the matching polynomial of any graph can be determined by the recursion relation (4.7) and the condition $M_{N_n} = x^n$ for all $n \geq 0$.*

Further relations between the matching polynomials of graphs and their subgraphs are given by the following theorem:

THEOREM 4.7 (I. GUTMAN, H. HOSOYA [GUHO1]):   *Let $v_1, \ldots, v_n$ be the vertices of G. Let $G - v_i$ be the subgraph of G obtained by deleting the vertex v_i. Then*

$$\sum_{i=1}^{n} M_{G-v_i}(x) = \sum_{k=0}^{m} (-1)^k (n - 2k) p(G, k) x^{n-2k-1}$$

$$= \frac{d}{dx} M_G(x). \qquad (4.9)$$

THEOREM 4.8 (E. J. FARRELL, S. A. WAHID [FaWa3],[Wah1]): *Let $e_1, \ldots, e_n$ be the edges of G. Let $G - e_i$ be the subgraph of G obtained by the deletion of the edge e_i. Then*

$$\sum_{i=1}^{m} M_{G-e_i}(x) = \sum_{k=0}^{m} (-1)^k (m-k) p(G,k) x^{n-2k}$$

$$= \frac{d}{dy}(y^{m-n/2} M_G(\sqrt{y})),$$
(4.10)

where $y = x^{(n-2k)/(m-k-1)}$.

Combining (4.9) and (4.10) we get

$$2 \sum_{i=1}^{m} M_{G-e_i}(x) = x \sum_{i=1}^{n} M_{G-v_i}(x) + (2m-n) M_G(x).$$
(4.11)

THEOREM 4.9 (C. D. GODSIL[God7]): *Let u and v be two distinct vertices of G. Let $\mathcal{P}_{uv}$ be the set of all paths connecting u and v. Then*

$$M_{G-u} M_{G-v} - M_G M_{G-u-v} = \sum_{P \in \mathcal{P}_{uv}} (M_{G-P})^2.$$
(4.12)

A result analogous to Theorem 4.9 is also known for the characteristic polynomial (see Theorem 3.8 in Chapter 3). Theorem 4.7 and Theorem 4.9 have the following corollary:

COROLLARY 4.9.1 (O. J. HEILMANN, E. H. LIEB [HeLi2]): *Let $\mathcal{P}_u$ be the set of all paths of G with initial vertex u. Then*

$$M'_G M_{G-u} - M_G M'_{G-u} = \sum_{P \in \mathcal{P}_u} (M_{G-P})^2$$

(where the prime denotes, as usual, differentiation).

THEOREM 4.10 (O. J. HEILMANN, E. H. LIEB [HeLi2]): *Using the notation of Corollary 4.9.1,*

$$M_G(x) M_{G-u}(y) - M_G(y) M_{G-u}(x) = (x-y) \sum_{P \in \mathcal{P}_u} M_{G-P}(x) M_{G-P}(y)$$

Now let $|P|$ denote the number of vertices in the path P, and let $\mathcal{P}$ be the set of all paths.

COROLLARY 4.10.1: $M_G(x)M_{G-u}(x) =$
$-x\sum_{P\in\mathcal{P}_u}(-1)^{|P|}(M_{G-P}(x))^2$.

COROLLARY 4.10.2: $M_G(x)M'_G(x) = -x\sum_{P\in\mathcal{P}}(-1)^{|P|}(M_{G-P}(x))^2$.

A result that generalizes (4.12) can be found in [HELI2]:

THEOREM 4.11 (O. J. HEILMANN, E. H. LIEB [HELI2]): *Let H be a nonempty subgraph of G and u a vertex of $G - H$. Let $\mathcal{P}_{uH}$ be the set of all paths from u to a vertex in H for which that terminal vertex is the only one in the path in H. Then*

$$M_{G-u}M_{G-H} - M_G M_{G-H-u} = \sum_{P\in\mathcal{P}_{uH}} M_{G-P}M_{G-H-P}.$$

One of the most intriguing properties of M_G is that it has only real roots.

THEOREM 4.12 (O. J. HEILMANN, E. H. LIEB, H. KUNZ): *All roots of the matching polynomial are real.*

This result was first proved in [HELI1] and [KUN1]. The paper [HELI2] offers three different proofs. Further more or less independent proofs can also be found in [GOD1], [GOGU1], [GOGU2], and [GUT34]. Here is the idea of one of the proofs given in [HELI2]. Suppose that one of the roots $\mu_i = a$ is complex. Then M_G must have another root $\mu_j = \bar{a}$ (where $\bar{a}$ denotes the complex conjugate of a). Setting $x = a$ and $y = \bar{a}$ in Theorem 4.10 we conclude that

$$\sum_{P\in\mathcal{P}_u} |M_{G-P}(a)|^2 = 0$$

which is a contradiction.

The unimodality of the sequence $p(G, k)$, $k = 1, \ldots, m$ is an immediate consequence of Theorem 4.12. Let $C(m, k)$ denote the combination of m things k at a time.

COROLLARY 4.12.1:   *The sequence $p(G, k)/C(m, k)$ is log-concave, i. e., the numbers*

$$\frac{p(G, k)/C(m, k)}{p(G, k-1)/C(m, k-1)}, \quad k = 1, \ldots, m$$

form a nonincreasing sequence.

COROLLARY 4.12.2: $p(G,k)^2 \geq p(G,k-1) \cdot p(G,k+1)$

A. J. Schwenk [SCHW3] has also proved the unimodality of the sequence $p(G,k)$ without using the fact that the roots of M_G are real.

The collection of roots of $M_G(x)$ exhibits a striking similarity to the spectrum of G. Some of these similarities are reflected in the following theorems. Let the roots of M_G be denoted by $\mu_1(G) \geq \mu_2(G) \geq \cdots \geq \mu_n(G)$.

THEOREM 4.13 (O. J. HEILMANN, E. H. LIEB [HELI2]):   *If v is a vertex of G, then $\mu_i(G) \geq \mu_i(G-v) \geq \mu_{i+1}(G)$.*

PROOF: In order to prove the inequalities given above, we use formula (4.8) and proceed by induction on the number of vertices of G. For the sake of brevity we shall consider only the case where the roots of M_G, M_{G-v}, and $M_{G-u'-v}$ are distinct. After verifying the validity of Theorem 4.13 for $n = 3$ and $n = 4$ (which is easy), we assume that it also holds for all graphs with $n-1$ vertices. Since $G-v$ is such a graph, we have

$$\mu_i(G-v) > \mu_{i+1}(G-u'-v) > \mu_{i+1}(G-v)$$

for all $u' \neq v$. From (4.9) it is now easy to see that $M_G(x) > 0$ for sufficiently large values of x; furthermore $M_G(\mu_k(G-v))$ is positive for even k and negative for odd k. Since M_G is continuous, the interval $(\mu_{k+1}(G-v), \mu_k(G-v))$ must contain a root for $k = 1, 2, \ldots, n-2$ as must $(-\infty, \mu_{n-1}(G-v))$ and $(\mu_1(G-v), \infty)$. Since there are n intervals in all, the roots of M_{G-v} must interlace the roots of M_G and thus Theorem 4.13 holds for all graphs with n vertices. This argument also shows that all of the roots of the matching polynomial are real.

THEOREM 4.14 (I. GUTMAN [GUT39]):   *Let v be a vertex of G. If G is connected, then $\mu_1(G) > \mu_1(G-v)$ and $\mu_1(G) > \mu_2(G)$.*

THEOREM 4.15 (I. GUTMAN [GUT39]):   *Let e be an edge of G. Then $\mu_1(G) \geq \mu_1(G-e)$. If G is connected, then $\mu_1(G) > \mu_1(G-e)$.*

THEOREM 4.16 (O. J. HEILMANN, E. H. LIEB [HELI2], I. GUTMAN [GUT32]):   *If G is a connected graph with maximum degree $D \geq 2$ and is not isomorphic to $K_{1,D}$, then*

$$\frac{1}{2}\sqrt{D+1+\sqrt{(D-1)^2+4}} \leq \mu_1(G) < 2\sqrt{D-1}. \qquad (4.13)$$

Further, the bound on the right side of (4.13) is the best possible.

THEOREM 4.17 (C. D. GODSIL, I. GUTMAN [GoGu3]):   Let $\lambda_1(G)$ be the largest eigenvalue of G. Then $\mu_1(G) \leq \lambda_1(G)$. If G is connected then $\mu_1(G) = \lambda_1(G)$ only if G is a tree.

Theorems 4.3 and 4.4 indicate that the matching and characteristic polynomials are closely related. There are, however, even more subtle relationships, as the following theorem shows.

THEOREM 4.18 (C. D. GODSIL [God1]):   For every graph G there exists a forest $F = F(G)$ such that $M_G(x)$ divides $P_F(x)$.

To prove Theorem 4.18 we need some preparation. Let G be a rooted graph whose root is the vertex v. Then the *Godsil tree* of G is the tree $T(G, v)$ obtained in the following manner: consider all paths of G which start at its root. These are the vertices of $T(G, v)$; two vertices are adjacent if one is contained in the other maximally, *i. e.*, one path has one vertex fewer than the other path and is contained in it. Notice that $T(G, v)$ is completely determined by the component containing v.

As an example, consider the graph G_0 given at the beginning of this chapter. Its Godsil trees $T(G_0, v_1)$ and $T(G_0, v_2)$ are given in the following figure:

It is clear that the other Godsil trees of G_0 are isomorphic to either $T(G_0, v_1)$ or $T(G_0, v_2)$. The vertices of $T(G_0, v_1)$ are labeled so that the sequence of labels between the vertex 1 and a given vertex corresponds to the appropriate path in G_0. The same is true for the labeling of the vertices of $T(G_0, v_2)$. The vertex of $T(G_0, v_2)$ indicated by an arrow, for instance, corresponds to the path $(v_2, v_1, v_6, v_5, v_3)$ of G_0.

Godsil trees have the following properties:

(i) $T(G, v)$ is a tree,

(ii) if G is a tree, then G and $T(G, v)$ are isomorphic, and

(iii) if $v_1, \ldots, v_r$ are the vertices in G adjacent to v, then $T(G, v)$ is obtained by taking $T(G - v, v_1) + \cdots + T(G - v, v_r)$ and adjoining a new vertex v that is adjacent to each vertex corresponding to the single element path (v_i) in $T(G - v, v_i)$, $i = 1, \ldots, r$.

THEOREM 4.19 (C. D. GODSIL [God1]):   Let v be a vertex of the graph G. Then

$$\frac{M_G(x)}{M_{G-v}(x)} = \frac{M_{T(G,v)}(x)}{M_{T(G,v)-v}(x)}. \tag{4.14}$$

In addition, if G is connected, then $M_G(x)$ divides $M_{T(G,v)}(x)$.

Theorem 4.3 insures, of course, that $M_{T(G,v)} = P_{T(G,v)}$.

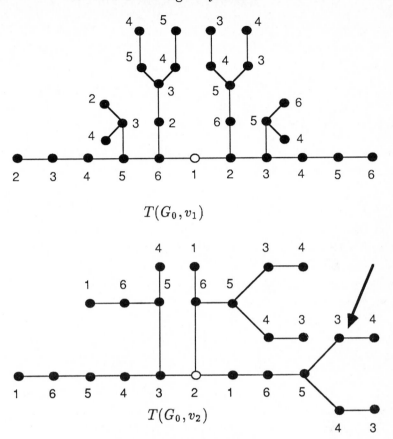

Figure 4.3 Godsil Trees

The forest $F(G)$ mentioned in the conclusion of Theorem 4.18 is now constructed as $F(G) = T(G, u_1) + \cdots + T(G, u_p)$ where $u_1, \ldots, u_p$ is a selection of vertices from G, one from each connected component. The resulting forest $F(G)$ is not unique; it depends on the choice of $u_1, \ldots, u_p$.

PROOF: Since $T(G, v)$ depends only on the component of G which contains the root v, it is sufficient to prove Theorem 4.19 for connected graphs. Because of property (ii), the theorem holds trivially for trees.

It is easy to verify the correctness of (4.14) for graphs with three or fewer vertices. Assume that (4.14) holds for all graphs with fewer than n vertices. We need only show that (4.14) holds for graphs with n vertices.

According to (4.8)

$$\frac{M_G}{M_{G-v}} = x - \frac{\sum_{e'} M_{G-u'-v}}{M_{G-v}}$$

and, from the induction hypothesis,

$$\frac{M_{G-v}}{M_{G-u'-v}} = \frac{M_{T(G-v,u')}}{M_{T(G-v,u')-u'}}.$$

Because of property (iii),

$$\frac{M_{T(G-v,u')}}{M_{T(G-v,u')-u'}} = \frac{M_{T(G,v)-v}}{M_{T(G,v)-u'-v}}.$$

Combining these three relations gives us

$$\frac{M_G}{M_{G-v}} = x - \frac{\sum_{e'} M_{T(G,v)-u'-v}}{M_{T(G,v)-v}}.$$

Theorem 4.19 now follows by an additional application of Equation (4.8).

PROOF OF THEOREM 4.18: We first show that if G is connected then M_G divides $M_{T(G,v)}$. We do this by induction on the number of vertices of G. The proof is immediate for connected graphs with fewer than four vertices. If M_{G-v} divides $M_{T(G-v,u')}$, then because of property (iii) it will also divide $M_{T(G,v)-v}$. Then, however, by Theorem 4.19 M_G must be a divisor of $M_{T(G,v)}$. Hence Theorem 4.18 is true for connected graphs. The validity for disconnected graphs is a straightforward consequence of Theorem 4.19.

COROLLARY 4.19.1: *All divisors of the polynomial $P_F(x)$ are matching polynomials.*

COROLLARY 4.19.2: *If G is connected, then $\mu_1(G) = \lambda_1(F(G))$.*

Theorem 4.12 also follows immediately from Theorem 4.19.

Theorem 4.18 is at present the most powerful result in the theory of matching polynomials. It allows the theory of matching polynomials to be viewed as part of the theory of graph spectra.

A related question has been posed [AIH5]: for any graph G, is it possible to construct a Hermitian matrix $H(G)$ such that M_G is its characteristic polynomial? Like the adjacency matrix, the (i, j) entry of $H(G)$ should be zero whenever the corresponding vertices are not adjacent. The existence of the matrix implies, of course, that the roots of the corresponding matching polynomial are real. The matrix $H(G)$ can be constructed for certain graphs including unicyclic graphs and cacti ([HEPA1], [SHNT1], [GRA1], [GRKT1]), but in general $H(G)$ does not exist [GUGM1].

The matching polynomial and/or their roots have been determined for many classes of graphs ([FAR1], [FAR7], [FAR8], [FAR9], [FAWA1], [FAWA2], [GUT10], [GUT25], [GUT28], [GUT32], [GUT33], [GUT41], [GUFW1], [GUPO1], [HOOH1], [WAH1]). While not all of the results will be presented here, some of the more interesting examples follow.

Let C_n, P_n, K_n denote, as usual, the circuit, the path, and the complete graph in the usual way, and let $K_{m,n}$ denote the complete bipartite graph on $m + n$ vertices. In addition, let T_n and U_n be the *Chebyshev functions* of the first and second kind, let H_n and H_n^* be the two standard forms of the *Hermite polynomial*, and let L_n and L_n^s be the *Laguerre polynomial* and generalized Laguerre polynomial.

THEOREM 4.20 (C. D. GODSIL, I. GUTMAN [GOGU2], I. GUTMAN, D M CVETKOVIĆ [GUCV1], H. HOSOYA [HOE1]):

$$M_{C_n}(x) = 2T_n(x/2) \text{ for } n \geq 3. \tag{4.15}$$

$$M_{P_n}(x) = 2(4 - x^2)^{-\frac{1}{2}} U_{n+1}(x/2) \text{ for } n \geq 0. \tag{4.16}$$

$$M_{K_n}(x) = 2^{n/2} H_n\left(\frac{x}{\sqrt{2}}\right) = H_n^*(x) \text{ for } n \geq 1. \tag{4.17}$$

$$M_{K_{n,n}}(x) = (-1)^n L_n(x^2) \text{ for } n \geq 1. \tag{4.18}$$

$$M_{K_{m,n}}(x) = (-1)^m x^{n-m} L_m^{n-m}(x^2) \text{ for } n > m \geq 1. \tag{4.19}$$

We shall prove that (4.17) holds. The application of (4.8) to an arbitrary vertex of K_n yields

$$M_{K_n}(x) = x M_{K_{n-1}}(x) - (n-1) M_{K_{n-2}}(x).$$

Also, $M_{K_1}(x) = x$, and $M_{K_2}(x) = x^2 - 1$. On the other hand, the Hermite polynomial $H_n^*(x)$ satisfies exactly the same recursion relation and initial conditions, and hence the relation (4.17) is obvious.

Further results of this type can be found in [GOD2], [HELI2], [ZAS3], and [ZAS4]. Since it is known that for $n > 25$ Hermite polynomials are not solvable by radicals, we immediately get the following conclusion:

COROLLARY 4.20.1 (C. D. GODSIL [GOD5]): *The matching polynomial of K_n is not solvable by radicals for $n > 25$.*

COROLLARY 4.20.2 (C. D. GODSIL [GOD5]): *The characteristic polynomial of the Godsil tree of K_n (which is unique) is not solvable by radicals for $n > 25$.*

COROLLARY 4.20.3 (C. D. GODSIL [GOD5]): *Let p_n be the proportion of trees on n vertices which have characteristic polynomials not solvable by radicals. Then $p_n \to 1$ as $n \to \infty$, and hence the characteristic polynomials of almost all trees are not solvable by radicals.*

We note that Corollary 4.20.2 and Corollary 4.20.3 answer an open question originally posed in [CVDSA1] (page 267).

Let $\overline{G}$ denote the complement of the graph G.

THEOREM 4.21 (T. ZASLAVSKY [ZAS2]): *M_G determines $M_{\overline{G}}$. In particular,*

$$M_{\overline{G}}(x) = \sum_{k=0}^{m} p(G,k) M_{K_{n-2k}}(x). \tag{4.20}$$

PROOF: The edges of K_n can be partitioned into those which belong to G and those which belong to $\overline{G}$. The number of selections of k independent edges in K_n such that at least j of them belong to G is $p(G,j)p(K_{n-2j}, k-j)$. On the other hand, $p(\overline{G},k)$ is equal to the number of selections of k independent edges in K_n such that none of them belong to G. By the inclusion–exclusion principle we obtain

$$p(\overline{G}, k) = \sum_{j=0}^{k}(-1)^j p(G, k)p(K_{n-2j}, k - j)$$

whose immediate consequence is Theorem 4.21.

COROLLARY 4.21.1: *If* $M_G = M_H$ *then* $M_{\overline{G}} = M_{\overline{H}}$ *and* $M_{\overline{G+pK_1}} = M_{\overline{H+pK_1}}$ *for all* $p \geq 1$.

COROLLARY 4.21.2 (C. D. GODSIL [GOD3]):

$$p(\overline{G}, k) = \frac{1}{(n - 2k)!\sqrt{2\pi}} \int_{-\infty}^{\infty} M_G(x)M_{K_{n-2k}}(x)e^{-x^2/2}\,dx \qquad (4.21)$$

In particular, the number of 1-factors of $\overline{G}$ *is given by*

$$\frac{1}{\sqrt{2\pi}} \int_{-\infty}^{\infty} M_G(x)e^{-x^2/2}\,dx.$$

Let B be a bipartite graph with an (m, n)-bipartition with $m \leq n$, and let $\overline{B}$ be the bipartite complement of B.

THEOREM 4.22 (I. GUTMAN [GUT32]): M_B *determines* $M_{\overline{B}}$. *In particular,*

$$M_{\overline{B}}(x) = \sum_{k=0}^{m} p(B, k)M_{K_{m-k,n-k}}(x). \qquad (4.22)$$

COROLLARY 4.22.1:

$$p(\overline{B}, k) = \frac{1}{(a - k)!(b - k)!} \int_{-\infty}^{\infty} M_B(\sqrt{x})M_{K_{m-k,n-k}}(\sqrt{x})e^{-x}\,dx \qquad (4.23)$$

Corollary 4.21.2 and Corollary 4.22.1 come from Theorem 4.20 using the fact that M_{K_n} and $M_{K_{m,n}}$ are orthogonal polynomials.

Theorem 4.22 also appears in [GOD3].

Let G_1 and G_2 be two graphs and let $G_1 \nabla G_2$ denote the complete product.

THEOREM 4.23 (I. GUTMAN [GUT32]): M_{G_1} and M_{G_2} determine $M_{G_1 \triangledown G_2}$. In particular,

$$M_{G_1 \triangledown G_2}(x) = \sum_k p(\overline{G_1} + \overline{G_2}, k) M_{K_{n_1+n_2-2k}}(x) \qquad (4.24)$$

where

$$p(\overline{G_1} + \overline{G_2}, k) =$$

$$\sum_{u,s,t} (-1)^{s+t} p(G_1, s) p(G_2, t) p(K_{n_1-2s}, u - s) p(K_{n_2-2t}, k - u - t).$$

$$(4.25)$$

COROLLARY 4.23.1: If $M_{G_1} = M_{G_2}$ and G is any graph, then $M_{G_1 \triangledown G} = M_{G_2 \triangledown G}$.

Suppose G is a graph with n vertices and K is a rooted graph; then $G[K]$ is obtained by taking n copies of K and identifying each root with a different vertex of G. Further, let L denote the graph obtained from K by deleting its root.

THEOREM 4.24 (C. D. GODSIL, I. GUTMAN [GoGu3]):

$$M_{G[K]}(x) = \left(M_L(x)\right)^n M_G\left(\frac{M_K(x)}{M_L(x)}\right) \qquad (4.26)$$

THEOREM 4.25 (I. GUTMAN [GUT32]): Let $G \circ L$ be the corona of G and L. Then

$$M_{G \circ L}(x) = \left(M_L(x)\right)^n M_G\left(\frac{x - M_L'(x)}{M_L(x)}\right) \qquad (4.27)$$

where $M_L'(x)$ is the first derivative of $M_L(x)$.

Section 4.2: The Matching Polynomial of Weighted Graphs

In this section we extend the definition of the matching polynomial to include graphs with weighted edges and loops. The following definition has proved to be reasonable in [AIH1], [GUT34], [GUPO1], and [GUMT1]. Let G be a graph whose vertices and edges are v_i, $i = 1, \ldots, n$ and e_j, $j = 1, \ldots, m$; G_W is obtained from G by associating a weight $w(e_j)$ to each edge e_j and a weight $w(v_i)$ to each loop at the vertex v_i. Suppose that e is an edge with u and v as end points; then the weighted matching polynomial of G_W is defined recursively by

$$M_{G_W}(x) = M_{G_W - e}(x) - w(e)^2 M_{G_W - u - v}(x) \qquad (4.28)$$

and

$$M_{G_W}(x) = M_{G'_W}(x) - w(v)M_{G_W - v}(x) \qquad (4.29)$$

where G'_W is formed by deleting the loop at the vertex v.

Note how equation (4.28) compares with (4.7). The weighted matching polynomial is identical to the matching polynomial if $w(v) = 0$ for any loop and $w(e) = 1$ for any edge. The advantage of this definition is that it immediately gives the following theorem:

THEOREM 4.26: (CF. THEOREM 4.3) *Let $M_{G_W}(x)$ and $P_{G_W}(x)$ be the matching and characteristic polynomials of G_W. Then, if G is a forest,*

$$M_{G_W}(x) = P_{G_W}(x). \qquad (4.30)$$

THEOREM 4.27 (I. GUTMAN [GUT34]): *If $w(e)$ and $w(v)$ are real for every edge e and vertex v, then the roots of the weighted matching polynomial of G_W are real.*

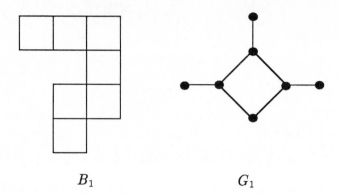

$$B_1 \qquad\qquad\qquad G_1$$

Figure 4.4 A Board and its Corresponding Graph

Section 4.3: The Rook Polynomial

Let N be the set of positive integers. A *board* B is a finite subset of $N \times N$. We may consider a board as a subset of an s by t chessboard. We say that the cell b_{ij} belongs to the board B if $(i, j) \in B$. The index i denotes the row while j denotes the column in which the cell b_{ij} is located. The board B_1, for example, is composed of the cells b_{11}, b_{12}, b_{13}, b_{23}, b_{32}, b_{33}, b_{42}:

We say that the cells b_{ij} and b_{kl} are independent if $i \neq k$ and $j \neq l$. In other words, if b_{ij} and b_{kl} are independent, then two rooks placed on b_{ij} and b_{kl} will not attack each other.

The number of ways that k independent cells can be chosen from B (*i.e.*, in which k rooks can be placed on B) is denoted by $r(B, k)$. By definition let $r(B, 0) = 1$ and $r(B, 1) = m$, the number of cells in B.

For the board B_1 defined above, $r(B_1, 0) = 1$, $r(B_1, 1) = 7$, $r(B_1, 2) = 11$, $r(B_1, 3) = 3$, $r(B_1, k) = 0$ for $k \geq 4$. In particular, it is possible to place three nonattacking rooks on B_1 in three distinct ways, *viz.*,

The traditional definition of the *rook polynomial* as given in [RIO1] is $\sum_k r(B, k) x^k$. For our discussion, however, we will define the rook polynomial of B to be

$$R_B(x) = \sum_{k=0}^{m} (-1)^k r(B, k) x^{s+t-2k} \qquad (4.31)$$

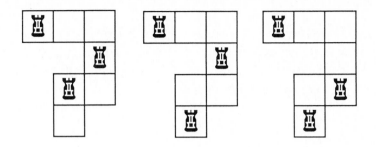

Figure 4.5 Nonattacking rooks on B_1

Rook theory (and within it the theory of the rook polynomial) has been developed independently from graph theory ([RIO1]). Nevertheless, the following simple result holds:

THEOREM 4.28 (C. D. GODSIL, I. GUTMAN [GOGU4]): *Let* $\mathcal{G}(s,t,m)$ *be the set of all labelled bipartite graphs on* $s+t$ *vertices having* m *edges and no isolated vertices. Let* $\mathcal{B}(s,t,m)$ *be the set of all boards with* s *rows,* t *columns, and* m *cells. Then for all* $s,t,m \geq 0$ *there exists a bijection* f *of* $\mathcal{B}(s,t,m)$ *into* $\mathcal{G}(s,t,m)$ *such that if* $G = f(B)$ *then* $p(G,k) = r(B,k)$ *for all* $k \geq 0$.

COROLLARY 4.28.1: *If* $G = f(B)$, *then* $R_B(x) = M_G(x)$.

In other words, every rook polynomial is a matching polynomial of a bipartite graph, and the matching polynomial of every bipartite graph without isolated vertices is a rook polynomial of some board.

The bipartite graph G_1 in Figure 4.4, for example, corresponds to the board B_1. The vertices of G_1 are labeled so that u_i and v_j are adjacent if and only if the cell b_{ij} is contained in B_1. We also have $R_{B_1}(x) = M_{G_1}(x) = x^7 - 7x^5 + 11x^3 - 3x$.

Theorem 4.28 thus imbeds the theory of rook polynomials into the theory of the matching polynomial. Several of the results presented in this chapter (Theorem 4.5, Theorem 4.22, as well as the equations (4.15), (4.16), (4.18), (4.19)) have also been proved under the guise of rook theory [RIO1]. It has been independently conjectured in [GOJW1] and proven in [NIJ1] that $R_B(x)$ has real roots.

Section 4.4: The Independence Polynomial

Instead of considering independent edges, one can study the selection of independent vertices. Hence let $i(G, k)$ be the number of ways in which k independent (*i.e.*, mutually nonadjacent) vertices can be selected from G. In addition define $i(G, 0) = 1$ for all graphs G. The *independence polynomial* of the graph G is then defined to be

$$I_G(x) = \sum_{k=0}^{n} (-1)^k i(G, k) x^{2n-2k}. \qquad (4.32)$$

An immediate consequence of this definition is the evaluation of the independence polynomial of a line graph.

THEOREM 4.29: *Let G be a graph with n vertices and m edges, and let $L(G)$ be its line graph. Then*

$$x^{2m} M_G(x) = x^{2n} I_{L(G)}(x). \qquad (4.33)$$

If G is not a line graph, then $I_G(x)$ need not have real roots [GUHA1]. Consequently the analogy between the matching and independence polynomials (equations 4.1 and 4.32) is not far reaching. A number of elementary properties of the independence polynomial are collected in [GUHA1].

Another generalization to the theory of matching polynomials would be to select other independent objects such as triangles or stars. It has been shown, however, that no matter what the object is, all polynomials whose coefficients are equal to the number of selections of some independent object are, in fact, independence polynomials [GUHA1].

Section 4.5: The F-Polynomial

E. J. Farrell [FAR5] has introduced a family of graph polynomials and named them F-polynomials. Let $\mathcal{F}$ be a set of connected graphs, and with each element $F \in \mathcal{F}$ associate an independent indeterminant. An F-cover of G is a spanning subgraph of G in which each component belongs to F. The *weight of an F-cover* is the monomial which is the product of the indeterminants associated with elements of that cover. The *F-polynomial* is then the sum over all possible covers of the respective weights.

A special case of an F-polynomial is the *circuit polynomial* [FAR4], [FAR5]. In this case, the set of circuits includes K_1 and K_2 as circuits with one and two vertices. A weight w_k is associated with each circuit with k vertices and $w = (w_1, w_2, \ldots, w_n)$. The circuit polynomial of a graph with n vertices is then denoted by $C(G; w)$.

The circuit polynomial for the bipartite graph G_0 defined at the beginning of this chapter, for example, is $C(G_0, w_1, w_2, w_3, w_4, w_5, w_6) = w_1^6 + 8w_1^4 w_2 + 14w_1^2 w_2^2 + 2w_2^3 + 2(w_1^3 + 3w_1 w_2)w_3 + w_1^2 w_4 + 2w_1 w_5 + w_6$.

THEOREM 4.30 (E. J. FARRELL [FAR4], [FAR5]):

$$P_G(x) = C(G; x, -1, -2, \ldots, -2). \qquad (4.34)$$

$$M_G(x) = C(G; x, -1, 0, \ldots, 0). \qquad (4.35)$$

Some further results on circuit polynomials can be found in [FAR10], [FAGR1], and [FAGR2] . E. J. Farrell has also considered several other special cases of the F-polynomial. They are not, however, within the context of this book. For a review of these properties see [FAR12] and [FAR13] .

Section 4.6: The Permanental Polynomial

Let A be the adjacency matrix of the graph G. The determinant of $\lambda I - A$ is the characteristic polynomial, while the permanent of the same matrix is called the *permanental polynomial* of the graph G. We shall denote it by $P_G^*(x)$. The theory of the permanental polynomial has many analogies with that of the characteristic polynomial. The coefficients, for example, can be calculated using an appropriate reformulation of the Sachs theorem from [CvDS1], p. 32 ([BoJó2], [BoJó5], [MeRW1]). Several other elementary results on the permanental polynomial can be found in [BoJó2], [BoJó3], [BoJó5], [MeRW1], [KaTG1].

THEOREM 4.31 (M. BOROWIECKI, T JÓZWIAK [BoJó3], [BoJó5]): *A graph on n vertices is bipartite if and only if $P_G^*(-x) = (-1)^n P_G^*(x)$.*

THEOREM 4.32 (M. BOROWIECKI, T. JÓZWIAK [BoJó3], [BoJó5], R. MERRIS, K. R. REBMANN, W. WATKINS [MwRW1]): *If G is a forest with n vertices and m edges, then*

$$P_G^*(x) = \sum_{k=0}^{m} p(G, k) x^{n-2k}. \tag{4.36}$$

COROLLARY 4.32.1: *Two forests have equal permanental polynomials if and only if they are cospectral.*

M. Borowiecki and T Józwiak have demonstrated in [BoJó3] that for every $n \geq 11$ there exists a pair of nonisomorphic, connected, cospectral graphs on n vertices which are not trees with the same permanental polynomial. Further results on the permanental polynomial can be found in [BaPa1]. For properties of the permanent of the adjacency matrix see [LeHP1].

Section 4.7: Polynomials and the Admittance Matrix

Let G be a graph, A its adjacency matrix, and D its degree matrix, i.e., the diagonal matrix with the vertex degrees of G for diagonal entries. The *admittance matrix* of G is $C = D - A$ (see [CvDSA1], p. 27). It is sometimes called the *Laplacian matrix* of G. We denote by $C_G(x)$ the characteristic polynomial of this matrix. Let

$$C_G(x) = x^n + c_1 x^{n-1} + \cdots + c_n \tag{4.37}$$

If $n \leq 10$, then no pair of nonisomorphic trees exist with n vertices having the same C_G polynomial; on the other hand, such pairs of trees do exist for all $n \geq 11$ [DiKZ1].

A. K. Kel'mans has given the basic properties of the coefficients of C_G (see [CvDSA1], p. 38). The following analogous result has been formulated recently for the permanental polynomial of the admittance matrix. Let

$$\operatorname{per}(xI - C) = x^n + p_1 x_{n-1} + \cdots + p_n. \tag{4.38}$$

Let V be the vertex set and E be the edge set of G. Define $[V, E]$ to be the set of ordered pairs $(v, e) \in V \times E$ such that v and e are incident. Two pairs from $[V, E]$ are said to overlap if they are the same in one coordinate. A *distinguished subset* of $[V, E]$ consists of pairwise nonoverlapping elements.

For $e \in E$, let $[V - e, E] = \{(v, e) \in [V, E] \mid v \notin e\}$. For $t = 1, 2, \ldots, [\frac{k}{2}]$ define $c_{k,t}$ to be the sum over all sets $\{e_1, \ldots e_t\}$ of independent edges of the total number of $k - 2t$ element distinguished subsets of $\bigcap_{j=1}^{t}[V - e_j, E]$. If $t = \frac{k}{2}$, then $c_{k,t}$ is just the number of collections of $\frac{k}{2}$ independent edges.

THEOREM 4.33 (R. MERRIS [MER2]): *Let T be a tree. Then*

$$p_k = c_k + (-1)^k \sum_{t=1}^{[k/2]} 2^t c_{k,t}. \tag{4.39}$$

The permanent of the admittance matrix has been examined by R. A. Brualdi [BRU1]. He proved, among other things, that if $\operatorname{per}(G)$ denotes the permanent of the admittance matrix of a graph G then the following theorem holds.

THEOREM 4.34 (R. A. BRUALDI [BRU1]):   Let B be a connected bipartite graph with an (m, n)-bipartition. Then

$$(2m - 1)(2n - 1) + 1 \leq \text{per}(G) \leq \text{per}(K_{m,n}) \qquad (4.40)$$

THEOREM 4.35 (R. A. BRUALDI [BRU1]):   Let T be a tree with n vertices. Then

(i) $\text{per}(K_{1,n-1}) \leq \text{per}(T) \leq \text{per}(P_n)$ $\qquad\qquad\qquad$ (4.41)

(ii) if T has a k-matching then $3^{k-2}(6n - 8k + 2) \leq \text{per}(T)$ $\qquad$ (4.42)

(iii) If T is not a star and has vertex degrees $d_1, \ldots, d_n$
$\quad\quad$ then $d_1 d_2 \ldots d_n (3n - 3)/(n - 2) \leq \text{per}(T)$. $\qquad\qquad$ (4.43)

The left hand inequality of (4.41) also holds for any bipartite graph with n vertices.

Section 4.8: The Distance Polynomial

The *distance matrix* of a graph G has rows and columns that correspond to the vertices of G with the (i, j) entry being equal to the distance between vertices i and j. the diagonal has only 0 entries, and entries of 1 appear in exactly the same positions as the adjacency matrix. So in one sense the distance matrix contains at least as much information as the adjacency matrix. On the other hand, given the adjacency matrix A of a graph, the distance between different vertices i and j is the smallest k such that $A_{i,j}^k$ is nonzero, so in fact the two matrices contain exactly the same information.

If the diameter of the graph is 2, then the distance matrix, D, satisfies $D = 2J - A - 2I$. Thus, for example, if the graph is the complete bipartite graph $K_{m,n}$, then $D + 2I$ has only two different rows so that the rank is 2. The trace of $D + 2I$ is $2(m + n)$ and the trace of $(D + 2I)^2$ is $4m^2 + 2mn + 4n^2$.

PROPOSITION 4.36: *The characteristic polynomial of the distance matrix of $K_{m,n}$ is*

$$(x + 2)^{m+n-2}(x^2 - 2(m + n - 2)x + 4(m - 1)(n - 1) - mn).$$

The special case of $K_{1,n}$ is given in [KRTR1] along with that of the cycle C_n.

If G is a regular graph of diameter 2, then the eigenvalues of the distance matrix D are evident since $D = 2J - A - 2I$. If G has n vertices and degree r, then the dominant eigenvalue of D is $2n - r - 2$. If λ_i is a subdominant eigenvalue of G then $-\lambda_i - 2$ is an eigenvalue of D.

The particular case of a strongly regular graph is the following:

PROPOSITION 4.37: *If G is a strongly regular with n vertices and eigenvalues r, λ_2, and λ_3, then the eigenvalues of the distance matrix are $2n - r - 2$, $-\lambda_2 - 2$, and $-\lambda_3 - 2$.*

COROLLARY 4.37.1: *Two regular graphs with the same degree and diameter 2 have cospectral distance matrices if and only if they have cospectral adjacency matrices.*

The distance matrix of a tree has interesting properties. Suppose the n vertices are ordered so that v_1 has degree 1 and is adjacent to v_2. Subtracting the the second row from the first and the second column from the first makes all entries in the first row and column equal to 1 except for the diagonal entry which is -2. Reordering the vertices if necessary to make v_2 have degree 1 and be adjacent to v_3 in the subtree determined by $\{v_2, \ldots, v_n\}$, subtract the third row and column from the second row and column respectively. Continuing in the same manner with the rest of the tree, the matrix B is obtained in which the last row and column has every entry equal to 1 except for the diagonal entry which is 0; the remainder of the matrix is of the form $-2I_{n-1}$. $B + 2I$ has rank 2 so that -2 is an eigenvalue of B with multiplicity $n-2$. Since the trace of B is $-2(n - 1)$ and the trace of B^2 is $6(n - 1)$, product of the remaining two eigenvalues is $-n + 1$. Hence the determinant of B is $(-2)^{n-2}(-n + 1)$.

THEOREM 4.38 (M. EDELBERG, M. R. GAREY, R. L. GRAHAM [EDGG] OF [CVDSA1]):   *The determinant of the distance matrix of a graph depends only on its blocks. In particular, if T is a tree with distance matrix $D(T)$, then*

$$\det D(T) = (-1)^{n-1}(n-1)2^{n-2}.$$

The following corollary follows immediately:

COROLLARY 4.38.1:   *If T is a tree with n vertices with distance matrix $D(T)$,*

$$\det D(T) = \begin{cases} < 0 & \text{if } n \text{ odd,} \\ 0 & \text{if } n = 1, \text{ and} \\ > 0 & \text{otherwise.} \end{cases}$$

Thus if T_n is a tree with n vertices, then $\det D(T_n)$ is an alternating sequence for $n = 2, 3, \ldots$, which implies that there is only one positive eigenvalue ([EDGG] of [CVDSA1]).

COROLLARY 4.38.2:   *The distance matrix of a tree has exactly one positive eigenvalue.*

Trees with cospectral distance matrices exist. The smallest has 17 vertices ([MCKA] of [CVDSA1]). All known trees with cospectral distance matrices have cospectral adjacency matrices. Whether or not this is always true is unknown.

There is an interesting application of distance matrices to an *addressing problem* for communication networks. Suppose a graph has n vertices; assign to each vertex an address which is an m-tuple with coordinates equal to 0, 1, or $*$. The distance between two addresses is the number of coordinates for which one address takes on the value 0 and the other takes on the value 1. We wish to make the assignments so that the distance between addresses is identical to the distance between the corresponding vertices in the graph. Then $N(G)$ is defined as the minimum value of m for which such an assignment can be made (and obviously one can make at least one assignment if m is large enough). P. Winkler [WIN1] has given an upper bound and R. L. Graham and H. O. Pollack have given a lower bound (see [CVDSA1], p.263).

THEOREM 4.39: *Let G be a graph with n vertices whose distance matrix has n_+ positive eigenvalues and n_- negative eigenvalues. Then*

$$\max \{n_-, n_+\} \leq N(G) \leq n - 1.$$

There is interest in determining the exact value of $N(G)$. In general this is very hard to do, but in some cases Theorem 4.39 provides an exact value.

COROLLARY 4.39.1 (R. L. GRAHAM, H. O. POLLACK): *If T is a tree with n vertices, then*

$$N(T) = n - 1.$$

PROOF: The cases of $n = 1$ and $n = 2$ are trivial. For $n > 2$, by Corollary 4.38.2, we have $n_+ = 1$, and since $\det T \neq 0$, it must be that $n_- = n - 1$, and hence Theorem 4.39 yields the result.

COROLLARY 4.39.2: *If $G = K_n$, then $N(G) = n - 1$.*

Further results of this type can be found in [LIN1].

Section 4.9: Miscellaneous Results

Some other graph polynomials have been defined which, in special cases, reduce to the characteristic polynomial. These include *Krishnamurthy's polynomial* [KRI1], the *tree polynomial* [FAR2], the *path polynomial* [FAR11], [FAR14], the *μ-polynomial* [GUPO2], [FAGU1], the *sextet polynomial* [GUT3], *Schneider's polynomial* [SCHN1] [GOD7] and the *pseudocharacteristic polynomial* [MER1].

Some further graph polynomials of interest in theoretical chemistry are presented in [KNTR1] and [GUT48].

Chapter 5

Applications to Chemistry and Other Branches of Science

In the present chapter we shall review some two hundred papers published since 1978 (together with a few older publications not mentioned in [CvDSa1]) which are concerned with the application of graph spectra to the natural sciences. The overwhelming majority of these papers deal with chemical problems and are published in chemical journals.

It is necessary to remind the readers of this chapter that investigations in theoretical chemistry (and to some extent also in theoretical physics) require mathematical results of somewhat different kind from those preferred by mathematicians. Thus many papers dealing with chemical applications of graph spectra contain only mathematical results (that is, theorems) of minor mathematical interest or contain no such results at all. Often chemical papers simply elaborate (or even rediscover) previously known mathematical results. Sometimes they offer only empirical observations on or approximate relations between certain graph spectral quantities. Nevertheless, such papers may be (and some certainly are) very useful and of great significance for the appropriate scientific field.

The present chapter will not assume any previous knowledge of theoretical chemistry or physics. Our main concern will be the mathematical content of the papers reviewed. Chemical and physical details will be omitted, or at most mentioned only briefly. The interested reader should consult the original literature or some of the many existing monographs and reviews: [AuYSG], [Cve3], [Dmi1], [Dmi2], [Dmi3], [GrGT1], [Gut10], [Gut12], [Gut40], [Gut48], [Kin3],

[MAL3], [TRI3], [TRI4], [AMGT1], [QUA1], [WILS1]. It is by no means possible to list here all papers in which graph eigenvalues have been either used or mentioned. Because of the limited space, only a selected number of results and research topics can be discussed at satisfactory length. Being aware that the selection is somewhat subjective, the authors still hope that the material of this chapter is a representative cross section of the current application of graph spectra in the natural sciences.

Section 5.1: On Hückel Molecular Orbital Theory

The *Hückel molecular orbital (HMO) theory* is nowadays the most important field of theoretical chemistry where graph eigenvalues occur. HMO theory deals with unsaturated *conjugated molecules*. In these molecules two kinds of electrons are distinguished, the σ-*electrons* and the π-*electrons*. Only the π-electrons are of interest in HMO theory; it is believed that they are responsible for the most important chemical and physical properties of conjugated molecules.

HMO theory describes the π-electrons of a conjugated molecule by means of an approximate *Schrödinger equation*

$$H\psi = E\psi \tag{5.1}$$

where H is the *Hamiltonian operator*, ψ the *wave function* (also called the *molecular orbital*) and E the *energy* of a π-electron. The HMO Hamiltonian can be written as

$$H = \alpha I + \beta A \tag{5.2}$$

where I is a unit matrix and A can be interpreted as the adjacency matrix of a certain graph called the *molecular (or Hückel) graph*. In the case of conjugated hydrocarbons, the molecular graph has no loops or multiple edges. Otherwise the molecular graph possesses weighted edges or loops. Hence we shall focus our attention on the HMO theory of *conjugated hydrocarbons*.

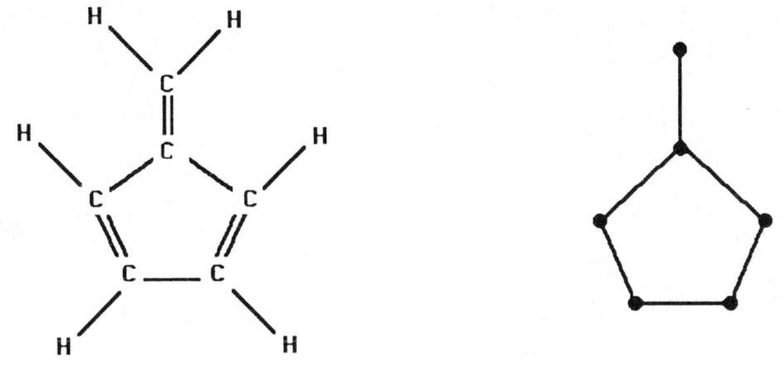

.Fulvene   Molecular Graph of Fulvene

Figure 5.1 Molecular Graphs

The formal definition of a molecular graph is given in [CvDSA1], p. 229, and an example is given below:

Let the molecular graph have n vertices (so that the corresponding conjugated hydrocarbon has n carbon atoms). Then because of (5.2) the equation (5.1) has n solutions

$$E_j = \alpha + \lambda_j \beta \qquad (5.3)$$

and

$$\psi_j = (c_{j1}, c_{j2}, \ldots, c_{jn}) \qquad (5.4)$$

for $j = 1, 2, \ldots, n$, where λ_j and $(c_{j1}, c_{j2}, \ldots, c_{jn})$ are the eigenvalues and eigenvectors of the molecular graph.

For the remainder of this chapter we shall assume that the graph eigenvalues are labeled in nonincreasing order and that the eigenvectors are real, normalized and mutually orthogonal.

The parameters α and β are constants in the HMO theory. It is usual and convenient to formally set $\alpha = 0$ and $\beta = 1$.

The number of π-electrons which are described by the wave function ψ_j is denoted by g_j with $g_j \in \{0, 1, 2\}$. In the most common cases,

$$g_j = 2 \text{ if } \lambda_j > 0 \text{ and } g_j = 0 \text{ if } \lambda_j < 0. \qquad (5.5)$$

In the general case (especially when $\lambda_j = 0$) the numbers g_j are determined in a somewhat more complicated manner (see [MARo1], [MARo2]).

The following are the most frequently considered HMO quantities. The *total π-electron energy* is defined as

$$E = \sum_{j=1}^{n} g_j \lambda_j. \tag{5.6}$$

The *charge* on the atom u is defined as

$$q_u = 1 - \sum_{j=1}^{n} g_j (c_{ju})^2, \tag{5.7}$$

where C_{ju} is the component of the j-th eigenvector of the molecular graph corresponding to the vertex u. The bond order between atoms u and v is defined as

$$p_{uv} = \sum_{j=1}^{n} g_j c_{ju} c_{jv}. \tag{5.8}$$

The basic properties of q_u and p_{uv} have been established in [CoLo1].

The number of vertices of the molecular graph is usually even. Then the energy of the *highest occupied molecular orbital (HOMO)* is $\lambda_{n/2}$. The energy of the *lowest unoccupied molecular orbital (LUMO)* is $\lambda_{n/2+1}$. The *HOMO-LUMO separation* is $\lambda_{n/2} - \lambda_{n/2+1}$.

It is of chemical interest to know whether the molecule considered has *nonbonding molecular orbitals*. Such an orbital corresponds (via (5.3) and (5.4)) to graph eigenvectors whose eigenvalues are equal to zero. There exist nonbonding molecular orbitals if det $A = 0$. This is one reason to investigate the determinant of the adjacency matrix of a molecular graph. The algebraic multiplicity of the number zero in the spectrum of a graph G is denoted by η.

The *topological resonance energy* is defined as

$$\text{TRE} = \sum_{j=1}^{n} g_j (\lambda_j - \mu_j), \tag{5.9}$$

where $\mu_j, j = 1, 2, \ldots n$, are the roots of the matching polynomial of the molecular graph in nonincreasing order.

Section 5.2: The Characteristic Polynomial

The calculation of the characteristic polynomial of the molecular graph remains one of the first applications of graph theory to HMO theory. The papers [AIH1], [GRPT1], [RIMD1], and [TRI2] all use variations of the Sachs Theorem ([CVDSA1], p. 32). Different matrix-theoretic results including the Cayley-Hamilton theorem [RAN5], Frame's method [BAL2], [BAL3] and others [ELB2], [ELB3], [GUSH1], [KIA3], [JIA2] have been applied to $P_G(\lambda)$. T. Au-Chin and K. Yuan-Sun [AUYU1] and F. Zhang [ZHA1], [ZHA2] have rediscovered the recursion formula of [CVDSA1] (p. 78) for computing P_G. M. Randić proposed the following three step calculation scheme to obtain P_G [RAN2]: (i) determine the n characteristic polynomials of the vertex-deleted subgraphs of G; (ii) determine the first derivative of P_G by summing these polynomials; (iii) determine P_G by integration using the (separately calculated) determinant of A. This algorithm is recommended for molecular graphs with more than ten vertices; its advantage over existing methods (which seems to be fairly uncertain) remains to be verified.

The methods for the calculation of P_G when G is a tree [BAL1], when G has vertices of degree one [BARA1] and when G has bridges [KACH1], [KACH2], [GUT25], [GUT30] are based on known results of spectral graph theory. In [RAN4], [HORA1] and [KIR1], P_G is expanded in terms of P_{P_k}, the characteristic polynomial of a path with k edges. Another method for the computation of P_G from the graph eigenvalues is given in [MKVP1].

Section 5.3: Cospectral Molecular Graphs

The existence of molecular graphs having the same spectrum continues to attract the attention of theoretical chemists (a review is given in [AMGT1]). Several novel series of cospectral nonisomorphic graphs have been constructed [GUT27], [HEI2], [JIA3], [LIWZ1], [SCHHE1], [JIA3].

Section 5.4: The Spectrum and the Automorphism Group of Molecular Graphs

The use of group theory based on the consideration of the *symmetry of the molecule* is a well established technique in theoretical chemistry (see [CvDSa1], p. 146). The group of symmetry operations of a conjugated molecule is a subgroup of the automorphism group of the corresponding molecular graph (see [WiKG] given in [CvDSa1]). Group-theoretical methods for factoring the characteristic polynomial of symmetric molecular graphs have been further elaborated in [Ama1], [Ama2], [AuYu1], [AuYu2], [Dav1], [Dix3], [GPTy1], [Hal2], [Kin1], [Liu1], [McCl1], [McCl2], [NoWW1], [TeCh1], [WaLz1], [Wil1], [YaWa1], [ZhLW1], [ZhLW2].

In [Ama1], for example, graphs with "twofold symmetry" are considered. A graph G is said to have twofold symmetry if its adjacency matrix can be written in the form

$$A = \begin{pmatrix} B_1 & B_2 \\ B_2 & B_1 \end{pmatrix} \tag{5.10}$$

where B_1 and B_2 are symmetric square matrices. The spectrum of G is then the union of the eigenvalues of the matrices $B_1 + B_2$ and $B_1 - B_2$.

Section 5.5: The Energy of a Graph

The total π-electron energy of a conjugated molecule is defined by (5.6). If the conditions (5.5) hold, then

$$E = \sum_{j=1}^{n} |\lambda_j|. \tag{5.11}$$

With this in mind we define the energy of a graph, $E(G)$, to be the sum of the absolute value of the eigenvalues of G. This allows the consideration of $E(G)$ for graphs other than molecular graphs [Gut12].

Numerous results about the energy of a graph have been obtained. For previous results concerning $E(G)$ see [Gut10], [Gut12], and [CvDSa1], pp. 237, 238, and 259.

It is known [CVDSA1], p. 82 that the characteristic polynomial of a bipartite graph can be written as

$$P_G(\lambda) = \sum_{k=0}^{[n/2]} (-1)^k b(G,k)\lambda^{n-2k}. \tag{5.12}$$

It is easy to show that $b(G,k) \geq 0$ for all $k = 0, 1, \ldots, [n/2]$.

THEOREM 5.1 (I. GUTMAN [GUT9], [GUT12] IN [CVDSA1]): *Let G be a bipartite graph and $P_G(\lambda)$ have the form (5.12). Then $E(G)$ is a monotonically increasing function of each $b(G,k)$ for $k = 1, 2, \ldots [n/2]$.*

PROOF: Let $R(x)$ and $Q(x)$ be two polynomials of degree n whose roots are $r_1, r_2, \ldots, r_n$ and $q_1, q_2, \ldots, q_n$ respectively. It is known that

$$p.v. \int_{-\infty}^{\infty} \ln\Big(\frac{R(ix)}{Q(ix)}\Big) dx = \pi \sum_{j=1}^{n} (|Re(r_j)| - |Re(q_j)|)$$

where p.v. stands for the proper value. Setting $R(x) = P_G(x)$ and $Q(x) = x^n$ and using (5.12), we arrive after appropriate transformations at

$$E(G) = \frac{1}{\pi} p.v. \int_{-\infty}^{\infty} x^{-2} \ln\Big(\sum_{k=0}^{[n/2]} b(G,k)x^{2k}\Big) dx$$

from which follows Theorem 5.1 immediately.

THEOREM 5.2 (I. GUTMAN [GUT9] IN [CVDSA1]): *Let T be a tree and F a forest on n vertices. Then for all $k \geq 1$,*

$$p(K_{1,n-1}, k) \leq p(T,k) \leq p(P_n, k), \text{ and} \tag{5.13}$$

$$p(nK_1, k) \leq p(F,k) \leq p(P_n, k), \tag{5.14}$$

where $p(G,k)$ denotes the number of k-matchings of the graph G.

COROLLARY 5.2.1:   *If T is a tree and F is a forest on n vertices, then*

$$E(K_{1,n-1}) \leq E(T) \leq E(P_n), \text{ and} \qquad (5.15)$$

$$E(nK_1) \leq E(F) \leq E(P_n). \qquad (5.16)$$

PROOF: For any tree T with n vertices, $p(T,1) = n - 1$. Hence (5.13) holds for $k = 1$. Since all edges of $K_{1,n-1}$ are mutually incident, it is clear that $p(K_{1,n-1}, k) = 0$ for $k > 1$. Therefore the left hand inequality of (5.13) is true for all k.

The right hand inequality is proven by induction on the number of vertices. For $n \leq 3$ the result is trivially true, and for $n = 4$ the result is easily verified. Now suppose that (5.13) is true for all trees with fewer than n vertices. Let T be a tree with n vertices, u a vertex of degree one, and v adjacent to u. Then by the induction hypothesis $p(T-u,k) \leq p(P_{n-1},k)$ and $p(T-u-v,k-1) \leq p(P_{n-2},k-1)$. Hence $p(T-u,k) + p(T-u-v,k-1) \leq p(P_{n-1},k) + p(P_{n-2},k-1)$. By Theorem 8.5 $p(T-u,k) + p(T-u-v,k-1) = p(T,k)$ and $p(P_{n-1},k) + p(P_{n-2},k-1) = p(P_n,k)$. Therefore the right hand inequality of (5.13) also holds for all k.

The left hand side of (5.14) is obvious since $p(nK_1,k) = 0$ for all $k \geq 1$. Thus Theorem 5.2 is proved.

A number of additional inequalities of this type can be found in [GUT12] from [CVDSA1], [ZHA3], [ZHLA1], and [ZHLA2].

THEOREM 5.3 (I. GUTMAN, [GUT49]):   *Let G be a graph with n vertices, m edges, t triangles, and q quadrilaterals. Let the spectrum of G have n_+ positive eigenvalues and n_- negative eigenvalues. Then*

$$12t\sqrt{\frac{n_+ n_-}{(n_+ + n_-)q - 4m^2}} \leq E(G)$$

$$\leq \sqrt{2m(n_+ + n_-) - (n_+ - n_-)^2\left(\frac{q}{2m} - \frac{9t^2}{m^2}\right)}. \qquad (5.17)$$

THEOREM 5.4 (L. TÜRKER[TÜR3]): *Let G be a bipartite graph with n vertices, m edges, and with the characteristic polynomial given by (5.12). In addition, let $\nu = \binom{[n/2]}{2}$. Then*

$$E(G) \le 2(\sqrt{4\nu b(G,2)} + m)^{1/2}.$$

PROOF: Note that for bipartite graphs $\sum_{i=1}^{[n/2]} \lambda_i = E(G)/2$ and $\sum_{i=1}^{[n/2]} \lambda_i^2 = m$. Therefore $\sum_{i<j}^{[n/2]} \lambda_i \lambda_j = E(G)^2/8 - m/2$ In addition, $b(G,2) = \sum_{i<j}^{[n/2]} \lambda_i^2 \lambda_j^2$. Theorem 5.4 now follows from these two equations and the fact that $\frac{1}{n} \sum r_i^2 \ge (\frac{1}{n} \sum r_i)^2$ for any real numbers $r_1, \dots, r_n$.

L. Türker has also shown that $4(\sqrt{4\nu b(G,2)} + m) \le 2mn$.

Additional upper and lower bounds for $E(G)$ are given in [GUT2], [GUT9], [GUT43], and [GUTN1]. Other properties of $E(G)$ are examined in [BOST1], [BOST2], [BOME1], [GUT5], [GUT31], [GUGR1], [GUPE1], [HAL1], [HAL3], and [JITH1]. These papers contain mainly approximate and/or empirical formulas for $E(G)$.

Section 5.6: S– and T– Isomers

Let G be a graph and u and v two of its vertices. Let G' be isomorphic to G with u' and v' vertices corresponding to u and v respectively. Construct the graph S by adding an edge joining u and u' and an edge joining v and v'. Similarly, construct the graph T from G and G' by joining u to v' and v to u'.

THEOREM 5.5 (O. E. POLANSKY, M. ZANDER [POZA1]):

$$P_T(x) - P_S(x) = (P_{G-u}(x) - P_{G-v}(x))^2, \text{ and} \qquad (5.18)$$

$$M_T(x) - M_S(x) = (M_{G-u}(x) - M_{G-v}(x))^2. \qquad (5.19)$$

PROOF: Apply the recurrence relation for the characteristic polynomial given on p. 78 of [CVDSA1] to the edges $\{u, u'\}$ and $\{v, v'\}$ of S and to the edges $\{u, v'\}$ and $\{v, u'\}$ of T. Observe that all cycles of S which contain $\{u, u'\}$ also contain $\{v, v'\}$. The equation (5.8) follows by straightforward calculation and the proof of equation (5.9) is easy; it is similar and based on an application of Theorem 4.5.

THEOREM 5.6 (A. GRAOVAC, I. GUTMAN, O. E. POLANSKY [GRGP1]): *If G is bipartite, then*

$$E(S) \geq E(T),\qquad\qquad(5.20)$$

but if G is not bipartite, then either $E(S) < E(T)$ or $E(S) > E(T)$ may occur.

The following result was observed in [POZA1] and proved in [GUGP1] and [GRGP2].

THEOREM 5.7: *Let $\lambda_i(S)$ and $\lambda_i(T)$ be the eigenvalues of S and T respectively in nonincreasing order, and let n be the number of vertices of G. Then for all $j = 1, 2, \ldots, n$,*

$$\lambda_{2j-1}(S) \geq \lambda_{2j-1}(T) \geq \lambda_{2j}(T) \geq \lambda_{2j}(S).\qquad(5.21)$$

The inequalities given in Theorem 5.7 have rather interesting and far reaching chemical consequences. They are the starting point for a number of theoretical and experimental investigations. In the chemical literature the inequalities (5.21) are known under the name *TEMO*, (an acronym for topological effect on molecular orbitals). For more details along these lines see [POL3], [POL4], [POL5], [POL6], [POMA1], and [POZA1].

It is also interesting that the Cauchy inequalities ([CVDSA1], p. 19) have recently been rediscovered by W. T. Dixon [DIX1]. For further application of the Cauchy inequalities, see [DIX2] and [GUTR1].

Section 5.7: Circuits and the Energy of a Graph

Let G be a graph and C a circuit. Then define $ef(G, C)$ by

$$ef(G, C) = \frac{1}{\pi} \int_{-\infty}^{\infty} \ln \left| \frac{P_G(ix)}{P_G(ix) + 2P_{G-C}(ix)} \right| dx. \qquad (5.22)$$

This definition, is given in [BoGu1] as a measure of the effect of C on $E(G)$.

THEOREM 5.8 (I. GUTMAN [GUT21]):   *If G is bipartite and the number of vertices of C is divisible by four, then $ef(G, C) < 0$.*

Other papers on this and related topics are [GUT26], [GuPo2], [HER1], and [GuHE1]. W. C. Herndon [HER1] has shown that the roots of the polynomial $P_G + 2P_{G-C}$ need not be real. This causes certain difficulties in the theory of $ef(G, C)$ [GuHE1].

Section 5.8: Charge and Bond Order

If the molecular graph is bipartite and conditions (5.5) are satisfied, then the charge q_u of equation (5.7) is equal to zero for all u (This important result was already known in 1940. See the reference [CoRu] in [CvDSa1]). If the molecular graph is not bipartite, then not much is known about q_u.

Let P_n be, as usual, the path with vertices $v_1, v_2, \ldots, v_n$, and let H_1 and H_2 be arbitrary graphs. Form G_n by joining v_1 to a vertex of H_1 and v_n to a vertex of H_2.

THEOREM 5.9 (I. GUTMAN [GUT7]):   *Let q_{v_i} be the charge on the atom corresponding to v_i in G_n, and let exactly one of the graphs H_1 or H_2 be bipartite. Then $q_{v_i} \cdot q_{v_{i+1}} < 0$ for $i = 1, 2, \ldots, n - 1$.*

In other words, the charges alternate in sign along the path P_n in G_n. Some related spectral properties of the graph G_n are deduced in [GUT18].

THEOREM 5.10 (I. GUTMAN [GUT37]):   *Let G be a simple bipartite graph, and let G_h be obtained from G by attaching a loop of weight $h \neq 0$ to any vertex. Then the charges corresponding to any two adjacent vertices of G_h have opposite signs.*

Further investigations about the charge on the atom have been given in [GUT23] and [GUPO2]. In [MARO2] q_u has been viewed as a mathematical index, characterizing the vertices of a nonbipartite graph.

In addition to the bond order p_{uv} defined by (5.8), there are several other quantities related to bond order considered in the chemical literature. Their comparative study is given in [GUT6] and [GUT8]. Some investigations of the dependence of p_{uv} on the structure of the molecular graph (not containing, however, any true mathematical theorems) are given in [GUT4], [GUT13], [GUPO2], [HOHO1], and [HOMU1].

Section 5.9: HOMO-LUMO Separation

The HOMO-LUMO separation is, in the most common chemical case, equal to the difference between the least positive and greatest negative eigenvalue of a molecular graph. In the case of bipartite graphs (with an even number of vertices), the HOMO-LUMO separation is just twice the smallest nonnegative eigenvalue.

THEOREM 5.11 (C. D. GODSIL [GOD8], [GOD9]):   *If F is a forest with n vertices containing a 1-factor, then $\lambda_{n/2}(F) \geq \lambda_{n/2}(P_n)$.*

Other known bounds for the least positive eigenvalue of a bipartite graph [GUT35] are not as sharp. Various approximate formulas have been given for the HOMO-LUMO separation in [GRGU3], [GURO1], [DYKA1], [DYKA2], [DYKA3], [KICH1], [TÜR2]. The question of its dependence on the structure of the molecular graph has been considered in [HAL2], [GUT29], and [GRTR1] without a great deal of success.

The knowledge of n_+, n_-, and η, the number of positive, negative, and zero eigenvalues respectively of a molecular graph is also of chemical importance. In the past a number of papers appeared dealing with the determination of η (see [CVDSA1], pp. 322-326). A paper not cited in [CVDSA1] is one by R. B. Mallion [MAL1], reporting the number of zeros in the spectrum of a circuit (which, of course, is a well known result – see [CVDSA1], p. 72). Another similar publication is [ELBA1] in which conditions for the existence of zero as an eigenvalue in the spectrum of a circuit, a path, or a wheel are established.

The numbers of n_+, n_-, and η have been used to classify conjugated hydrocarbons in [HEEL1] and [KPRT1].

Recently Sinanoglu [SIN1], [SIN2] developed a theory for predicting the stabilities, distortions and reactions of organic and inorganic molecules which goes far beyond the simple Hückel molecular orbital model. In this theory the numbers n_+, n_-, and η play a crucial role. Particularly important are the transformations that leave these numbers unchanged. Several such transformations have been obtained as, for example, when all the edges incident to a vertex are given an arbitrary nonzero weight [SIN1]. For some earlier work along these lines, see [GUT3] in [CvDSA1].

Section 5.10: The Determinant of the Adjacency Matrix

There are two reasons to investigate the determinant of the adjacency matrix of a molecular graph. First, if this determinant is zero, then $\eta > 0$, a useful bit of chemical information in itself. Second, for certain molecular graphs (*i.e.*, benzenoid molecules) det A is closely related to the number of 1-factors, a quantity of great chemical importance (see [CvDSA1], pp. 239-244).

In [GRGU1], [GRGU2], [KIA1], [KIA2], [JIA1], [YAN1], and [GUT48] various recursive techniques have been described for the calculation of det A. The next two theorems give such techniques.

Let G be a graph and u and v two adjacent vertices of G. Let the vertex u be adjacent to $v, u_1, \ldots, u_s$ and v be adjacent to $u, v_1, \ldots, v_t$. For any adjacent pair of vertices x and y, let $w(xy)$ denote the weight of the edge joining them.

THEOREM 5.12 (Y. JIANG [KIA1], [JIA1]): *Suppose the edge $\{u,v\}$ of the graph G is not contained in a triangle, and suppose that H is formed by deleting u and v from G and joining u_i to v_j, $i = 1, \ldots, s$, $j = 1, \ldots, t$. In addition let $w(u_i v_j) = -w(uu_i) \cdot w(vv_j)/w(uv)$. Then*

$$\det A(G) = w(uv)^2 \det A(H). \qquad (5.23)$$

If the edge $\{u,v\}$ belongs to a triangle, then equation (5.23) remains valid if a loop of weight $2w(xu) \cdot w(xv)/w(uv)$ is attached to each vertex x adjacent to both u and v in G.

THEOREM 5.13 (I. GUTMAN [GUT48]): *Let G be a bipartite graph with n vertices, and let $\delta(G) = \sqrt{(-1)^{n/2} \det A(G)}$. If uv is an edge of G, then either*

$$\delta(G) = \delta(G - uv) + \delta(G - u - v), \text{ or} \qquad (5.24)$$

$$\delta(G) = \delta(G - uv) - \delta(G - u - v), \text{ or} \qquad (5.25)$$

$$\delta(G) = -\delta(G - uv) + \delta(G - u - v). \qquad (5.26)$$

All three relations may hold for different edges in the same graph.

PROOF: If the number of vertices of G is odd, then $\delta(G) = \delta(G - uv) = \delta(G - u - v) = 0$ and Theorem 5.13 holds trivially. Consider, therefore, the case when n is even. Let C be the set of circuits of G containing uv. Let P be the set of all paths in G connecting the vertices u and v. It is known ([CVDSA1], p. 78)

$$P_G = P_{G-uv} - P_{G-u-v} - 2 \sum_{C \in C} P_{G-C}$$

and, from Theorem 3.8,

$$P_{G-u} \cdot P_{G-v} - P_G \cdot P_{G-u-v} = \Big(\sum_{P \in P} P_{G-P} \Big)^2.$$

Since u and v are adjacent,

$$\sum_{P \in P} P_{G-P} = P_{G-u-v} + \sum_{C \in C} P_{G-C}.$$

Combining the above three equations one gets

$$4(P_{G-u} \cdot P_{G-v} - P_G \cdot P_{G-u-v}) = (P_G \cdot P_{G-uv} - P_{G-u-v})^2$$

which can be transformed into

$$4P_{G-u} \cdot P_{G-v} =$$

$$(\sqrt{P_G} - \sqrt{P_{G-uv}} - i\sqrt{P_{G-u-v}}) \cdot (\sqrt{P_G} - \sqrt{P_{G-uv}} + i\sqrt{P_{G-u-v}}) \cdot$$
$$\cdot (\sqrt{P_G} + \sqrt{P_{G-uv}} - i\sqrt{P_{G-u-v}}) \cdot (\sqrt{P_G} + \sqrt{P_{G-uv}} + i\sqrt{P_{G-u-v}}).$$

For a bipartite graph H with n vertices, $P_H(0) = (-1)^n \det A(H)$ and $\det A(H) = 0$ whenever n is odd. The evaluation of the polynomials in the equation given above at zero yields the new equation

$$(\delta(G) - \delta(G - uv) - \delta(G - u - v)) \cdot (\delta(G) - \delta(G - uv) + \delta(G - u - v)) \cdot$$
$$(\delta(G) + \delta(G - uv) - \delta(G - u - v)) \cdot (\delta(G) + \delta(G - uv) + \delta(G - u - v))$$
$$= 0$$

from which follows Theorem 5.13.

The inverse of the adjacency matrix was proposed as a means for the ordering of molecular graphs [ElBH1].

A method for the calculation of A^{-1} based on the Coates formula (see [CvDSa1], p. 47) is given in [Tür1].

Section 5.11: The Magnetic Properties of Conjugated Hydrocarbons

J. Aihara [Aih6], [Aih4] has developed a theory of London (or *ring current*) *magnetic susceptibility* of conjugated molecules. According to this theory, if G is the molecular graph and $C_1, \dots, C_t$ are the circuits contained in G, then the *London susceptibility* χ_G is given by

$$\chi_G = \sum_{j=1}^{t} \chi_j \tag{5.27}$$

where χ_j is the susceptibility contribution coming from the j-th circuit, and it can be shown that

$$\chi_j = \sigma_j \sum_{i=1}^{n} \frac{P_{G-C_j}(\lambda_i)}{P_G'(\lambda_i)} \tag{5.28}$$

In formula (5.28) the constant σ_j is a parameter depending on the geometry of the molecule, P'_G is the first derivative of P_G, and $\lambda_1, \ldots, \lambda_n$ are the graph eigenvalues. This formula holds only if the eigenvalues of G are all simple; if this is not the case, then appropriate adjustments can be made as in [AIHO1] and [HOAI1]. From the knowledge of χ_i one can calculate the ring currents [AIHO2].

THEOREM 5.14 (N. MIZOGUCHI [MIZ1]): *If the right hand side of (5.28) exists and the graph G is bipartite, then $\frac{\chi_i}{\sigma_j} < 0$ whenever the number of vertices of C_i is divisible by four and $\frac{\chi_i}{\sigma_j} > 0$ otherwise.*

PROOF: Using arguments similar to those in Theorem 5.1, one shows that

$$\sum_{i=1}^{n} \frac{P_{G-C}(\lambda_i)}{P'_G(\lambda_i)} = -\frac{1}{\pi} p.v. \int_{-\infty}^{\infty} \frac{P_{G-C}(ix)}{P_G(ix)} dx.$$

Applying (5.12) we get

$$\frac{P_{G-C}(ix)}{P_G(ix)} = i^{|C|} \frac{\sum_k b(G-C,k)x^{n-|C|-2k}}{\sum_k b(G,k)x^{n-2k}}$$

where $|C|$ is the number of vertices of the circuit C. The theorem now follows from $b(G-C,k) \geq 0$ and $b(G,k) \geq 0$ for all k.

J. Aihara has also discussed the relationships between χ_G and the topological resonance energy in [AIH8] and [AIH9].

Section 5.12: The Topological Resonance Energy

The concept of topological resonance energy, equation (5.9), was introduced independently by J. Aihara [AIH2] and I. Gutman *et al.* [GUMT2]. TRE was interpreted as a measure of aromaticity of a conjugated molecule. Since 1976 the theory of TRE, with more or less success, has been applied to a great number of conjugated molecules; the number of published papers on TRE approaches one hundred. For more details on this matter, see [TRI1], [ILST1], and [TRI4]. In this section we shall present only a few exact results about TRE.

By TRE(G) we shall mean that the TRE corresponds to the conjugated system whose molecular graph is G.

As an immediate consequence of Theorem 4.3, we have that $TRE(G) = 0$ if G is a tree.

THEOREM 5.15 (I. GUTMAN [GUT20]): *If G is unicyclic and C is its circuit, then $TRE(G) < 0$ if the number of vertices of C is divisible by four and $TRE(G) > 0$ otherwise.*

THEOREM 5.16 (I. GUTMAN, M. MILUN, N. TRINAJSTIĆ [GUMT2]): *If the relations (5.5) hold, then*

$$TRE(G) = \frac{1}{\pi} \int_{-\infty}^{\infty} \ln \frac{P_G(ix)}{M_G(ix)}\, dx. \qquad (5.29)$$

COROLLARY 5.16.1: *If G is unicyclic, then $TRE(G) = ef(G, C)$ where $ef(G, C)$ is as defined in (5.22).*

THEOREM 5.17 (I. GUTMAN [GUT41]): *If G is a hexagonal system then $TRE(G)$ is positive.*

More information about hexagonal systems is given in the next section.

Methods for calculating TRE are developed in [HEEL1] and [MOTR1]. An approximate method for the calculation of TRE of graphs with weighted edges and loops is given in [HEPA1] and further developed in [GUT36] and [GUT44]. Some other properties of TRE are examined in [GUT22], [GUT26], [GUMO1], [GUMO2], and [GUMO3].

Section 5.13: Some Spectral Properties of Hexagonal Systems

Graphs representing *benzenoid hydrocarbons* (sometimes called *benzenoid graphs, hexagonal animals*, or *hexagonal systems*) are of great importance in chemistry. In fact, hexagonal systems form the most frequently examined class of molecular graphs.For their basic properties see Chapter 8.2 in [CVDSA1] and the reviews [GUT40] and [GUT42]. Various spectral quantities of a large number of hexagonal systems have been tabulated in [GUPM1], [GUPM2], and [YASH1].

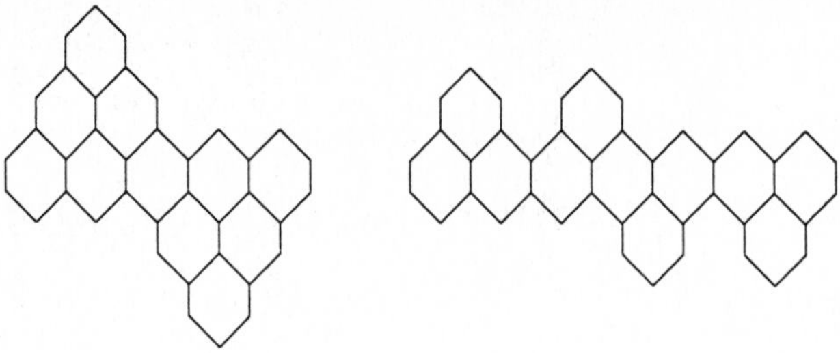

$$H_1 \qquad\qquad\qquad H_2$$

Figure 5.2 Two Hexagonal Systems

The graphs H_1 and H_2 of Figure 5.2 are examples of hexagonal systems.

Hexagonal systems are bipartite graphs. Consider, therefore, a hexagonal system H as a bipartite graph on $n_1 + n_2$ vertices. Then $n_1 \neq n_2$ is sufficient for the existence of zeros in the spectrum of H. This condition is not, however, necessary; in fact H_1 and H_2 are the smallest hexagonal systems having $n_1 = n_2$ and (two) zeros in the spectrum.

Explicit combinatorial expressions for the first few coefficients of the characteristic polynomial of hexagonal systems were obtained by Dias [DIA1]. The same author conjectured [DIA2] that hexagonal systems with n vertices where n is divisible by 4 have -1 in their spectra. Counterexamples have shown the conjecture to be false [GUKR1].

THEOREM 5.18 (I. GUTMAN [GUT41]): *Let G be a hexagonal system and let its characteristic and matching polynomials have respective coefficients of $b(G,k)$ and $p(G,k)$ for x^k. Then*
(a) $b(G,k) = 0$ if and only if $p(G,k) = 0$, and
(b) $b(G,k) \geq p(G,k)$ for all k.

Note that Theorem 5.17 is a corollary of conclusion (*b*) of Theorem 5.18 and Theorem 5.16.

The spectra and/or characteristic polynomials of some hexagonal systems are determined in [DESO1] and [HOOH1]. Let L_h and A_h denote the following hexagonal systems with h hexagons:

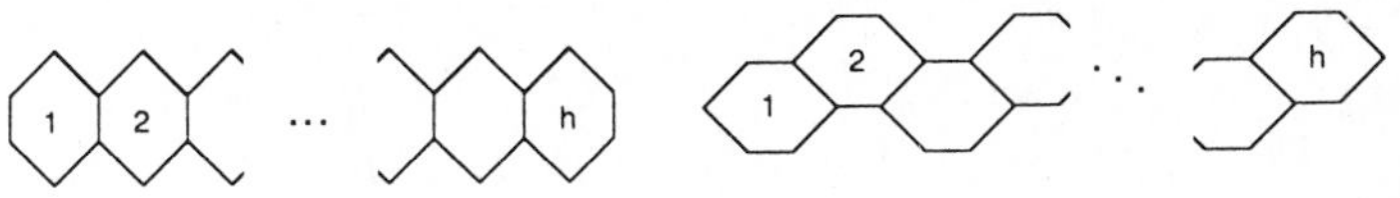

$$L_h \qquad\qquad A_h$$

Figure 5.3 Some Special Animals

THEOREM 5.19 (H. HOSOYA, N. OHKAMI [HoOH1]):   *For $h \geq 4$*

$$P_{L_h}(x) + P_{L_{h-4}}(x) - (x^4 - 5x^2 + 4)(P_{L_{h-1}}(x) + P_{L_{h-3}}(x)) +$$
$$+ (2x^4 - 6x^2 + 6)P_{L_{h-2}}(x) = 0 \qquad (5.31)$$

where $P_{L_0}(x) = x^2 - 1$, and for $h \geq 5$

$$P_{A_h}(x) + P_{A_{h-5}}(x) - (x^4 - 5x^2 + 4)(P_{A_{h-1}}(x) + P_{A_{h-4}}(x)) +$$
$$+ (2x^4 - 5x^2 + 3)(P_{A_{h-2}}(x) + P_{A_{h-3}}(x)) = 0 \qquad (5.32)$$

where $P_{A_0}(x) = x^2 - 1$.

In [HoOH1] it is in fact demonstrated that (5.31) and (5.32) hold for a class of graphs containing the graphs L_h and A_h in Figure 5.3 as subgraphs.

In [GUT38] a result on the number of 1-factors in hexagonal systems is deduced using the theory of graph spectra.

$$* \quad * \quad * \quad * \quad *$$

In this section we shall mention some further classes of molecular graphs which have attracted the particular attention of theoretical chemists.

A graph is called a *Möbius graph* if one of its edges has weight -1. HMO properties of Möbius graphs are considered in [GRTR1], [GRTR2], [GUT17], [POL1], and [POL2].

HMO properties of graphs with weighted edges and loops are examined in a large number of the publications already cited. Some additional papers in this area include [GUT24], [GUT45], [GUT47], [RIMA1].

Spectra of various particular graphs or particular classes of graphs have been used in many investigations in HMO theory, but a review of all of these results is beyond the scope of this book.

Section 5.14: Miscellaneous HMO Results

Some additional reactivity indices for conjugated hydrocarbons have been studied by means of the theory of graph spectra in [GUT 15] and [GUT16]. Electronic properties of *conjugated polymers* were investigated in [GRPTY1] and [NOWW1].

An application of the transfer matrix method for the construction of the eigenvectors of some molecular graphs is elaborated in [JIIF1]. Another method, in which Chebyshev polynomials are used, is given in [KAS1]. The detemination of doubly degenerate eigenvectors of bipartite molecular graphs is considered in [SAM1].

In [SIME1] the product $G \times K_2$ of the molecular graph G and K_2 and the possible chemical application of its spectrum are considered (for the product definition see [CVDSA1], p. 65). In [MESI1] a generalization of the product $G \times K_2$ is proposed.

Section 5.15: Molecular Orbital Approaches Other than the HMO Model

There have been many attempts to use an HMO-like Hamiltonian similar to equation (5.2) for describing the electrons of other types of molecules. All approaches along these lines use graph spectra, although often only implicitly. The work of Sinanoglu [SIN1], [SIN2] has already been mentioned in section 5.5.

D. M. P. Mingos in [MIN1] and [MIN2] proposed an HMO-like theory of *organometallic complexes*. S. J. Cyvin in [CYV1] and [CYV2] developed a quantum-mechanical model of conjugated hydrocarbons having a triple bond. In [HEI1] and [HEER1] the *ionization potentials* of saturated hydrocarbons are described by means of a model based on the eigenvalues of certain graphs.

R. B. King and D. H. Rouvray in [KIRO1] and [KIRO2] rationalized the chemical bonding in *polyhedral boranes*, *carboranes* and *metal clusters* by using the spectrum of K_n. J. Aihara applied the TRE model to polyhedral boranes in [AIH3] and to *inorganic clusters* in [AIH7]. In the latter case, the molecular graph was again K_n.

O. Bilek and P. Kadura in [BiKa1] and [Kad1], and L. Salem and C. Leforestier in [SaLe1] applied the HMO model for the calculation of the electronic properties of finite *crystal lattices*.

It seems that none of the papers cited in this section contain novel mathematical results. Nevertheless, they do show that the application of the theory of graph spectra goes much beyond the scope of the Hückel molecular orbital model of unsaturated conjugated hydrocarbons.

Section 5.16: Applications of Graph Eigenvalues in Physics and Chemistry Other than Molecular Orbital Models

In this section we refer to a number chemical and physical investigations which have only one thing in common — they all use graph spectra.

In a number of publications the index of a molecular graph is considered as a measure of the *branching* of the carbon-atom skeleton of a hydrocarbon molecule (see [CvG4] in [CvDSa1]). An iterative method for the calculation of the index of a molecular graph is described in [PoGu1].

Let G be a graph and F a 1-factor. H. Joela proposes a theory in [Joe1] in which is constructed a graph $H(G)$ whose vertices are the edges of F with two vertices in $H(G)$ adjacent if, as edges of F, there exists a third edge incident to both of them. The index of $H(G)$ is then calculated.

Let G be a graph, A its adjacency matrix, and u a vertex. M. Randić in [Ran1] defines a *walk-based code* as the sequence $A^k_{uu}, k = 2, 3, \ldots$ and considers it as a property of the vertex u. Two vertices are called *isocodal* if their walk-based codes coincide. It has been shown in [KnMS1] and [RaWG1] that

(a) there exist trees with several nonequivalent but isocodal vertices (here two vertices are said to be equivalent if they belong to the same orbit of the automorphism group),

(b) there exist pairs of cospectral trees no two vertices of which are isocodal, and

(c) there exist pairs of trees having isocodal vertices that are not cospectral.

The problem of counting spanning trees of molecular graphs by means of graph spectra is considered in [MAL2] and [GUME1].

In [TAHP1] a model of *isomerisation reactions* is given in which the spectrum of K_n occurs. In [CJLF1], [CHLI1], and [YALY1] systems of linear equations occurring in chemical kinetics are solved by use of the Coates flow graph (see [CVDSA1], p. 47).

There is also an application of the the characteristic polynomial of graphs in mechanical engineering; the problem of constructing the characteristic polynomials from smaller parts in cinematic chains is considered in [JAHA1].

In a statistical model of polymer molecules, the eigenvalues of the admittance matrix (see [CVDSA1], p. 27 and section 8.7) play a role. In *polymer statistics* this matrix is called the *Kirchhoff matrix*. As an example, the mean square of the radius of gyration of a polymer molecule is proportional to $\sum \frac{1}{\mu_i}$ where the sum is taken over all nonzero eigenvalues of the admittance matrix. The same eigenvalues are used for the calculation of the relaxation times, the distribution function of the inertial tensor, the partition function, etc. For more details of this theory see [EIC1], [EIC2], [EIC3], [EIMA1], [FOR1], [GUT14], and the numerous papers cited therein.

Section 5.17: Graph Eigenvalues in Geography and the Social Sciences.

In certain geographic studies, [GOU1] and [TIN1], the eigenvalues and eigenvectors of the adjacency matrix of *transportation networks* have been studied in order to obtain information about their connectedness. They conclude that the more highly connected a network is, the larger is its index λ_1. In addition, the greater the degree of a vertex, the greater is the corresponding component in the eigenvector of λ_1. Badly connected networks exhibit the reverse properties for their dominant eigenvalue and eigenvector. An analogous interpretation of the second and subsequent eigenvalues of the adjacency matrix is also attempted in [HAY1] and [TIN2], but with less success. For a review of the geographical applications of graph spectra, see [CLHA1] and [CLHO1].

Finally, it should be noted that the eigenvalues of weighted digraphs play a certain role in the social sciences (see [RoBr1] and [Rob1]). The scope of this research, however, is not within the ambit of this book.

Chapter 6
Spectra of Infinite Graphs

There have been several attempts to extend the well developed theory of graph spectra from the finite to the infinite case. These include the works of L. Collatz [COL1], A. Torgašev [TOR1], and B. Mohar [MOH3]. Each of these approaches has assets and liabilities, and we shall first outline each approach.

L. Collatz [COL1] deals with graphs with periodic structure. With this approach a graph is considered to be imbedded in $\Re^n$. A finite subgraph K is called a *kernel* of G if there exist independent vectors $v_1, \ldots, v_n$ in $\Re^n$ such that a translation by v_i takes vertices of K to vertices of G and edges of K to edges of G.

Suppose that $u_1, \ldots, u_N$ are the vertices of K. A vertex u is said to be of *type i* if there exists a translation which is a linear combination of $v_1, \ldots, v_n$ and takes u to u_i. It is easy to see that for any i and j, $1 \leq i, j \leq N$, there is a constant a_{ij} such that each vertex of type i is adjacent to a_{ij} vertices of type j. The matrix $A = (a_{ij})$ is the adjacency matrix of a finite graph H which is called a *divisor* of G in the same spirit in which divisors are defined for finite graphs (see Chapter 4 of [CVDSA1]). A kernel is called a *minimal kernel* if none of its proper subgraphs is a kernel of the original graph. The corresponding divisor is called a *minimal divisor*, and the *spectrum of an infinite graph* is defined to be the eigenvalues of all of its minimal divisors.

The approach of A. Torgašev deals with a $0 - a$ adjacency matrix $A = (a_{ij})$ where $a_{ij} = a^{i+j-2}$ if vertices v_i and v_j are adjacent and 0 otherwise (a is a constant with $0 < a < 1$). The spectrum of a graph is the discrete operator spectrum of this matrix. The disadvantage of this definition is that the resulting spectrum is not independent of the labeling of the vertices. On the other hand, many of the important spectral properties of finite graphs do extend to the infinite case when defined this way, and they are independent of the labeling of the vertices and the choice of the constant a. As will be seen later in this chapter, the results are general enough to justify this approach which, at first glance, seems to be so unnatural.

The approach of B. Mohar uses the 0-1 adjacency matrix and the *general operator spectrum* (of both bounded and unbounded operators). In general, the spectrum consists of more than just eigenvalues. The class of graphs considered is limited to *locally finite graphs* and to graphs with the *unique extension property* (to be defined later) which contain, for example, all graphs with *uniformly bounded vertex degrees*. This approach has the advantage of defining a spectrum that is a *graph invariant*. The spectrum is, however, neither discrete nor real in general. Nonetheless this approach is a direct generalization and does have some interesting consequences.

Just as the main tool used with finite graphs is linear algebra, the main tool used with infinite graphs is functional analysis and in particular the spectral theory of operators in infinite dimensional Hilbert spaces. Hence the spectral theory of infinite graphs becomes a mixed area situated between functional analysis and graph theory. All the tools from the general spectral theory of operators can be found in the standard textbooks on functional analysis, for example [STON1] or [TAY1].

There have been no new results concerning the Collatz approach since [CVDSA1] was published, and hence it will not be discussed here. Sections 6.1 to 6.9 will discuss the Torgasev approach while the remainder of the chapter will be devoted to the Mohar approach.

Section 6.1: General Properties

In this chapter we shall describe infinite graphs. By this, unless stated otherwise, we shall mean graphs with no loops or multiple edges and with a vertex set $V = \{v_i \mid i \in N\}$. Usually the vertices are indexed by the natural numbers. The adjacency matrix $A = A(G) = (a_{ij})$ of the graph G is an infinite matrix defined by

$$a_{ij} = \begin{cases} a^{i+j-2}, & \text{if } v_i \text{ and } v_j \text{ are adjacent} \\ 0 & \text{otherwise,} \end{cases}$$

where a is a fixed positive constant satisfying $0 < a < 1$. Hence A is 0 on the diagonal.

Let H be a fixed separable infinite-dimensional complex Hilbert space and $\{e_i\}$ any orthonormal basis. H might be, for example, the sequence space ℓ^2, and $\{e_i\} = \{\delta_{ij}\}$ where δ_{ij} is the Kronecker delta. Then $A(G)$ is the matrix representation of an operator on H with respect to the basis $\{e_i\}$; we call this operator the adjacency operator of the graph. This operator is symmetric and bounded, $i.\,e.$, $\|A\| < \infty$. Moreover, since

$$n(A)^2 = \sum_{i,j} a_{ij}^2 = \sum_{i,j} a^{2i+2j-4} < \infty,$$

it is a Hilbert-Schmidt operator, too, and hence must be a compact selfadjoint operator.

The spectrum of the graph is defined to be the *spectrum* $\sigma(A)$ of the adjacency operator $A = A(G)$ and is denoted $\sigma(G)$. It obviously depends both on the choice of the constant a and on the labeling of the graph. But, since the adjacency operator is selfadjoint and represents a Hilbert-Schmidt operator, the spectrum is always real and discrete, $i.\,e.$, it consists only of zero and a sequence of eigenvalues. All of the eigenvalues are real, but zero, in general, is not an eigenvalue. The sequence of all nonzero eigenvalues ordered by decreasing absolute value tends to zero if it is infinite. If the eigenvalues are $\lambda_1, \lambda_2, \lambda_3, \ldots$ with $\lambda_1 \geq |\lambda_2| \geq |\lambda_3| \cdots > 0$, then each λ_i has finite multiplicity and the spectrum is $\sigma(G) = \{\lambda_1, \lambda_2, \lambda_3, \ldots\} \cup \{0\}$.

When it is necessary to stress the fact that the spectrum $\sigma(G)$ depends on the constant a and the labeling of the vertices, we denote the spectrum by $\sigma(G; a)$.

The largest eigenvalue $\lambda_1 = \|A\|$ is called the *spectral radius* of G, and is usually denoted by $r(G) = r(G; a)$. Since $|\lambda_i| \leq \lambda_1$ for all i, the spectrum of a graph lies in the interval $[-r(G), r(G)]$.

If x is an arbitrary vector in H, then $x = \sum_{i=1} x_i e_i$ which we also denote as $(x_1, x_2, \ldots)'$. The vector x is *positive* if $x_i > 0$ for all i and is *negative* if $x_i < 0$ for all i, and its *norm*, $\|x\|$ is defined by

$$\|x\| = (\sum_{i=1}^{\infty} |x_i|^2)^{1/2}.$$

As in the finite case, a Hilbert-Schmidt operator A is called *reducible* (in the matrix sense) if there is a nonempty proper subset of the orthonormal basis whose linear span is invariant under A. And, exactly as in the finite case, $A(G)$ is *irreducible* if and only if the graph G is connected. The following theorem from [TOR1] summarizes these results.

THEOREM 6.1 (A. TORGAŠEV [TOR1]):   *The spectrum $\sigma(G; a)$ of a connected infinite graph G consists of the zero and a sequence of real eigenvalues $\lambda_1, \lambda_2, \lambda_3, \ldots$ with $\lambda_1 \geq |\lambda_2| \geq |\lambda_3| \ldots > 0$. Each nonzero λ_i has finite multiplicity and, if there are an infinite number of eigenvalues, $\lambda_i \to 0$ as $i \to \infty$. The spectrum $\sigma(G; a) = \{\lambda_1, \lambda_2, \lambda_3, \ldots\} \cup \{0\}$ lies in the interval $[-r(G), r(G)]$ where $r(G) = \lambda_1$ is the spectral radius of G. The largest eigenvalue λ_1 is simple, and if $-\lambda_1$ is an eigenvalue, then it is simple, too. If $r_0 = r(K_\infty)$ is the spectral radius of the infinite complete graph, then $r(G) \leq r_0$.*

M. Petrović ([PET4], [PET6]) has generalized these results to disconnected graphs and derived other properties to be described in section 6.2.

A Hilbert-Schmidt operator is called *nuclear* if its sequence $\{|\lambda_i|\}$ satisfies $\sum_i |\lambda_i| < \infty$.

THEOREM 6.2 (M. PETROVIĆ, [PET6]):   *For any labeling of the vertices of an infinite graph G, and for any value of a, $0 < a < 1$, the adjacency operator $A(G)$ is nuclear. In addition, $\sum_i \lambda_i = 0$.*

Section 6.2: Spectral Properties of some Classes of Infinite Graphs

In this section we shall look at several classes of infinite graphs. First we look at the infinite complete graph K_∞ and the *complete bipartite graph* K_{N_1,N_2}. We assume here that N_1 and N_2 correspond to the vertices and that $N_1 \cup N_2 = N$.

THEOREM 6.3 (A. TORGAŠEV [TOR1]):   *The spectrum of the infinite complete graph consists of 0 and an infinite sequence of simple eigenvalues $\xi_0 > 0$, $\xi_1 < \xi_2 < \cdots < 0$ such that $\xi_n \in (-a^{2n-2}, -a^{2n})$. In addition,*

$$\xi_0 = \sum_{i=1}^{\infty} |\xi_i|.$$

Hence we see that for any $a \in (0,1)$ we have the negative part of the spectrum of K_∞ contained in $(-1,0)$.

THEOREM 6.4 (A. TORGAŠEV [TOR1]):   *If $G = K_{N_1,N_2}$ is an infinite complete bipartite graph, then*

$$\sigma(K_{N_1,N_2}) = \{\pm r, 0\},$$

where $r = (\sum_{i \in N_1} a^{2i-2})^{1/2}(\sum_{i \in N_2} a^{2i-2})^{1/2}$; $\lambda = 0$ is also an eigenvalue of G.

Now we consider connected bipartite graphs.

THEOREM 6.5 (A. TORGAŠEV, M. PETROVIĆ, [TOR1], [PET6]):
  (i) *An infinite graph is bipartite if and only if its spectrum is symmetric with respect to 0, i.e., if $\lambda \in \sigma(G; a)$ then $-\lambda \in \sigma(G; a)$ with the same multiplicity.*
 (ii) *An infinite connected graph with spectral radius r is bipartite if and only if $-r$ is an eigenvalue.*

M. Petrović has also described the spectrum of all *complete multipartite graphs*. Suppose that $N_1, \ldots, N_m$ ($m \leq \infty$) is a partition of the positive integers. Let $K(N_1, \ldots, N_m)$ be the complete multipartite graph with $N_1, \ldots, N_m$ as *characteristic sets*. Now for each $i \leq m$ define c_i by the equation

$$c_i = \sum_{j \in N_i} a^{2j-2}, \quad i = 1, \ldots, m.$$

This sequence obviously satisfies $\sum_{i=1}^{m} c_i = 1/(1 - a^2)$.

THEOREM 6.6 (M. PETROVIĆ [PET2]):   Let $G = K(N_1, \ldots, N_m)$, $(m \leq \infty)$ be a connected infinite complete m-partite graph. Then for each $a \in (0,1)$ we have

(i) G has exactly m nonzero eigenvalues with exactly one of them positive. All eigenvalues distinct from 0 and $-c_i$ are simple and determined by the equation

$$\sum_{i=1}^{m} \frac{c_i}{\lambda + c_i} = 1.$$

(ii) $\lambda = -c_i$ is an eigenvalue with multiplicity $p - 1$ where p is the number of times λ appears in the sequence $\{c_i, i = 1, 2, \ldots, m\}$.

(iii) $\lambda = 0$ is an eigenvalue of infinite multiplicity unless $G = K_\infty$. In that case 0 is not an eigenvalue.

THEOREM 6.7 (M. PETROVIĆ [PET2]):   For a given value of $a \in (0,1)$, a (possibly disconnected) infinite graph has only one positive eigenvalue if and only if it is a complete multipartite graph.

The last two theorems are generalizations of similar theorems for finite graphs (see [CVDSA1]).

THEOREM 6.8 (M. PETROVIĆ [PET2]):   Let $G = K(N_1, \ldots, N_m)$, $(m \leq \infty)$ be a connected infinite complete m-partite graph. The index $r = r(G)$ satisfies $r \leq (m - 1)/m(1 - a^2)$ with equality if and only if $c_1 = \cdots = c_m$.

THEOREM 6.9 (A. TORGAŠEV [TOR1]):   Let H be an infinite connected graph (with a given labeling of the vertices), and let $G = L(H)$ be its line graph. If $\lambda(G; a)$ is the least eigenvalue of G, then $\lambda(G; a) > -2$.

Further generalizations of well known results for finite graphs are given by M. Petrović.

THEOREM 6.10 (M. PETROVIĆ [PET6]):   Let G be a graph whose (possibly infinite number of) connected components are $G_1, G_2, \ldots$ where each G_i (finite or infinite) possesses the induced labeling from G. Then

$$\sigma(G; a) = \bigcup_i \sigma(G_i; a) \cup \{0\}.$$

In addition, G is connected if and only if the spectral radius is a simple eigenvalue with a positive eigenvector.

Hence, as in the finite case, for many problems it is sufficient to consider only connected graphs.

THEOREM 6.11 (M. PETROVIĆ [PET2]): *Let G be an infinite graph with a given labeling of the vertices, and let $\lambda_1^+ \geq \lambda_2^+ \geq \ldots > 0$ and $\lambda_1^- \leq \lambda_2^- \leq \ldots < 0$ be the positive and negative eigenvalues of G. Suppose G_0 is an induced subgraph of G (including the labelings) with eigenvalues $\mu_1^- \leq \mu_2^- \leq \ldots < 0$ and $\mu_1^+ \geq \mu_2^+ \geq \ldots > 0$. Then $\lambda_n^+ \geq \mu_n^+$ and $\lambda_n^- \leq \mu_n^-$.*

Section 6.3: The Characteristic Function of an Infinite Graph

M. Petrović has introduced the concept of the characteristic function of an *infinite digraph* or graph. In our context we shall restrict out attention to graphs.

By Theorem 6.2 the adjacency matrix is always nuclear, and hence the concept of the *characteristic determinant* $D_A(\lambda)$ can be used. It is defined by

$$D_A(\lambda) = \prod_i (1 - \lambda\lambda_i)$$

(see [GOKR1], p.199).

Thus the *characteristic function* of an infinite graph can be defined by $f_G(\lambda) = D_A(\lambda)$ where $\{\lambda_i\}$ is the sequence of all nonzero eigenvalues of the graph G. It is an entire analytic function satisfying $f_G(0) = 1$, and hence may be expanded around zero to get $f_G(\lambda) = \sum_n c_n \lambda^n$ with $c_0 = 1$.

The equation $f_G(\lambda) = 0$ is called the *characteristic equation* of G, and its nonzero roots are exactly the reciprocals $\tilde{\lambda} = 1/\lambda_i$, $(\lambda_i \in \sigma(G; a) \setminus 0)$. We then have $|\tilde{\lambda}_i| \to \infty$ if the sequence of eigenvalues is infinite.

Since the matrix $A(G)$ is always a von Koch matrix (see [KOC1] or [GOKR1], p. 214), the coefficients c_n of the characteristic function will be determined by

$$c_n = (-1)^n \sum_{i_1 < \cdots < i_n} D_n(i_1, \ldots, i_n), \quad (n \in N)$$

where $D_n(i_1,\ldots,i_n)$ is the determinant of the principal submatrix of A with rows and columns $i_1,\ldots,i_n$.

It can also be proved that $c_1 = -\text{tr}(A) = -\sum_i \lambda_i$, and hence, by Theorem 6.2 we have $c_1 = 0$. Thus for any graph G we write

$$f_G(\lambda) = 1 + \sum_{i=2}^{\infty} c_n \lambda_n.$$

From this starting point many properties of the characteristic polynomial can be generalized to the characteristic function of infinite graphs. We shall note some of the major ones; the reader is referred to [PET6] for more details.

(1) If $\{G_i \mid i \in I\}$ is the set of connected components of G, then

$$f_G(\lambda) = \prod_{i \in I} f_{G_i}(\lambda).$$

(2) If G is any undirected graph without loops or multiple edges, then

$$c_n = \sum (-1)^{p(U)} 2^{c(U)} \quad (n = 2, 3, \ldots)$$

where the summation is taken over all basic figures with n vertices. This is a direct extension of the finite case described in [CvDSa1], p. 35.

(3) The length of the shortest odd circuit of G equals the index of the first nonzero odd coefficient.

From Theorem 6.5 we note that the characteristic function of an arbitrary bipartite graph is even. Consequently all the odd coefficients of $f_G(\lambda)$ are zero. It is known that this is a characterization of bipartite graphs [PET6].

From (2) the following characteristic functions can be calculated.

(4) If $G = K_\infty$, then

$$f_G(\lambda) = 1 - \sum_{i=2}^{\infty} (n-1) \frac{a^{n(n-1)}}{(1-a^2)\cdots(1-a^{2n})}.$$

(5) If $G = T(\infty)$, the one way infinite path labeled such that v_1 is of degree 1 and v_i is adjacent to v_{i-1} and v_{i+1} for $i > 1$, then $c_{2n+1} = 0$ for all $n \geq 0$ and

$$f_G(\lambda) = 1 + \sum_{i=2}^{\infty} (-1)^n \frac{a^{2n(2n-1)}}{(1-a^4)\cdots(1-a^{4n})}.$$

Other results of this type may be found in [PET6].

Section 6.4: Graphs with a Finite Spectrum

A graph G has a finite spectrum if it has only a finite number of nonzero eigenvalues. Since the multiplicity of any nonzero eigenvalue is finite, it is equivalent to say that G has a finite number of distinct eigenvalues. In this case zero is an eigenvalue of infinite multiplicity; moreover, the codimension of the corresponding eigenspace is finite ([TOR1]). Hence the whole spectrum $\sigma(G; a)$ of such a graph consists only of eigenvalues.

Infinite graphs with finite spectrum are important not only because of their similar structure to that of finite graphs, but also because the spectrum of $\sigma(G; a)$, being determined by an algebraic equation, can be explicitly found.

Let the equivalence relation $\sim$ be defined on the set $V(G)$ by letting $u \sim v$ if u and v have the same neighbours in G. Note then u and v are nonadjacent. Suppose that the quotient $V(G)/\sim = \{V_1, V_2, \ldots\}$. Then each V_i consists only of isolated vertices, and if there is one edge between V_i and V_j, then there is every possible edge between V_i and V_j. The subsets V_i are called the *characteristic subsets* of G. A graph is of *type k* if there are k characteristic subsets. If k is finite, then G is said to be of *finite type*. The *canonical graph* of G is the quotient graph; G is called *canonical* if it is equal to its own quotient graph.

If G is a complete multipartite graph $K(N_1, \ldots, N_m)$, for example, the canonical graph is K_m.

The following theorem relates graphs with a finite spectrum with graphs of finite type.

THEOREM 6.12 (A. TORGAŠEV [TOR2]):
 (i) *A connected graph G has a finite spectrum if and only if it has finite type.*
 (ii) *A graph of finite type and its canonical graph have the same number of nonzero eigenvalues (including multiplicities).*
 (iii) *All the nonzero eigenvalues are determined by the equation*

$$\det(B - \lambda D) = 0 \qquad (6.1)$$

where B is the 0-1 adjacency matrix of the canonical graph and D is the diagonal matrix with the i-th diagonal entry equal to $1/A_i$, $A_i = \sum_{j \in V_i} a^{2j-2}$.

(iv) *If G is connected, of type k, and the number of nonzero eigenvalues is t, then $k \leq 2^t - 1$. If it is disconnected, then $k \leq t(2^t - 1)$.*

The graphs of finite type with five or fewer distinct nonzero eigenvalues are determined in [TOR2] and [TOR3]. The basic method is to reduce the problem to one of finite graphs by using (ii) in Theorem 6.12.

Section 6.5: Operations on Infinite Graphs

In this section we consider some unary and n-ary operations on infinite graphs, and investigate when such an operation gives a graph with a finite spectrum, or, equivalently (by Theorem 6.12) a graph of finite type.

We shall look at the effects of changing the constant a, of the relabelings of the graph, of taking line graphs and total graphs, and at the induced subgraphs. In some cases an infinite spectrum results while for others necessary and sufficient conditions for a finite spectrum to occur are given.

We say that a graph G has a *t-finite spectrum* if it has exactly t non-zero eigenvalues (including multiplicities). The next theorem shows that this definition is well defined.

THEOREM 6.13 (A. TORGAŠEV [TOR5]):   *If G has a t-finite spectrum for some constant a and labeling, then G has a t-finite spectrum for all $a \in (0,1)$ and all labelings.*

THEOREM 6.14 (A. TORGAŠEV [TOR5]):   *If G has a t-finite spectrum and G_0 is an induced subgraph of G, then G_0 has a t_0-spectrum for some t_0 satisfying $0 \leq t_0 \leq t$.*

THEOREM 6.15 (A. TORGAŠEV [TOR5]):   *If G has a t-finite spectrum, the its complement $\overline{G}$ has an infinite spectrum.*

THEOREM 6.16 (A. TORGAŠEV [TOR5]):   *The disjoint union $G = G_1 \cup G_2$ has a finite spectrum if and only if G_1 and G_2 have finite spectra. Further, if t_1 and t_2 are the types of G_1 and G_2, then $t = t_1 + t_2$ is the type of G.*

The *complete product* $G_1 \nabla G_2$ is obtained by taking copies of G_1 and G_2 and joining every vertex in G_1 with each vertex in G_2.

THEOREM 6.17 (A. TORGAŠEV [TOR5]):   *The complete product of* G_1 *and* G_2 *has a finite spectrum if and only if* G_1 *and* G_2 *do. In that case, if* t_1 *and* t_2 *are the types of* G_1 *and* G_2 *and* t *is the type of* G *then*

$$t_1 + t_2 \le t \le t_1 + t_2 + 2.$$

The *Cartesian product* of two graphs G_1 and G_2 has the Cartesian product $V(G_1) \times V(G_2)$ as its vertex set with two vertices joined if they are adjacent in one coordinate and equal in the other.

THEOREM 6.18 (A. TORGAŠEV [TOR5]):   *The Cartesian product of* G_1 *and* G_2 *has a finite spectrum if and only if* G_1 *and* G_2 *do. In that case, if* t_1 *and* t_2 *are the types of* G_1 *and* G_2 *and* t *is the type of* G *then* $t = t_1 t_2$.

The sum, total, and line graphs are defined for infinite graphs in exactly the same manner as for finite graphs (see [CVDSA1], pp. 61, 64).

THEOREM 6.19 (A. TORGAŠEV [TOR5]):   *The sum, total, and line graphs always have infinite spectra.*

Further results of this type for other n-ary operations can be found in [TOR1], [TOR4], [TOR5], [PET3], [PET4], and [PET5].

Section 6.6: The Automorphism Group of an Infinite Graph

If A is the adjacency matrix of a connected infinite graph (which depends, in general, on the constant a and the labeling of the vertices), then any permutation matrix P that satisfies $AP = PA$ is called an automorphism of G. We identify this permutation matrix with the permutation ω of the vertex set which it represents. If P is an automorphism then is can be seen that it preserves adjacency.

It is also easy to see that a permutation ω of $V(G)$ is an automorphism if and only if $\omega(i) + \omega(j) = i + j$ for any pair of adjacent vertices v_i and v_j. Hence the permutation group $\Gamma(G)$ depends on the labeling of the vertices but is independent of the choice of the constant a.

THEOREM 6.20 (A. TORGAŠEV [TOR6]):   *Any connected infinite graph satisfies $|\Gamma(G)| \leq 2$.*

It is then natural to ask is when the automorphism group is non-trivial. Let ω be an automorphism and let $t = \omega(1) - 1$. If $t = 0$ then ω is the identity. Otherwise, since G is connected, we have a partition of the vertices into N_1, vertices with $\omega(i) - i = t$, and N_2, those with $\omega(i) - i = -t$, and this makes G bipartite. Hence ω maps N_1 onto N_2 and vice versa. Indeed, if $i \in N_1$ and $\omega(i) = j$, then $j - i = t$ and $\omega(j) = j - t = i$ and ω is involutional (as is also implied by Theorem 6.20). This implies that there exists an integer d such that $N_1 = \{kd + r \mid k$ a nonnegative even integer, and $r = 1, \ldots d\}$, and $N_2 = \{kd + r \mid k$ a positive odd integer, and $r = 1, \ldots d\}$. This is also sufficient for $\Gamma(G)$ to be nontrivial.

As an application of this result, the connected infinite graphs with a small number of nonzero eigenvalues are considered in [TOR6]. If the number of eigenvalues is odd, then the automorphism group is trivial. If there are two or four eigenvalues, a complete characterization of graphs with nontrivial automorphism group can be given

In [TOR10] the automorphism group of infinite generalized line graphs are considered.

THEOREM 6.21 (A. TORGAŠEV [TOR10]):   *The infinite two way path is the only connected infinite generalized line graph having a nontrivial automorphism group.*

Section 6.7: Infinite Generalized Line Graphs

It is known (see Chapter 1) that any finite connected graph with more than 36 vertices satisfies $\lambda(G) \geq -2$ if and only if it is a generalized line graph, abbreviated GLG. This characterization can be extended to infinite graphs. Let H be a connected finite or infinite graph, and let $\bar{k} = (k_1, k_2, \ldots)$ be a sequence of cardinals with $k_i \leq \aleph_0$ for each i, and let $CP(k_i)$ be the corresponding cocktail party graphs. If H is finite, assume that at least one k_i is infinite. The infinite generalized line graph $G = L(H; \bar{k})$ is then defined in the same way as in the finite case. Obviously G is connected.

On the other hand, let G be any connected infinite graph. We shall say that the sequence G_m $(m \in N)$ of finite induced subgraphs of G is a *good sequence* if each G_m is a GLG with $G_m = L(H_m, ; \bar{k}_m)$ where (1) $\cup_m G_m = G$, and (2) H_m is a subgraph of H_{m+1} and $k_{m_i} \leq k_{m_{i+1}}$.

THEOREM 6.22 (A. TORGAŠEV [TOR10]): *A connected infinite graph is a GLG if and only if it has a good sequence of its induced subgraphs.*

THEOREM 6.23 (A. TORGAŠEV [TOR10]): *Two connected infinite $GLGs$ are isomorphic if and only if their root graphs and underlying cocktail party graphs are isomorphic.*

Let G be a connected infinite graph, $r(G; a)$ its spectral radius, and $\lambda(G; a)$ its least eigenvalue. It is known [TOR7] that $r(G; a)$ is a strictly increasing function of a and $\lambda(G; a)$ is a decreasing function of a. Hence we can define the limiting values $r(G) = \sup\{r(G; a) \mid 0 < a < 1\}$, and $\lambda(G) = \inf\{\lambda(G; a) \mid 0 < a < 1\}$.

In [TOR7] the following relations are proved:

$$r(G) = \sup\{r(F) \mid F \text{ a finite induced subgraph of } G\}, \qquad (6.2)$$

$$\lambda(G) = \inf\{\lambda(F) \mid F \text{ a finite induced subgraph of } G\}. \qquad (6.3)$$

By Theorem 4.1 ([MOH3]) and the relation (6.2) we have that $r(G)$ is finite if and only if G has uniformly bounded degrees. The corresponding question for $\lambda(G)$ is an open question.

Notice that (6.2) and (6.3) imply that $r(G)$ and $\lambda(G)$ are invariants of a graph and are independent of the choice of the constant a or the labeling of the vertices.

Now consider graphs with $\lambda(G) \geq -2$, or, equivalently, $\lambda(G; a) \geq -2$ for each $a \in (0, 1)$. By (6.3) this is equivalent to $\lambda(F) > -2$ for all finite induced subgraphs F contained in G.

From Theorem 6.22 and Theorem 6.23 the following result can, with some effort, be derived.

THEOREM 6.24 (A. TORGAŠEV [TOR7], [TOR10]): *A connected infinite graph G has $\lambda(G) \geq -2$ if and only if it is a generalized line graph.*

It is interesting to note that all $GLGs$ satisfy the strict inequality $\lambda(G; a) > -2$ for all $a \in (0, 1)$. This was proved for line graphs in [TOR1]. Hence we have the following theorem.

THEOREM 6.25 (A. TORGAŠEV [TOR7]):   *A connected infinite graph G has $\lambda(G; a) > -2$ for each $a \in (0, 1)$ if and only if it is a generalized line graph.*

Notice that Theorem 6.24 is proved in [TOR7] by a consistent usage of the *infinite root systems* in infinite dimensional Hilbert spaces.

We conclude this section by describing all infinite connected graphs satisfying $\lambda(G) > -2$.

By Theorem 6.24 any such graph is a generalized line graph. We should note that both the methods used and results obtained are in contrast to those of the finite case ([DOCV1]). There are two conditions concerning infinite trees that are important:

(1) H is an infinite tree which has a sequence of paths whose lengths become arbitrarily long, and

(2) H does not satisfy (1), has exactly one vertex of infinite degree, and the degrees of the other vertices are bounded.

THEOREM 6.26 (A. TORGAŠEV [TOR13]):   *A connected infinite graph G satisfies $\lambda(G) > -2$ if and only if $G = L(H; k)$ where either*

(i) *$G = L(H)$ where H is a tree satisfying condition (2) above, or*

(ii) *$G = L(H; 1, 0, 0, \ldots)$ where H is a tree satisfying (2) above and the cocktail party graph $CP(1)$ corresponds to a vertex of finite degree*

Section 6.8: The D-spectrum of Infinite Graphs

We know from Theorem 6.25 that, in contrast to the finite case, any infinite generalized line graph satisfies $\lambda(G; a) > -2$. So it is reasonable to ask if there is a different approach to the spectrum of an infinite graph which more closely approximates the finite case. The *D-spectrum* from [TOR4] is one such approach to this problem.

Let $D = \text{diag}(1, a^2, a^4, \ldots)$ be the diagonal operator with respect to a fixed basis $\{e_i\}$ of a separable Hilbert space. This operator is obviously a selfadjoint Hilbert-Schmidt one. The complex number λ is called a *D-eigenvalue* of G if there is a nonzero vector $x = \sum_i x_i e_i$ with $\sum_i |x_i|^2 < \infty$ satisfying the equation $A(G)x = \lambda D x$. The vector x is then called the *D-eigenvector* corresponding to the eigenvalue λ. It is clear that $\lambda = 0$ is a *D*-eigenvalue if and only if it is a usual eigenvalue.

The set of all distinct D-eigenvalues of G is denoted by $\sigma_D(G) = \sigma_D(G; a)$ and is called the D-spectrum of G. It is always real and, in general, depends on the choice of the constant a and the labeling of the vertices of G. The multiplicities of the eigenvalues can be computed, but they are not essential for the discussion here.

Some examples are:

(1) If $G = K_\infty$ then $\sigma_D(G) = \{-1\}$.

(2) If G is the complete bipartite graph $K(N_1, N_2)$, then $\sigma_D(G) = 0$.

(3) If $G = T(\infty)$, the infinite one way path, then $\sigma_D(G) = \emptyset$.

At present there is no known infinite graph whose D-spectrum actually depends on the labeling of the vertices. Indeed, for graph of finite type we can give the following theorem.

THEOREM 6.27 (A. TORGAŠEV [TOR4]): *If G is a connected infinite graphs of finite type, then*

(i) $0 \in \sigma_D(G)$,

(ii) $\sigma_D(G)$ *is independent of the choice of constant a and the labeling of the vertices, and*

(iii) $\sigma_D(G)$ *has at most m elements where m is the number of characteristic subsets of G that are finite.*

The D-spectrum for line graphs is very much like the usual spectrum for finite graphs.

THEOREM 6.28 (A. TORGAŠEV [TOR4]): *If G is a graph and $\lambda \in \sigma_D(L(G))$, then $\lambda \geq -2$.*

The D-spectrum allows a direct generalization of a result of M. Doob ([CVDSA1], p. 169) for finite graphs (although the proof is somewhat different and more complicated).

THEOREM 6.29 (A. TORGAŠEV [TOR4]): *A graph G has $-2 \in \sigma_D(L(G))$ if and only if it has at least one even cycle or two distinct odd cycles.*

Hence the property of having $-2 \in \sigma D(L(G))$ is independent of the choice of the constant a and the labeling of the vertices.

In [TOR4] $\sigma_D(G)$ is calculated for the line graphs of almost all complete multipartite graphs (only the case when all of the characteristic parts of G are finite is missing). The following theorem is typical.

THEOREM 6.30 (A. TORGAŠEV [TOR4]):

(i) If $G = K_\infty$, then $\sigma_D(L(G)) = \{-2\}$ for any labeling of the vertices.
(ii) If $G = K(N_1, N_2)$,then
    (a) $\sigma_D(L(G)) = \{-2\}$ if both N_1 and N_2 are infinite,
    (b) $\sigma_D(L(G)) = \{-2, a_1 - 2\}$ if $|N_1| = a_1 > 1$, and
    (c) $\sigma_D(L(G)) = \{-1\}$ if $|N_1| = 1$.

The D-spectrum of generalized line graphs has also been investigated.

THEOREM 6.31 (A. TORGAŠEV [TOR7]):   For any generalized line graph G and each $\lambda \in \sigma_D(G)$ we have $\lambda \geq -2$. The inequality is strict if and only if either

(i) $G = L(H)$ where H is an infinite tree or unicyclic with an odd cycle, or
(ii) $G = L(H; 1, 0, 0, \ldots)$ where H is an infinite tree.

We note that this is a direct generalization of the result for finite graphs ([DOCV1]). Also, the property of -2 being in the D-spectrum is independent of the choice of constant a and labeling of the vertices for generalized line graphs.

The D-spectrum is usually easier to compute than is the other spectra of infinite graphs that are defined in this chapter.

Section 6.9: Graphs with Uniformly Bounded Spectra

For any positive constant L, one may attempt to find all infinite connected graphs such that $\sigma(G; a)$ lies in the interval $[-L, L]$ for every choice of constant $a \in (0, 1)$. This class of graphs will be denoted by $M(L)$ as in [TOR8]. This property is obviously equivalent to $r(G) \leq L$. The set of graphs satisfying this property for some L is called a set of graphs with *uniformly bounded spectra*. In this case the degrees of the vertices are also finite and uniformly bounded.

THEOREM 6.32 (A. TORGAŠEV [TOR8]): *Any graph in $M(L)$ has (for any constant a) an infinite spectrum.*

By equation (6.2) it is easy to see that $r(G) \leq L$ is equivalent to $r(F) \leq L$ for all finite induced subgraphs F of G. Hence for certain values of L it is possible to appeal to known theorems about finite graph spectra, as is done in the following case. Let $T(n_1, n_2, \ldots, n_t)$ be the tree formed by taking t paths with $n_1, n_2, \ldots n_t$ edges each and identifying one vertex of degree one from each. Thus $T(n_1, n_2, \ldots, n_t)$ has one vertex of degree t, $\sum_i (n_i - 1)$ vertices of degree two and t vertices of degree one.

THEOREM 6.33 (A. TORGAŠEV [TOR8]):

(i) *The only infinite connected graphs with $r(G) \leq 2$ are $T(\infty)$, $T(\infty, \infty)$, and $T(1, 1, \infty)$. Each of these graphs has $r(G) = 2$.*

(ii) *The only connected infinite graphs with $2 < r(G) < \sqrt{(2 + \sqrt{5})}$ are $T(1, t, \infty)$, $(t < \infty)$ $T(1, \infty, \infty)$, and $T(2, 2, \infty)$.*

Each of the graphs in Theorem 6.33 is bipartite; the maximal L such that $M(L)$ contains only bipartite has been determined in [TOR8] to be $3\sqrt{2}/2$ and this bound is attained by $T(\infty, \infty, \infty)$.

Section 6.10: Another Approach to the Spectrum of an Infinite Graph

A different approach to the spectrum of infinite graphs has been given by B. Mohar in [MOH3]. In this section we shall describe his results. The concepts and properties of general bounded and unbounded symmetric operators in a Hilbert space are standard and are available in any standard textbook on Functional analysis (e. g., [TAY1]).

All graphs to be considered here are assumed to be countable, unoriented and locally finite, i. e., the degree of each vertex is finite.

The adjacency matrix $A = A(G) = (a_{ij})$ of the graph G is the infinite matrix where a_{ij} is the number of edges joining the vertices v_i and v_j.

Now let $\{e_i\}$ be an orthonormal basis in the Hilbert space ℓ^2, and let A_0 be the operator whose matrix representation in this basis is (a_{ij}).

The domain $D(A_0)$ of A_0 is a dense subspace of ℓ^2, and A_0 is symmetric and closable.

Its closure $A(G) = \bar{A}_0$ is called the *adjacency operator* of G. It is a closed symmetric operator in a dense domain of ℓ^2 but, in general, is not selfadjoint. It is not necessarily bounded, either, but if it is bounded, then it is selfadjoint. Hence it is important to know when $A(G)$ is bounded.

If the adjacency operator is selfadjoint, then the graph has what Mohar calls the *unique extension property*. It is known [MoOm1] that not every locally finite graph has the unique extension property. The spectrum can be divided into three parts: the *point spectrum*, the continuous spectrum, and the *residual spectrum* which are denoted by $\sigma_p(G)$, $\sigma_c(G)$, and $\sigma_r(G)$ respectively. The elements of $\sigma_p(G)$ are also called the *eigenvalues* of G. If G has the unique extension property, then $\sigma_r(G) = \emptyset$ and the spectrum is real.

The spectrum does not depend on the particular labeling of the vertices of G since unitary equivalent operators have the same spectrum. The same is true for $\sigma_p(G)$, $\sigma_c(G)$, and $\sigma_r(G)$. Hence this spectral definition is a graph invariant and is also a natural generalization of the definition for finite graphs.

In general, $\sigma(G)$ is a closed subset of the complex numbers, but, unless the adjacency operator is selfadjoint, it may not be real and it may be unbounded. We can only say that $\sigma_p(G)$ and $\sigma_c(G)$ are real and that $\sigma_r(G)$ is either empty or contains all nonzero imaginary numbers. The problem of determining those graphs G for which $\sigma_r(G)$ is empty is still open. This problem is equivalent to characterizing graphs with the unique extension property.

Here are some examples:

Let $T(\infty)$ and $T(\infty, \infty)$ be the one way and two way infinite paths. Then in both cases the adjacency operator is selfadjoint, the spectrum coincides with the interval $[-2, 2]$, and the point spectrum is empty.

THEOREM 6.34 (B. MOHAR [MOH3]):   *Let $G_1, G_2, \ldots$ be (finitely or infinitely many) connected components of the graph G. Then*

$$\sigma(G) = \overline{\cup_i \sigma(G_i)}, \text{ and}$$

$$\sigma_p(G) = \cup_i \sigma_p(G_i)$$

THEOREM 6.35 (B. MOHAR [MOH3]): *The adjacency operator is bounded if and only if the degrees of the vertices are uniformly bounded by some constant M. In this case $\|A\| \leq M$, $\sigma(G) \subseteq [-M, M]$, and $A(G)$ is a selfadjoint operator defined on the whole space ℓ^2. The operator A is compact if and only if the number of edges of G is finite.*

The *essential spectrum* of G is the union of $\sigma_c(G)$ and all the eigenvalues of infinite multiplicity.

THEOREM 6.36 (B. MOHAR [MOH3]): *Adding or deleting a finite number of edges does not change the essential spectrum of G.*

THEOREM 6.37 (B. MOHAR [MOH3]): *Let G be a bipartite graph. Then both the point and continuous spectra are symmetric with respect to zero.*

The spectral radius of G is the spectral radius of $A(G)$, and it is denoted by $r(G)$. Since A is symmetric, we have

$$r(G) = \|A\| = \sup\{< Ax, x > | \ x \in D(A), \|x\| = 1\},$$

where $\|A\| = \infty$ if A is unbounded.

THEOREM 6.38 (B. MOHAR [MOH3]):

$$r(G) = \sup\{r(F) \mid F \text{ a finite induced subgraph of } G\}$$

and if A is bounded then $r(G) \in \sigma(G)$.

The spectrum of several products of graphs can be computed; these are analogous with the finite case given in [CVDSA1], p. 65, p. 70.

Let $G_1, G_2, \ldots, G_n$ be given graphs and B a set of n-tuples on the symbols 0 and 1 not containing $(0, \ldots, 0)$. The NEPS graph with basis B has been defined in Chapter 3. If the vertex degrees of the the graphs G_i are uniformly bounded, then the adjacency operator of the NEPS graph with basis B is equal to the closure

$$A = \overline{\sum_{\beta \in B} A_1^{\beta_1} \otimes \cdots \otimes A_n^{\beta_n}}$$

where $A_i = A(G_i)$ and $\beta = (\beta_1, \ldots, \beta_n)$. In this case the spectrum of the graph is equal to the algebraic sum

$$\sigma(A) = \sum_{\beta \in B} \sigma(G_1)^{\beta_1} \cdots \sigma(G_n)^{\beta_n}.$$

This result appeared in [MoOm1].

Further investigations in this direction are contained in [MoOm2]. Let G be a connected locally finite graph, and let $D = (d_{ij})$ be a matrix with integer coefficients. D is called a *front divisor* of G if there is a partition of $V(G)$ into sets $V_1, V_2, V_3, \ldots$, such that (1) for each i, j and each $v \in V_i$ there are exactly d_{ij} edges emanating from v and having terminal vertex in V_j, and (2) for each i the set V_i is finite.

For finite graphs it is known that if D is a front divisor of G then $\sigma(G)$ contains the spectrum of D (see [CvDSa1], Chapter 4).

THEOREM 6.39 (B. MOHAR, M. OMLADIČ [MoOm2]): *Let G be a locally finite graph with bounded vertex degrees and $D = (d_{ij})$ its front divisor with the corresponding partition $V(G) = V_1 \cup V_2 \cup \cdots$. If $\tilde{D}$ is the matrix with (i, j) entry satisfying $D_{ij} = (|V_i|/|V_j|)^{1/2} \cdot d_{ij}$, then $\sigma(\tilde{D}) \subseteq \sigma(D)$.*

In [MoOm2] some sufficient conditions for the equality $\sigma(\tilde{D}) = \sigma(D)$ are given. In particular, if G is a distance-regular graph with parameters p^i_{jk} as described in Chapter 2, then the matrix $P_k(k = 0, 1, 2 \ldots)$ given by $(P_k)_{ij} = p^i_{jk}$ is a front divisor of the $G^{(k)}$, the graph with the same vertex set as G and two vertices adjacent if and only if they are at distance k in G.

THEOREM 6.40 (B. MOHAR, M. OMLADIČ [MoOm2]): *Let G be a locally finite distance-regular graph, let n_i be the number of vertices which are at distance i from any fixed vertex, and $(\tilde{P}_k)_{ij} = (n_i/n_j)^{1/2} p^i_{jk}$. Then for each k the spectrum of the matrix $\tilde{P}_k$ is equal to the spectrum $\sigma(G^{(k)})$ of the graph $G^{(k)}$.*

Using the above theorem the spectrum of all infinite locally finite graphs is determined in [MoOm2].

This approach has the advantage of being a more direct generalization of the spectrum of a finite graph than the approach used in [Tor1] since it deals with the 0-1 adjacency matrix. In addition, the spectrum is a graph invariant. However, the spectrum is, in general, considerably more complicated and is neither discrete nor real; the graphs must be locally finite in order to make worthwhile computations. Moreover, the well known spectral theorems for finite graphs usually do not carry over to the infinite case in a straightforward way. This would seem to indicate that new approaches and ideas are needed.

Spectra of Graphs with Seven Vertices

This section contains data on all graphs with seven vertices. Similar data for graphs with five or fewer vertices may be found in [CvDSa1], and for graphs with six vertices in [CvPe1]. The graphs are ordered lexicographically by their spectral moments.

The data was computed using the interactive program GRAPH developed at the University of Belgrade. Stirling University allowed the authors to use their computing facilities to produce these tables.

Number	Eigenvalues Characteristic polynomial coefficients and spectral moments	Graph
6 - 1	1.84776 1.41421 0.76537 0.00000 -0.76537 -1.41421 -1.84776 1 0 -6 0 10 0 -4 0; 0 12 0 32 0 96 0	
6 - 2	1.93185 1.41421 0.51764 0.00000 -0.51764 -1.41421 -1.93185 1 0 -6 0 9 0 -2 0; 0 12 0 36 0 120 0	
6 - 3	1.96962 1.28558 0.68404 0.00000 -0.68404 -1.28558 -1.96962 1 0 -6 0 9 0 -3 0; 0 12 0 36 0 126 0	
6 - 4	2.00000 1.00000 1.00000 0.00000 -1.00000 -1.00000 -2.00000 1 0 -6 0 9 0 -4 0; 0 12 0 36 0 132 0	
6 - 5	2.00000 1.41421 0.00000 0.00000 0.00000 -1.41421 -2.00000 1 0 -6 0 8 0 0 0; 0 12 0 40 0 144 0	
6 - 6	2.05288 1.20864 0.56997 0.00000 -0.56997 -1.20864 -2.05288 1 0 -6 0 8 0 -2 0; 0 12 0 40 0 156 0	
6 - 7	2.10100 1.25928 0.00000 0.00000 0.00000 -1.25928 -2.10100 1 0 -6 0 7 0 0 0; 0 12 0 44 0 180 0	
6 - 8	2.13578 1.00000 0.66215 0.00000 -0.66215 -1.00000 -2.13578 1 0 -6 0 7 0 -2 0; 0 12 0 44 0 192 0	
6 - 9	2.17533 1.12603 0.00000 0.00000 0.00000 -1.12603 -2.17533 1 0 -6 0 6 0 0 0; 0 12 0 48 0 216 0	
6 - 10	2.28825 0.87403 0.00000 0.00000 0.00000 -0.87403 -2.28825 1 0 -6 0 4 0 0 0; 0 12 0 56 0 288 0	
6 - 11	2.44949 0.00000 0.00000 0.00000 0.00000 0.00000 -2.44949 1 0 -6 0 0 0 0 0; 0 12 0 72 0 432 0	
7 - 12	2.00000 1.24698 1.24698 -0.44504 -0.44504 -1.80194 -1.80194 1 0 -7 0 14 0 -7 -2; 0 14 0 42 0 140 14	
7 - 13	2.10100 1.25928 1.00000 0.00000 -1.00000 -1.25928 -2.10100 1 0 -7 0 13 0 -7 0; 0 14 0 46 0 182 0	
7 - 14	2.15154 1.26848 0.61803 0.42063 -0.89579 -1.61803 -1.94486 1 0 -7 0 13 -2 -6 2; 0 14 0 46 10 176 84	

Number	Eigenvalues Characteristic polynomial coefficients and spectral moments	Graph
7 - 15	2.19869 1.24698 0.71354 0.00000 -0.44504 -1.80194 -1.91223 1 0 -7 0 12 -2 -3 0 ; 0 14 0 50 10 200 98	
7 - 16	2.21432 1.00000 1.00000 0.00000 -0.53919 -1.67513 -2.00000 1 0 -7 0 12 -2 -4 0 ; 0 14 0 50 10 206 98	
7 - 17	2.18890 1.41421 0.45685 0.00000 -0.45685 -1.41421 -2.18890 1 0 -7 0 11 0 -2 0 ; 0 14 0 54 0 236 0	
7 - 18	2.25619 1.18994 0.61803 0.00000 -0.35650 -1.61803 -2.08963 1 0 -7 0 11 -2 -2 0 ; 0 14 0 54 10 236 98	
7 - 19	2.23607 1.41421 0.00000 0.00000 0.00000 -1.41421 -2.23607 1 0 -7 0 10 0 0 0 ; 0 14 0 58 0 266 0	
7 - 20	2.26382 1.27932 0.48831 0.00000 -0.48831 -1.27932 -2.26382 1 0 -7 0 10 0 -2 0 ; 0 14 0 58 0 278 0	
7 - 21	2.27639 1.18593 0.64159 0.00000 -0.64159 -1.18593 -2.27639 1 0 -7 0 10 0 -3 0 ; 0 14 0 58 0 284 0	
7 - 22	2.32437 1.14718 0.53037 0.00000 -0.53037 -1.14718 -2.32437 1 0 -7 0 9 0 -2 0 ; 0 14 0 62 0 320 0	
7 - 23	2.33441 1.00000 0.74196 0.00000 -0.74196 -1.00000 -2.33441 1 0 -7 0 9 0 -3 0 ; 0 14 0 62 0 326 0	
7 - 24	2.35830 1.19935 0.00000 0.00000 0.00000 -1.19935 -2.35829 1 0 -7 0 8 0 0 0 ; 0 14 0 66 0 350 0	
7 - 25	2.37608 1.00000 0.59519 0.00000 -0.59519 -1.00000 -2.37608 1 0 -7 0 8 0 -2 0 ; 0 14 0 66 0 362 0	
7 - 26	2.44949 1.00000 0.00000 0.00000 0.00000 -1.00000 -2.44949 1 0 -7 0 6 0 0 0 ; 0 14 0 74 0 434 0	
7 - 27	2.23321 1.56429 0.67291 -0.36473 -1.00000 -1.27240 -1.83328 1 0 -7 -2 13 6 -5 -2 ; 0 14 6 46 40 182 224	
7 - 28	2.25327 1.64485 0.23266 0.00000 -1.00000 -1.20331 -1.92747 1 0 -7 -2 12 6 -2 0 ; 0 14 6 50 40 206 224	
7 - 29	2.29701 1.49330 0.63999 -0.46314 -1.00000 -1.00000 -1.96716 1 0 -7 -2 12 6 -4 -2 ; 0 14 6 50 40 218 238	

Number	Eigenvalues Characteristic polynomial coefficients and spectral moments	Graph
7 - 30	2.34292 1.41421 0.47068 0.00000 -1.00000 -1.41421 -1.81361 1 0 -7 -2 12 4 -4 0 ; 0 14 6 50 50 218 322	
7 - 31	2.36234 1.24698 0.82578 -0.44504 -0.67964 -1.50848 -1.80194 1 0 -7 -2 12 4 -5 -2 ; 0 14 6 50 50 224 336	
7 - 32	2.37988 1.41421 0.29137 0.00000 -0.75102 -1.41421 -1.92022 1 0 -7 -2 11 4 -2 0 ; 0 14 6 54 50 248 336	
7 - 33	2.38937 1.36680 0.39436 0.00000 -1.00000 -1.18523 -1.96531 1 0 -7 -2 11 4 -3 0 ; 0 14 6 54 50 254 336	
7 - 34	2.41421 1.00000 1.00000 -0.41421 -1.00000 -1.00000 -2.00000 1 0 -7 -2 11 4 -5 -2 ; 0 14 6 54 50 266 350	
7 - 35	2.43828 1.13856 0.61803 0.00000 -0.82025 -1.61803 -1.75660 1 0 -7 -2 11 2 -4 0 ; 0 14 6 54 60 260 434	
7 - 36	2.36865 1.52616 0.00000 0.00000 -0.78771 -1.00000 -2.10710 1 0 -7 -2 10 6 0 0 ; 0 14 6 58 40 278 252	
7 - 37	2.43091 1.32691 0.30114 0.00000 -1.00000 -1.00000 -2.05896 1 0 -7 -2 10 4 -2 0 ; 0 14 6 58 50 290 350	
7 - 38	2.46760 1.18826 0.38671 0.00000 -0.60434 -1.52743 -1.91080 1 0 -7 -2 10 2 -2 0 ; 0 14 6 58 60 290 448	
7 - 39	2.47449 1.11432 0.52415 0.00000 -0.76148 -1.38906 -1.96243 1 0 -7 -2 10 2 -3 0 ; 0 14 6 58 60 296 448	
7 - 40	2.53744 0.84926 0.61803 0.00000 -0.48909 -1.61803 -1.89761 1 0 -7 -2 9 0 -2 0 ; 0 14 6 62 70 332 560	
7 - 41	2.54501 1.00000 0.43941 0.00000 -0.83021 -1.00000 -2.15421 1 0 -7 -2 8 2 -2 0 ; 0 14 6 66 60 374 476	
7 - 42	2.56155 1.00000 0.00000 0.00000 0.00000 -1.56155 -2.00000 1 0 -7 -2 8 0 0 0 ; 0 14 6 66 70 362 574	
7 - 43	2.59440 0.91588 0.00000 0.00000 0.00000 -1.38826 -2.12202 1 0 -7 -2 7 0 0 0 ; 0 14 6 70 70 404 588	
7 - 44	2.68133 0.64207 0.00000 0.00000 0.00000 -1.00000 -2.32340 1 0 -7 -2 4 0 0 0 ; 0 14 6 82 70 530 630	

Number	Eigenvalues Characteristic polynomial coefficients and spectral moments	Graph
8 - 45	2.34292 1.00000 1.00000 0.47068 -1.00000 -1.81361 -2.00000 1 0 -8 0 17 -4 -10 4 ; 0 16 0 60 20 268 196	
8 - 46	2.35830 1.19935 1.00000 0.00000 -1.00000 -1.19935 -2.35829 1 0 -8 0 15 0 -8 0 ; 0 16 0 68 0 352 0	
8 - 47	2.37720 1.27389 0.80194 0.00000 -0.55496 -1.65109 -2.24698 1 0 -8 0 15 -2 -5 0 ; 0 16 0 68 10 334 112	
8 - 48	2.44785 1.17619 0.65618 0.00000 -0.26463 -1.83237 -2.18322 1 0 -8 0 14 -4 -2 0 ; 0 16 0 72 20 364 224	
8 - 49	2.46506 1.00000 0.61803 0.50955 -0.70002 -1.61803 -2.27460 1 0 -8 0 14 -4 -5 2 ; 0 16 0 72 20 382 210	
8 - 50	2.49386 1.00000 0.76233 0.00000 -0.27139 -1.68701 -2.29779 1 0 -8 0 13 -4 -2 0 ; 0 16 0 76 20 412 224	
8 - 51	2.44949 1.41421 0.00000 0.00000 0.00000 -1.41421 -2.44949 1 0 -8 0 12 0 0 0 ; 0 16 0 80 0 448 0	
8 - 52	2.47367 1.25235 0.55911 0.00000 -0.55911 -1.25235 -2.47367 1 0 -8 0 12 0 -3 0 ; 0 16 0 80 0 466 0	
8 - 53	2.52434 1.00000 0.79229 0.00000 -0.79229 -1.00000 -2.52434 1 0 -8 0 11 0 -4 0 ; 0 16 0 84 0 520 0	
8 - 54	2.53959 1.24520 0.00000 0.00000 0.00000 -1.24520 -2.53958 1 0 -8 0 10 0 0 0 ; 0 16 0 88 0 544 0	
8 - 55	2.57793 1.16372 0.00000 0.00000 0.00000 -1.16372 -2.57794 1 0 -8 0 9 0 0 0 ; 0 16 0 92 0 592 0	
8 - 56	2.58874 1.00000 0.54629 0.00000 -0.54629 -1.00000 -2.58874 1 0 -8 0 9 0 -2 0 ; 0 16 0 92 0 604 0	
8 - 57	2.61313 1.08239 0.00000 0.00000 0.00000 -1.08239 -2.61313 1 0 -8 0 8 0 0 0 ; 0 16 0 96 0 640 0	
8 - 58	2.62285 0.84424 0.63867 0.00000 -0.63867 -0.84424 -2.62285 1 0 -8 0 8 0 -2 0 ; 0 16 0 96 0 652 0	
8 - 59	2.64575 1.00000 0.00000 0.00000 0.00000 -1.00000 -2.64575 1 0 -8 0 7 0 0 0 ; 0 16 0 100 0 688 0	

Number	Eigenvalues Characteristic polynomial coefficients and spectral moments	Graph
8 - 60	2.67624 0.91527 0.00000 0.00000 0.00000 -0.91527 -2.67624 1 0 -8 0 6 0 0 0 ; 0 16 0 104 0 736 0	
8 - 61	2.41421 1.41421 1.00000 -0.41421 -1.00000 -1.41421 -2.00000 1 0 -8 -2 17 6 -10 -4 ; 0 16 6 60 50 280 350	
8 - 62	2.47283 1.46260 0.61803 0.00000 -1.00000 -1.61803 -1.93543 1 0 -8 -2 16 4 -7 0 ; 0 16 6 64 60 310 448	
8 - 63	2.47283 1.46260 0.61803 0.00000 -1.00000 -1.61803 -1.93543 1 0 -8 -2 16 4 -7 0 ; 0 16 6 64 60 310 448	
8 - 64	2.48769 1.32357 0.88515 -0.25001 -0.85744 -1.65667 -1.93230 1 0 -8 -2 16 4 -8 -2 ; 0 16 6 64 60 316 462	
8 - 65	2.52064 1.28447 0.61803 0.27183 -1.23118 -1.61803 -1.84576 1 0 -8 -2 16 2 -9 2 ; 0 16 6 64 70 322 546	
8 - 66	2.37854 1.77946 0.18717 0.00000 -1.00000 -1.15012 -2.19505 1 0 -8 -2 15 8 -2 0 ; 0 16 6 68 40 328 238	
8 - 67	2.55535 1.19458 0.77989 0.00000 -0.89107 -1.71772 -1.92103 1 0 -8 -2 15 2 -7 0 ; 0 16 6 68 70 358 574	
8 - 68	2.55160 1.37196 0.51848 0.00000 -1.00000 -1.26585 -2.17620 1 0 -8 -2 14 4 -5 0 ; 0 16 6 72 60 394 476	
8 - 69	2.57723 1.29201 0.52195 0.00000 -0.66770 -1.72348 -2.00000 1 0 -8 -2 14 2 -4 0 ; 0 16 6 72 70 388 588	
8 - 70	2.54078 1.51249 0.22812 0.00000 -1.00000 -1.00000 -2.28139 1 0 -8 -2 13 6 -2 0 ; 0 16 6 76 50 424 378	
8 - 71	2.55937 1.38981 0.62620 -0.50487 -0.77466 -1.00000 -2.29585 1 0 -8 -2 13 6 -4 -2 ; 0 16 6 76 50 436 392	
8 - 72	2.57411 1.41421 0.27538 0.00000 -0.63787 -1.41421 -2.21161 1 0 -8 -2 13 4 -2 0 ; 0 16 6 76 60 424 490	
8 - 73	2.59616 1.18264 0.80194 -0.51572 -0.55496 -1.26308 -2.24698 1 0 -8 -2 13 4 -5 -2 ; 0 16 6 76 60 442 504	
8 - 74	2.60950 1.25979 0.44685 0.00000 -0.60838 -1.57105 -2.13669 1 0 -8 -2 13 2 -3 0 ; 0 16 6 76 70 430 602	

Number	Eigenvalues Characteristic polynomial coefficients and spectral moments	Graph
8 - 75	2.62001 1.13106 0.66341 0.00000 -0.83393 -1.39622 -2.18433 1 0 -8 -2 13 2 -5 0; 0 16 6 76 70 442 602	
8 - 76	2.61270 1.34352 0.28332 0.00000 -0.70646 -1.24306 -2.29002 1 0 -8 -2 12 4 -2 0; 0 16 6 80 60 472 504	
8 - 77	2.64455 1.17834 0.46959 0.00000 -0.65415 -1.40039 -2.23794 1 0 -8 -2 12 2 -3 0; 0 16 6 80 70 478 616	
8 - 78	2.64932 1.10966 0.58758 0.00000 -0.78671 -1.30591 -2.25393 1 0 -8 -2 12 2 -4 0; 0 16 6 80 70 484 616	
8 - 79	2.65544 1.21076 0.00000 0.00000 0.00000 -1.86620 -2.00000 1 0 -8 -2 12 0 0 0; 0 16 6 80 80 460 728	
8 - 80	2.66908 1.00000 0.61803 0.00000 -0.52398 -1.61803 -2.14510 1 0 -8 -2 12 0 -3 0; 0 16 6 80 80 478 728	
8 - 81	2.64797 1.27463 0.29216 0.00000 -0.86210 -1.00000 -2.35266 1 0 -8 -2 11 4 -2 0; 0 16 6 84 60 520 518	
8 - 82	2.69963 0.80194 0.76088 0.00000 -0.55496 -1.46051 -2.24698 1 0 -8 -2 11 0 -3 0; 0 16 6 84 80 526 742	
8 - 83	2.72456 0.88815 0.55292 0.00000 -0.46508 -1.39263 -2.30792 1 0 -8 -2 10 0 -2 0; 0 16 6 88 80 568 756	
8 - 84	2.34292 2.00000 0.47068 -1.00000 -1.00000 -1.00000 -1.81361 1 0 -8 -4 17 16 -2 -4; 0 16 12 60 80 268 448	
8 - 85	2.47635 1.84083 0.35373 -0.50231 -1.00000 -1.38228 -1.78632 1 0 -8 -4 16 14 -1 -2; 0 16 12 64 90 310 574	
8 - 86	2.52700 1.78601 0.33112 -0.68211 -1.00000 -1.00000 -1.96202 1 0 -8 -4 15 14 0 -2; 0 16 12 68 90 352 602	
8 - 87	2.64119 1.41421 0.72374 -0.58922 -1.00000 -1.41421 -1.77571 1 0 -8 -4 15 10 -6 -4; 0 16 12 68 110 388 840	
8 - 88	2.65825 1.48774 0.47984 -0.42896 -1.00000 -1.28528 -1.91158 1 0 -8 -4 14 10 -3 -2; 0 16 12 72 110 418 854	
8 - 89	2.68554 1.41421 0.33490 0.00000 -1.27133 -1.41421 -1.74912 1 0 -8 -4 14 8 -4 0; 0 16 12 72 120 424 952	

Number	Eigenvalues Characteristic polynomial coefficients and spectral moments	Graph
8 - 90	2.69353 1.32973 0.61803 -0.32973 -1.00000 -1.61803 -1.69353 1 0 -8 -4 14 8 -5 -2 ; 0 16 12 72 120 430 966	
8 - 91	2.67513 1.53919 0.00000 0.00000 -1.00000 -1.21432 -2.00000 1 0 -8 -4 13 10 0 0 ; 0 16 12 76 110 448 868	
8 - 92	2.70928 1.41421 0.19394 0.00000 -1.00000 -1.41421 -1.90321 1 0 -8 -4 13 8 -2 0 ; 0 16 12 76 120 460 980	
8 - 93	2.70928 1.41421 0.19394 0.00000 -1.00000 -1.41421 -1.90321 1 0 -8 -4 13 8 -2 0 ; 0 16 12 76 120 460 980	
8 - 94	2.72092 1.30085 0.58434 -0.39380 -1.00000 -1.25375 -1.95857 1 0 -8 -4 13 8 -4 -2 ; 0 16 12 76 120 472 994	
8 - 95	2.73205 1.00000 1.00000 -0.73205 -1.00000 -1.00000 -2.00000 1 0 -8 -4 13 8 -6 -4 ; 0 16 12 76 120 484 1008	
8 - 96	2.73205 1.41421 0.00000 0.00000 -0.73205 -1.41421 -2.00000 1 0 -8 -4 12 8 0 0 ; 0 16 12 80 120 496 1008	
8 - 97	2.74687 1.27511 0.54876 -0.50330 -1.00000 -1.00000 -2.06745 1 0 -8 -4 12 8 -3 -2 ; 0 16 12 80 120 514 1022	
8 - 98	2.76485 1.23947 0.32569 0.00000 -1.00000 -1.37455 -1.95546 1 0 -8 -4 12 6 -3 0 ; 0 16 12 80 130 514 1120	
8 - 99	2.76485 1.23947 0.32569 0.00000 -1.00000 -1.37455 -1.95546 1 0 -8 -4 12 6 -3 0 ; 0 16 12 80 130 514 1120	
8 - 100	2.77107 1.10271 0.69571 -0.50583 -0.65383 -1.45561 -1.95423 1 0 -8 -4 12 6 -4 -2 ; 0 16 12 80 130 520 1134	
8 - 101	2.78859 1.22013 0.24079 0.00000 -1.00000 -1.17901 -2.07050 1 0 -8 -4 11 6 -2 0 ; 0 16 12 84 130 556 1148	
8 - 102	2.80721 1.12212 0.29793 0.00000 -0.67829 -1.69093 -1.85804 1 0 -8 -4 11 4 -2 0 ; 0 16 12 84 140 556 1260	
8 - 103	2.81361 1.00000 0.52932 0.00000 -1.00000 -1.34292 -2.00000 1 0 -8 -4 11 4 -4 0 ; 0 16 12 84 140 568 1260	
8 - 104	2.83207 1.06166 0.30891 0.00000 -0.75722 -1.37057 -2.07485 1 0 -8 -4 10 4 -2 0 ; 0 16 12 88 140 604 1288	

Number	Eigenvalues Characteristic polynomial coefficients and spectral moments	Graph
8 - 105	2.83207 1.06166 0.30891 0.00000 -0.75722 -1.37057 -2.07485 1 0 -8 -4 10 4 -2 0 ; 0 16 12 88 140 604 1288	
8 - 106	2.85169 0.78304 0.61803 0.00000 -0.69131 -1.61803 -1.94341 1 0 -8 -4 10 2 -3 0 ; 0 16 12 88 150 610 1400	
8 - 107	2.83316 1.18798 0.00000 0.00000 -0.80406 -1.00000 -2.21707 1 0 -8 -4 9 6 0 0 ; 0 16 12 92 130 640 1204	
8 - 108	2.85577 1.00000 0.32164 0.00000 -1.00000 -1.00000 -2.17741 1 0 -8 -4 9 4 -2 0 ; 0 16 12 92 140 652 1316	
8 - 109	2.87167 0.84353 0.44220 0.00000 -0.61289 -1.46488 -2.07964 1 0 -8 -4 9 2 -2 0 ; 0 16 12 92 150 652 1428	
8 - 110	2.90321 0.80606 0.00000 0.00000 0.00000 -1.70928 -2.00000 1 0 -8 -4 8 0 0 0 ; 0 16 12 96 160 688 1568	
8 - 111	2.94388 0.66483 0.00000 0.00000 0.00000 -1.36837 -2.24034 1 0 -8 -4 6 0 0 0 ; 0 16 12 104 160 784 1624	
9 - 112	2.64119 1.00000 0.72374 0.41421 -0.58922 -1.77571 -2.41421 1 0 -9 0 17 -6 -5 2 ; 0 18 0 94 30 570 364	
9 - 113	2.68133 1.00000 0.64207 0.00000 0.00000 -2.00000 -2.32341 1 0 -9 0 16 -8 0 0 ; 0 18 0 98 40 594 504	
9 - 114	2.64575 1.00000 1.00000 0.00000 -1.00000 -1.00000 -2.64575 1 0 -9 0 15 0 -7 0 ; 0 18 0 102 0 690 0	
9 - 115	2.67406 1.13371 0.61803 0.00000 -0.25882 -1.61803 -2.54895 1 0 -9 0 15 -4 -2 0 ; 0 18 0 102 20 660 252	
9 - 116	2.75153 0.84108 0.61803 0.00000 0.00000 -1.61803 -2.59262 1 0 -9 0 13 -6 0 0 ; 0 18 0 110 30 756 378	
9 - 117	2.72375 1.17761 0.44090 0.00000 -0.44091 -1.17761 -2.72375 1 0 -9 0 12 0 -2 0 ; 0 18 0 114 0 822 0	
9 - 118	2.77517 1.13949 0.00000 0.00000 0.00000 -1.13949 -2.77517 1 0 -9 0 10 0 0 0 ; 0 18 0 122 0 918 0	
9 - 119	2.80921 0.88647 0.56789 0.00000 -0.56789 -0.88647 -2.80921 1 0 -9 0 9 0 -2 0 ; 0 18 0 126 0 984 0	

Number	Eigenvalues Characteristic polynomial coefficients and spectral moments	Graph
9 - 120	2.87662 0.85152 0.00000 0.00000 0.00000 -0.85152 -2.87662 1 0 -9 0 6 0 0 0; 0 18 0 138 0 1134 0	
9 - 121	2.92081 0.68474 0.00000 0.00000 0.00000 -0.68474 -2.92081 1 0 -9 0 4 0 0 0; 0 18 0 146 0 1242 0	
9 - 122	2.65544 1.21076 0.61803 0.61803 -1.61803 -1.61803 -1.86620 1 0 -9 -2 21 0 -16 6; 0 18 6 78 90 432 798	
9 - 123	2.64261 1.49391 0.61803 0.00000 -0.91023 -1.61803 -2.22628 1 0 -9 -2 19 4 -8 0; 0 18 6 86 70 492 616	
9 - 124	2.68718 1.14085 1.00000 0.00000 -1.00000 -1.63957 -2.18846 1 0 -9 -2 19 2 -11 0; 0 18 6 86 80 510 742	
9 - 125	2.70769 1.28568 0.74068 0.00000 -0.83440 -1.66092 -2.23874 1 0 -9 -2 18 2 -8 0; 0 18 6 90 80 546 756	
9 - 126	2.73320 1.15679 0.70765 0.25249 -0.93366 -1.75110 -2.16538 1 0 -9 -2 18 0 -9 2; 0 18 6 90 90 552 868	
9 - 127	2.67513 1.53919 0.41421 0.00000 -1.00000 -1.21432 -2.41421 1 0 -9 -2 17 6 -5 0; 0 18 6 94 60 582 518	
9 - 128	2.74387 1.26987 0.76271 -0.37691 -0.59875 -1.37439 -2.42640 1 0 -9 -2 16 4 -6 -2; 0 18 6 98 70 642 672	
9 - 129	2.77339 1.24698 0.48003 0.00000 -0.44504 -1.80194 -2.25342 1 0 -9 -2 16 0 -3 0; 0 18 6 98 90 624 910	
9 - 130	2.78100 1.12611 0.68660 0.00000 -0.59485 -1.70174 -2.29712 1 0 -9 -2 16 0 -5 0; 0 18 6 98 90 636 910	
9 - 131	2.80569 1.10649 0.61803 0.00000 -0.54410 -1.61803 -2.36807 1 0 -9 -2 15 0 -4 0; 0 18 6 102 90 684 924	
9 - 132	2.81361 1.00000 0.52932 0.41421 -1.00000 -1.34292 -2.41421 1 0 -9 -2 15 0 -7 2; 0 18 6 102 90 702 910	
9 - 133	2.82153 1.00000 0.65752 0.00000 -0.39196 -1.81898 -2.26811 1 0 -9 -2 15 -2 -3 0; 0 18 6 102 100 678 1050	
9 - 134	2.82910 0.82796 0.61803 0.46631 -0.78207 -1.61803 -2.34129 1 0 -9 -2 15 -2 -6 2; 0 18 6 102 100 696 1036	

Number	Eigenvalues Characteristic polynomial coefficients and spectral moments	Graph
9 - 135	2.73205 1.56155 0.00000 0.00000 -0.73205 -1.00000 -2.56155 1 0 -9 -2 14 8 0 0; 0 18 6 106 50 714 434	
9 - 136	2.84021 1.09057 0.56100 0.00000 -0.75189 -1.21026 -2.52964 1 0 -9 -2 13 2 -4 0; 0 18 6 110 80 792 826	
9 - 137	2.84559 1.18283 0.00000 0.00000 0.00000 -1.57294 -2.45548 1 0 -9 -2 13 0 0 0; 0 18 6 110 90 768 952	
9 - 138	2.81565 1.38385 0.00000 0.00000 -0.59013 -1.00000 -2.60937 1 0 -9 -2 12 6 0 0; 0 18 6 114 60 822 588	
9 - 139	2.85939 1.13533 0.35599 0.00000 -0.54087 -1.24988 -2.55997 1 0 -9 -2 12 2 -2 0; 0 18 6 114 80 834 840	
9 - 140	2.87649 1.00000 0.46768 0.00000 -0.42112 -1.39825 -2.52480 1 0 -9 -2 12 0 -2 0; 0 18 6 114 90 834 966	
9 - 141	2.89999 0.90977 0.50911 0.00000 -0.44382 -1.30580 -2.56925 1 0 -9 -2 11 0 -2 0; 0 18 6 118 90 888 980	
9 - 142	2.64119 1.73205 0.72374 -0.58922 -1.00000 -1.73205 -1.77571 1 0 -9 -4 21 14 -9 -6; 0 18 12 78 110 426 840	
9 - 143	2.75802 1.24698 1.09744 -0.44504 -1.19091 -1.66455 -1.80194 1 0 -9 -4 20 10 -13 -6; 0 18 12 82 130 504 1120	
9 - 144	2.76156 1.36333 1.00000 -1.00000 -1.00000 -1.00000 -2.12489 1 0 -9 -4 19 12 -11 -8; 0 18 12 86 120 546 1036	
9 - 145	2.71814 1.73792 0.37338 -0.42760 -1.00000 -1.20962 -2.19223 1 0 -9 -4 18 14 -2 -2; 0 18 12 90 110 546 896	
9 - 146	2.77571 1.58922 0.27626 0.00000 -1.00000 -1.64119 -2.00000 1 0 -9 -4 18 10 -4 0; 0 18 12 90 130 558 1134	
9 - 147	2.77571 1.58922 0.27626 0.00000 -1.00000 -1.64119 -2.00000 1 0 -9 -4 18 10 -4 0; 0 18 12 90 130 558 1134	
9 - 148	2.81622 1.36659 0.69269 -0.22558 -1.00000 -1.75552 -1.89440 1 0 -9 -4 18 8 -8 -2; 0 18 12 90 140 582 1274	
9 - 149	2.81622 1.36659 0.69269 -0.22558 -1.00000 -1.75552 -1.89440 1 0 -9 -4 18 8 -8 -2; 0 18 12 90 140 582 1274	

Number	Eigenvalues Characteristic polynomial coefficients and spectral moments	Graph
9 - 150	2.80983 1.50592 0.49270 -0.37584 -0.79929 -1.48985 -2.14348 1 0 -9 -4 17 10 -4 -2 ; 0 18 12 94 130 612 1176	
9 - 151	2.85004 1.10105 1.00000 -0.43144 -1.00000 -1.38238 -2.13726 1 0 -9 -4 17 8 -9 -4 ; 0 18 12 94 140 642 1316	
9 - 152	2.86081 1.25410 0.61803 0.00000 -1.11491 -1.61803 -2.00000 1 0 -9 -4 17 6 -8 0 ; 0 18 12 94 150 636 1414	
9 - 153	2.86081 1.25410 0.61803 0.00000 -1.11491 -1.61803 -2.00000 1 0 -9 -4 17 6 -8 0 ; 0 18 12 94 150 636 1414	
9 - 154	2.86620 1.00000 1.00000 -0.21076 -1.00000 -1.65544 -2.00000 1 0 -9 -4 17 6 -9 -2 ; 0 18 12 94 150 642 1428	
9 - 155	2.88263 1.10446 0.61803 0.24834 -1.32199 -1.61803 -1.91345 1 0 -9 -4 17 4 -10 2 ; 0 18 12 94 160 648 1526	
9 - 156	2.80027 1.62946 0.00000 0.00000 -1.00000 -1.15729 -2.27244 1 0 -9 -4 16 12 0 0 ; 0 18 12 98 120 642 1064	
9 - 157	2.86455 1.30880 0.63702 -0.28808 -1.00000 -1.31993 -2.20237 1 0 -9 -4 16 8 -6 -2 ; 0 18 12 98 140 678 1330	
9 - 158	2.88335 1.23026 0.57272 0.00000 -1.17331 -1.37088 -2.14214 1 0 -9 -4 16 6 -7 0 ; 0 18 12 98 150 684 1442	
9 - 159	2.87888 1.38935 0.26119 0.00000 -1.00000 -1.27220 -2.25722 1 0 -9 -4 15 8 -3 0 ; 0 18 12 102 140 714 1344	
9 - 160	2.88409 1.31829 0.55740 -0.43020 -0.71798 -1.35241 -2.25918 1 0 -9 -4 15 8 -4 -2 ; 0 18 12 102 140 720 1358	
9 - 161	2.90213 1.24879 0.44912 0.00000 -1.00000 -1.38991 -2.21012 1 0 -9 -4 15 6 -5 0 ; 0 18 12 102 150 726 1470	
9 - 162	2.92391 1.00000 0.73712 0.00000 -1.00000 -1.50970 -2.15133 1 0 -9 -4 15 4 -7 0 ; 0 18 12 102 160 738 1596	
9 - 163	2.89511 1.41421 0.00000 0.00000 -0.60271 -1.41421 -2.29240 1 0 -9 -4 14 8 0 0 ; 0 18 12 106 140 750 1372	
9 - 164	2.90101 1.36642 0.19108 0.00000 -1.00000 -1.13771 -2.32080 1 0 -9 -4 14 8 -2 0 ; 0 18 12 106 140 762 1372	

Number	Eigenvalues Characteristic polynomial coefficients and spectral moments	Graph
9 - 165	2.92765 1.09947 0.71763 -0.38430 -0.73647 -1.33852 -2.28546 1 0 -9 -4 14 6 -5 -2 ; 0 18 12 106 150 780 1512	
9 - 166	2.93346 1.21347 0.27256 0.00000 -0.56015 -1.72110 -2.13824 1 0 -9 -4 14 4 -2 0 ; 0 18 12 106 160 762 1624	
9 - 167	2.93610 1.17566 0.36745 0.00000 -0.66165 -1.64663 -2.17093 1 0 -9 -4 14 4 -3 0 ; 0 18 12 106 160 768 1624	
9 - 168	2.95372 1.00000 0.56588 0.00000 -0.63487 -1.88474 -2.00000 1 0 -9 -4 14 2 -4 0 ; 0 18 12 106 170 774 1750	
9 - 169	2.94331 1.21187 0.31775 0.00000 -1.00000 -1.12953 -2.34340 1 0 -9 -4 13 6 -3 0 ; 0 18 12 110 150 822 1526	
9 - 170	2.95577 1.16116 0.28036 0.00000 -0.59718 -1.53990 -2.26022 1 0 -9 -4 13 4 -2 0 ; 0 18 12 110 160 816 1652	
9 - 171	2.96073 1.06812 0.47926 0.00000 -0.83282 -1.38158 -2.29372 1 0 -9 -4 13 4 -4 0 ; 0 18 12 110 160 828 1652	
9 - 172	2.94272 1.32584 0.00000 0.00000 -0.84662 -1.00000 -2.42194 1 0 -9 -4 12 8 0 0 ; 0 18 12 114 140 858 1428	
9 - 173	2.96935 1.00000 0.72108 -0.48698 -0.79711 -1.00000 -2.40634 1 0 -9 -4 12 6 -4 -2 ; 0 18 12 114 150 882 1568	
9 - 174	2.97717 1.10838 0.28905 0.00000 -0.64501 -1.38870 -2.34091 1 0 -9 -4 12 4 -2 0 ; 0 18 12 114 160 870 1680	
9 - 175	3.00441 0.83526 0.51015 0.00000 -0.40540 -1.78251 -2.16192 1 0 -9 -4 12 0 -2 0 ; 0 18 12 114 180 870 1932	
9 - 176	3.00000 1.00000 0.41421 0.00000 -1.00000 -1.00000 -2.41421 1 0 -9 -4 11 4 -3 0 ; 0 18 12 118 160 930 1708	
9 - 177	3.01318 0.83103 0.55720 0.00000 -0.66043 -1.37751 -2.36347 1 0 -9 -4 11 2 -3 0 ; 0 18 12 118 170 930 1834	
9 - 178	3.02386 0.68295 0.61803 0.00000 -0.42430 -1.61803 -2.28251 1 0 -9 -4 11 0 -2 0 ; 0 18 12 118 180 924 1960	
9 - 179	3.01760 1.00000 0.31005 0.00000 -0.86968 -1.00000 -2.45797 1 0 -9 -4 10 4 -2 0 ; 0 18 12 122 160 978 1736	

Number	Eigenvalues Characteristic polynomial coefficients and spectral moments	Graph
9 - 180	3.03864 0.88642 0.00000 0.00000 0.00000 -1.58981 -2.33525 1 0 -9 -4 10 0 0 0;　　0 18 12 122 180 966 1988	
9 - 181	2.70928 2.00000 0.19394 -1.00000 -1.00000 -1.00000 -1.90321 1 0 -9 -6 19 24 5 -2;　　0 18 18 86 150 510 1106	
9 - 182	2.81361 1.73205 0.52932 -1.00000 -1.00000 -1.34292 -1.73205 1 0 -9 -6 19 20 -3 -6;　　0 18 18 86 170 558 1386	
9 - 183	2.77304 1.91889 0.00000 -0.43751 -1.00000 -1.36162 -1.89281 1 0 -9 -6 18 22 6 0;　　0 18 18 90 160 558 1260	
9 - 184	2.87564 1.68564 0.27915 -0.56394 -1.00000 -1.38752 -1.88896 1 0 -9 -6 17 18 1 -2;　　0 18 18 94 180 642 1568	
9 - 185	2.88474 1.63487 0.43412 -1.00000 -1.00000 -1.00000 -1.95372 1 0 -9 -6 17 18 -1 -4;　　0 18 18 94 180 654 1582	
9 - 186	2.92403 1.57602 0.31493 -0.52892 -1.00000 -1.33632 -1.94975 1 0 -9 -6 16 16 0 -2;　　0 18 18 98 190 702 1736	
9 - 187	2.94587 1.55204 0.29447 -0.71514 -1.00000 -1.00000 -2.07723 1 0 -9 -6 15 16 1 -2;　　0 18 18 102 190 750 1778	
9 - 188	2.98311 1.41421 0.14291 0.00000 -1.25106 -1.41421 -1.87496 1 0 -9 -6 15 12 -2 0;　　0 18 18 102 210 768 2016	
9 - 189	2.98723 1.36093 0.44701 -0.38100 -1.00000 -1.54719 -1.86697 1 0 -9 -6 15 12 -3 -2;　　0 18 18 102 210 774 2030	
9 - 190	2.99368 1.26253 0.65816 -0.60748 -1.00000 -1.35861 -1.94829 1 0 -9 -6 15 12 -5 -4;　　0 18 18 102 210 786 2044	
9 - 191	3.00000 1.00000 1.00000 -1.00000 -1.00000 -1.00000 -2.00000 1 0 -9 -6 15 12 -7 -6;　　0 18 18 102 210 798 2058	
9 - 192	3.01856 1.31440 0.16518 0.00000 -1.00000 -1.66285 -1.83529 1 0 -9 -6 14 10 -2 0;　　0 18 18 106 220 822 2184	
9 - 193	3.02228 1.24698 0.49132 -0.44504 -0.77446 -1.73914 -1.80194 1 0 -9 -6 14 10 -3 -2;　　0 18 18 106 220 828 2198	
9 - 194	3.02450 1.21544 0.53564 -0.35341 -1.00000 -1.47909 -1.94308 1 0 -9 -6 14 10 -4 -2;　　0 18 18 106 220 834 2198	

Number	Eigenvalues Characteristic polynomial coefficients and spectral moments	Graph
9 - 195	3.03439 1.32038 0.00000 0.00000 -0.80552 -1.54926 -2.00000 1 0 -9 -6 13 10 0 0 ; 0 18 18 110 220 864 2226	
9 - 196	3.05356 1.17790 0.27444 0.00000 -1.00000 -1.56950 -1.93641 1 0 -9 -6 13 8 -3 0 ; 0 18 18 110 230 882 2352	
9 - 197	3.05694 1.06611 0.61803 -0.40406 -0.78554 -1.61803 -1.93345 1 0 -9 -6 13 8 -4 -2 ; 0 18 18 110 230 888 2366	
9 - 198	3.05694 1.06611 0.61803 -0.40406 -0.78554 -1.61803 -1.93345 1 0 -9 -6 13 8 -4 -2 ; 0 18 18 110 230 888 2366	
9 - 199	3.05401 1.28267 0.00000 0.00000 -1.00000 -1.18818 -2.14849 1 0 -9 -6 12 10 0 0 ; 0 18 18 114 220 918 2268	
9 - 200	3.07040 1.16627 0.19850 0.00000 -1.00000 -1.34823 -2.08694 1 0 -9 -6 12 8 -2 0 ; 0 18 18 114 230 930 2394	
9 - 201	3.08424 1.04933 0.33321 0.00000 -0.85835 -1.68539 -1.92303 1 0 -9 -6 12 6 -3 0 ; 0 18 18 114 240 936 2520	
9 - 202	3.08613 1.00000 0.42801 0.00000 -1.00000 -1.51414 -2.00000 1 0 -9 -6 12 6 -4 0 ; 0 18 18 114 240 942 2520	
9 - 203	3.09178 1.00000 0.59916 -0.49069 -1.00000 -1.00000 -2.20025 1 0 -9 -6 11 8 -3 -2 ; 0 18 18 118 230 990 2450	
9 - 204	3.10019 1.04520 0.24400 0.00000 -0.81701 -1.47922 -2.09315 1 0 -9 -6 11 6 -2 0 ; 0 18 18 118 240 984 2562	
9 - 205	3.11315 0.88139 0.43334 0.00000 -0.74447 -1.78842 -1.89500 1 0 -9 -6 11 4 -3 0 ; 0 18 18 118 250 990 2688	
9 - 206	3.10278 1.14637 0.00000 0.00000 -1.00000 -1.00000 -2.24914 1 0 -9 -6 10 8 0 0 ; 0 18 18 122 230 1026 2478	
9 - 207	3.11397 1.07759 0.00000 0.00000 -0.56504 -1.46194 -2.16459 1 0 -9 -6 10 6 0 0 ; 0 18 18 122 240 1026 2604	
9 - 208	3.14479 0.84197 0.33313 0.00000 -0.76305 -1.34106 -2.21579 1 0 -9 -6 9 4 -2 0 ; 0 18 18 126 250 1092 2772	
9 - 209	3.15501 0.61803 0.52525 0.00000 -0.57262 -1.61803 -2.10764 1 0 -9 -6 9 2 -2 0 ; 0 18 18 126 260 1092 2898	

Number	Eigenvalues Characteristic polynomial coefficients and spectral moments	Graph
9 - 210	3.19258 0.61803 0.00000 0.00000 0.00000 -1.61803 -2.19258 1 0 -9 -6 7 0 0 0; 0 18 18 134 270 1188 3108	
9 - 211	3.20740 0.55452 0.00000 0.00000 0.00000 -1.47535 -2.28657 1 0 -9 -6 6 0 0 0; 0 18 18 138 270 1242 3150	
9 - 212	3.09787 1.45266 0.24722 -1.00000 -1.00000 -1.00000 -1.79774 1 0 -9 -8 13 18 3 -2; 0 18 24 110 270 930 2688	
9 - 213	3.10947 1.48434 0.00000 -0.67884 -1.00000 -1.00000 -1.91497 1 0 -9 -8 12 18 6 0; 0 18 24 114 270 966 2730	
9 - 214	3.17320 1.15038 0.40661 -0.53112 -1.00000 -1.45285 -1.74621 1 0 -9 -8 11 12 -1 -2; 0 18 24 118 300 1062 3178	
9 - 215	3.18766 1.14173 0.37585 -0.74644 -1.00000 -1.00000 -1.95880 1 0 -9 -8 10 12 0 -2; 0 18 24 122 300 1110 3234	
9 - 216	3.23607 0.61803 0.61803 0.00000 -1.23607 -1.61803 -1.61803 1 0 -9 -8 9 6 -4 0; 0 18 24 126 330 1188 3654	
9 - 217	3.24771 0.82822 0.27022 0.00000 -1.00000 -1.45508 -1.89107 1 0 -9 -8 8 6 -2 0; 0 18 24 130 330 1230 3710	
9 - 218	3.27307 0.85956 0.00000 0.00000 -1.00000 -1.00000 -2.13264 1 0 -9 -8 6 6 0 0; 0 18 24 138 330 1326 3822	
10 - 219	2.90321 0.80606 0.73205 0.00000 0.00000 -1.70928 -2.73205 1 0 -10 0 16 -8 0 0; 0 20 0 136 40 1040 560	
10 - 220	2.96176 1.00000 0.47749 0.00000 -0.47749 -1.00000 -2.96176 1 0 -10 0 11 0 -2 0; 0 20 0 156 0 1352 0	
10 - 221	3.02045 0.93643 0.00000 0.00000 0.00000 -0.93643 -3.02045 1 0 -10 0 8 0 0 0; 0 20 0 168 0 1520 0	
10 - 222	3.05923 0.80069 0.00000 0.00000 0.00000 -0.80069 -3.05923 1 0 -10 0 6 0 0 0; 0 20 0 176 0 1640 0	
10 - 223	3.16228 0.00000 0.00000 0.00000 0.00000 0.00000 -3.16228 1 0 -10 0 0 0 0 0; 0 20 0 200 0 2000 0	
10 - 224	2.90417 1.21546 0.61803 0.25823 -0.87751 -1.61803 -2.50035 1 0 -10 -2 20 0 -9 2; 0 20 6 120 100 866 1106	

Number	Eigenvalues Characteristic polynomial coefficients and spectral moments	Graph
10 - 225	2.96664 0.80194 0.70259 0.45229 -0.55496 -2.12152 -2.24698 1 0 -10 -2 19 -6 -5 2 ; 0 20 6 124 130 902 1540	
10 - 226	2.98054 1.00000 0.70649 0.00000 -0.44175 -1.66935 -2.57593 1 0 -10 -2 17 -2 -4 0 ; 0 20 6 132 110 1016 1302	
10 - 227	3.01179 0.87433 0.67691 0.00000 -0.25064 -1.74070 -2.57169 1 0 -10 -2 16 -4 -2 0 ; 0 20 6 136 120 1064 1456	
10 - 228	2.97509 1.31263 0.26248 0.00000 -0.57717 -1.23442 -2.73861 1 0 -10 -2 15 4 -2 0 ; 0 20 6 140 80 1124 910	
10 - 229	3.03032 0.86537 0.73205 0.00000 -0.57942 -1.31627 -2.73205 1 0 -10 -2 14 0 -4 0 ; 0 20 6 144 100 1196 1204	
10 - 230	3.09906 0.94370 0.00000 0.00000 0.00000 -1.20488 -2.83788 1 0 -10 -2 10 0 0 0 ; 0 20 6 160 100 1412 1260	
10 - 231	2.90321 1.41421 0.80606 0.00000 -1.41421 -1.70928 -2.00000 1 0 -10 -4 24 8 -16 0 ; 0 20 12 104 160 704 1568	
10 - 232	2.95945 1.43061 0.64244 -0.22965 -0.82105 -1.73787 -2.24393 1 0 -10 -4 21 8 -8 -2 ; 0 20 12 116 160 836 1666	
10 - 233	2.97941 1.32863 0.65199 0.00000 -1.00000 -1.76578 -2.19426 1 0 -10 -4 21 6 -10 0 ; 0 20 12 116 170 848 1792	
10 - 234	2.98206 1.29063 0.70992 0.00000 -1.07702 -1.67822 -2.22737 1 0 -10 -4 21 6 -11 0 ; 0 20 12 116 170 854 1792	
10 - 235	3.00000 1.00000 1.00000 0.00000 -1.00000 -2.00000 -2.00000 1 0 -10 -4 21 4 -12 0 ; 0 20 12 116 180 860 1932	
10 - 236	3.00083 1.13885 0.77051 0.17040 -1.20602 -1.69790 -2.17668 1 0 -10 -4 21 4 -13 2 ; 0 20 12 116 180 866 1918	
10 - 237	3.03531 1.00000 0.61803 0.46433 -1.27678 -1.61803 -2.22286 1 0 -10 -4 20 2 -13 4 ; 0 20 12 120 190 926 2072	
10 - 238	3.01329 1.09130 1.00000 -0.39180 -1.00000 -1.27192 -2.44087 1 0 -10 -4 19 8 -10 -4 ; 0 20 12 124 160 968 1736	
10 - 239	3.03407 1.17569 0.66371 0.00000 -0.85392 -1.72332 -2.29623 1 0 -10 -4 19 4 -8 0 ; 0 20 12 124 180 956 1988	

Number	Eigenvalues Characteristic polynomial coefficients and spectral moments	Graph
10 - 240	3.01760 1.41421 0.31005 0.00000 -0.86968 -1.41421 -2.45797 1 0 -10 -4 18 8 -4 0; 0 20 12 128 160 992 1736	
10 - 241	3.05266 1.15737 0.61803 0.00000 -0.83379 -1.61803 -2.37625 1 0 -10 -4 18 4 -7 0; 0 20 12 128 180 1010 2016	
10 - 242	3.05945 1.18663 0.46806 0.00000 -0.54180 -2.00000 -2.17234 1 0 -10 -4 18 2 -4 0; 0 20 12 128 190 992 2156	
10 - 243	3.06374 1.08656 0.64046 0.00000 -0.66886 -1.86065 -2.26126 1 0 -10 -4 18 2 -6 0; 0 20 12 128 190 1004 2156	
10 - 244	3.06871 1.00000 0.61803 0.27775 -1.00000 -1.61803 -2.34646 1 0 -10 -4 18 2 -9 2; 0 20 12 128 190 1022 2142	
10 - 245	3.02667 1.40068 0.49485 -0.37415 -1.00000 -1.00000 -2.54804 1 0 -10 -4 17 10 -4 -2; 0 20 12 132 150 1052 1638	
10 - 246	3.07486 1.00000 0.76597 0.00000 -1.00000 -1.38069 -2.46013 1 0 -10 -4 17 4 -8 0; 0 20 12 132 180 1076 2044	
10 - 247	3.07486 1.00000 0.76597 0.00000 -1.00000 -1.38069 -2.46013 1 0 -10 -4 17 4 -8 0; 0 20 12 132 180 1076 2044	
10 - 248	3.05584 1.37200 0.18571 0.00000 -0.79206 -1.27210 -2.54939 1 0 -10 -4 16 8 -2 0; 0 20 12 136 160 1100 1792	
10 - 249	3.08815 1.12798 0.50900 0.00000 -0.77671 -1.45769 -2.49073 1 0 -10 -4 16 4 -5 0; 0 20 12 136 180 1118 2072	
10 - 250	3.10278 1.14637 0.00000 0.00000 0.00000 -2.00000 -2.24914 1 0 -10 -4 16 0 0 0; 0 20 12 136 200 1088 2352	
10 - 251	3.08889 1.26464 0.22171 0.00000 -0.67098 -1.34449 -2.55977 1 0 -10 -4 15 6 -2 0; 0 20 12 140 170 1160 1960	
10 - 252	3.11980 1.14483 0.27311 0.00000 -0.57135 -1.39622 -2.57017 1 0 -10 -4 14 4 -2 0; 0 20 12 144 180 1220 2128	
10 - 253	3.13902 1.04953 0.00000 0.00000 0.00000 -1.72481 -2.46375 1 0 -10 -4 14 0 0 0; 0 20 12 144 200 1208 2408	
10 - 254	3.13776 1.09750 0.28101 0.00000 -0.61359 -1.28839 -2.61429 1 0 -10 -4 13 4 -2 0; 0 20 12 148 180 1280 2156	

Number	Eigenvalues Characteristic polynomial coefficients and spectral moments	Graph
10 - 255	3.15236 0.81032 0.65452 0.00000 -0.69376 -1.32942 -2.59403 1 0 -10 -4 13 2 -4 0 ; 0 20 12 148 190 1292 2296	
10 - 256	3.15633 1.00000 0.00000 0.00000 0.00000 -1.63090 -2.52543 1 0 -10 -4 13 0 0 0 ; 0 20 12 148 200 1268 2436	
10 - 257	3.17641 0.76648 0.52771 0.00000 -0.40836 -1.47131 -2.59094 1 0 -10 -4 12 0 -2 0 ; 0 20 12 152 200 1340 2464	
10 - 258	3.17210 1.00000 0.29980 0.00000 -0.78179 -1.00000 -2.69011 1 0 -10 -4 11 4 -2 0 ; 0 20 12 156 180 1400 2212	
10 - 259	3.20541 0.84444 0.00000 0.00000 0.00000 -1.38786 -2.66199 1 0 -10 -4 10 0 0 0 ; 0 20 12 160 200 1448 2520	
10 - 260	3.21432 0.73205 0.46081 0.00000 -0.67513 -1.00000 -2.73205 1 0 -10 -4 9 2 -2 0 ; 0 20 12 164 190 1520 2408	
10 - 261	2.91072 1.79940 0.61803 -0.79940 -1.00000 -1.61803 -1.91072 1 0 -10 -6 24 22 -7 -8 ; 0 20 18 104 190 710 1708	
10 - 262	2.99276 1.51517 0.86054 -0.81705 -1.00000 -1.64675 -1.90467 1 0 -10 -6 23 18 -12 -10 ; 0 20 18 108 210 800 2044	
10 - 263	3.03032 1.41421 0.86537 -0.57942 -1.31627 -1.41421 -2.00000 1 0 -10 -6 22 16 -12 -8 ; 0 20 18 112 220 860 2212	
10 - 264	2.98487 1.76222 0.28163 -0.43616 -1.00000 -1.43411 -2.15845 1 0 -10 -6 21 20 0 -2 ; 0 20 18 116 200 848 1932	
10 - 265	3.06644 1.22219 1.00000 -0.65222 -1.00000 -1.63641 -2.00000 1 0 -10 -6 21 14 -12 -8 ; 0 20 18 116 230 920 2394	
10 - 266	3.03967 1.64941 0.00000 0.00000 -0.88395 -1.80513 -2.00000 1 0 -10 -6 20 16 0 0 ; 0 20 18 120 220 908 2240	
10 - 267	3.05486 1.53396 0.53269 -0.51196 -1.00000 -1.44808 -2.16148 1 0 -10 -6 20 16 -5 -4 ; 0 20 18 120 220 938 2268	
10 - 268	3.10852 1.14379 0.84424 -0.15329 -1.34254 -1.73877 -1.86196 1 0 -10 -6 20 10 -12 -2 ; 0 20 18 120 250 980 2674	
10 - 269	3.00000 1.81361 0.00000 -0.47068 -1.00000 -1.00000 -2.34292 1 0 -10 -6 19 22 6 0 ; 0 20 18 124 190 932 1862	

Number	Eigenvalues Characteristic polynomial coefficients and spectral moments	Graph
10 - 270	3.11071 1.25644 0.75749 -0.37345 -1.24479 -1.34412 -2.16228 1 0 -10 -6 19 12 -9 -4 ; 0 20 18 124 240 1022 2590	
10 - 271	3.11551 1.43289 0.19691 0.00000 -1.00000 -1.56610 -2.17921 1 0 -10 -6 18 12 -3 0 ; 0 20 18 128 240 1046 2604	
10 - 272	3.11551 1.43289 0.19691 0.00000 -1.00000 -1.56610 -2.17921 1 0 -10 -6 18 12 -3 0 ; 0 20 18 128 240 1046 2604	
10 - 273	3.13658 1.24057 0.61803 -0.24057 -1.00000 -1.61803 -2.13658 1 0 -10 -6 18 10 -7 -2 ; 0 20 18 128 250 1070 2758	
10 - 274	3.13658 1.24057 0.61803 -0.24057 -1.00000 -1.61803 -2.13658 1 0 -10 -6 18 10 -7 -2 ; 0 20 18 128 250 1070 2758	
10 - 275	3.14884 1.17837 0.55254 0.00000 -1.09027 -1.78948 -2.00000 1 0 -10 -6 18 8 -8 0 ; 0 20 18 128 260 1076 2884	
10 - 276	3.10510 1.52251 0.31289 -0.57049 -1.00000 -1.00000 -2.37002 1 0 -10 -6 17 16 0 -2 ; 0 20 18 132 220 1088 2380	
10 - 277	3.13264 1.41421 0.14044 0.00000 -1.00000 -1.41421 -2.27307 1 0 -10 -6 17 12 -2 0 ; 0 20 18 132 240 1100 2646	
10 - 278	3.13765 1.34505 0.46863 -0.33273 -1.00000 -1.32526 -2.29334 1 0 -10 -6 17 12 -4 -2 ; 0 20 18 132 240 1112 2660	
10 - 279	3.14258 1.25143 0.67115 -0.54048 -1.00000 -1.21286 -2.31182 1 0 -10 -6 17 12 -6 -4 ; 0 20 18 132 240 1124 2674	
10 - 280	3.14930 1.28162 0.50427 -0.38406 -0.70649 -1.65054 -2.19411 1 0 -10 -6 17 10 -4 -2 ; 0 20 18 132 250 1112 2800	
10 - 281	3.14997 1.29552 0.34015 0.00000 -1.16795 -1.37544 -2.24225 1 0 -10 -6 17 10 -5 0 ; 0 20 18 132 250 1118 2786	
10 - 282	3.15587 1.10523 0.80194 -0.55496 -0.76814 -1.49296 -2.24698 1 0 -10 -6 17 10 -7 -4 ; 0 20 18 132 250 1130 2814	
10 - 283	3.16590 1.08250 0.71944 -0.25386 -0.88503 -1.68047 -2.14848 1 0 -10 -6 17 8 -7 -2 ; 0 20 18 132 260 1130 2940	
10 - 284	3.17741 1.00000 0.67836 0.00000 -1.00000 -1.85577 -2.00000 1 0 -10 -6 17 6 -8 0 ; 0 20 18 132 270 1136 3066	

Number	Eigenvalues Characteristic polynomial coefficients and spectral moments	Graph
10 - 285	3.16886 1.20828 0.55819 -0.30885 -1.00000 -1.30540 -2.32107 1 0 -10 -6 16 10 -5 -2 ; 0 20 18 136 250 1178 2842	
10 - 286	3.19095 1.05926 0.52883 0.00000 -0.90816 -1.71232 -2.15856 1 0 -10 -6 16 6 -6 0 ; 0 20 18 136 270 1184 3108	
10 - 287	3.20505 0.80898 0.61803 0.30594 -1.17599 -1.61803 -2.14398 1 0 -10 -6 16 4 -9 2 ; 0 20 18 136 280 1202 3234	
10 - 288	3.19493 1.11182 0.58149 -0.45666 -0.60700 -1.50764 -2.31693 1 0 -10 -6 15 8 -4 -2 ; 0 20 18 140 260 1232 3024	
10 - 289	3.20264 1.13179 0.30747 0.00000 -0.70678 -1.71706 -2.21807 1 0 -10 -6 15 6 -3 0 ; 0 20 18 140 270 1226 3150	
10 - 290	3.20720 1.19056 0.19231 0.00000 -0.82770 -1.38287 -2.37950 1 0 -10 -6 14 8 -2 0 ; 0 20 18 144 260 1280 3052	
10 - 291	3.21882 1.08870 0.31550 0.00000 -0.75826 -1.53791 -2.32685 1 0 -10 -6 14 6 -3 0 ; 0 20 18 144 270 1286 3192	
10 - 292	3.23163 0.84820 0.61803 0.00000 -0.80016 -1.61803 -2.27967 1 0 -10 -6 14 4 -5 0 ; 0 20 18 144 280 1298 3332	
10 - 293	3.21019 1.26071 0.00000 0.00000 -1.00000 -1.00000 -2.47090 1 0 -10 -6 13 10 0 0 ; 0 20 18 148 250 1328 2954	
10 - 294	3.22333 1.15259 0.19538 0.00000 -1.00000 -1.12747 -2.44383 1 0 -10 -6 13 8 -2 0 ; 0 20 18 148 260 1340 3094	
10 - 295	3.24555 0.87690 0.51451 0.00000 -0.76404 -1.51859 -2.35433 1 0 -10 -6 13 4 -4 0 ; 0 20 18 148 280 1352 3374	
10 - 296	3.25342 0.80194 0.51997 0.00000 -0.55496 -1.77339 -2.24698 1 0 -10 -6 13 2 -3 0 ; 0 20 18 148 290 1346 3514	
10 - 297	3.24142 1.00000 0.58249 -0.54082 -0.78362 -1.00000 -2.49948 1 0 -10 -6 12 8 -3 -2 ; 0 20 18 152 260 1406 3150	
10 - 298	3.27307 0.85956 0.00000 0.00000 0.00000 -2.00000 -2.13264 1 0 -10 -6 12 0 0 0 ; 0 20 18 152 300 1388 3696	
10 - 299	3.27256 0.90638 0.30660 0.00000 -0.65189 -1.36845 -2.46521 1 0 -10 -6 11 4 -2 0 ; 0 20 18 156 280 1460 3458	

Number	Eigenvalues Characteristic polynomial coefficients and spectral moments	Graph
10 - 300	3.00000 2.00000 0.00000 -1.00000 -1.00000 -1.00000 -2.00000 1 0 -10 -8 21 32 12 0 ; 0 20 24 116 240 860 2184	
10 - 301	3.09363 1.77819 0.18923 -0.67615 -1.00000 -1.54122 -1.84368 1 0 -10 -8 20 26 5 -2 ; 0 20 24 120 270 962 2674	
10 - 302	3.10710 1.78771 0.00000 -0.52616 -1.00000 -1.36865 -2.00000 1 0 -10 -8 19 26 8 0 ; 0 20 24 124 270 1004 2716	
10 - 303	3.18428 1.50876 0.41699 -0.69872 -1.00000 -1.47833 -1.93299 1 0 -10 -8 18 20 -1 -4 ; 0 20 24 128 300 1118 3220	
10 - 304	3.20326 1.36747 0.61803 -0.75394 -1.00000 -1.61803 -1.81679 1 0 -10 -8 18 18 -5 -6 ; 0 20 24 128 310 1142 3374	
10 - 305	3.16053 1.69230 0.00000 -0.69230 -1.00000 -1.00000 -2.16053 1 0 -10 -8 17 24 8 0 ; 0 20 24 132 280 1124 2968	
10 - 306	3.20999 1.46735 0.30070 -0.46052 -1.00000 -1.59463 -1.92289 1 0 -10 -8 17 18 0 -2 ; 0 20 24 132 310 1172 3402	
10 - 307	3.23703 1.39183 0.31663 -0.58699 -0.72518 -1.73521 -1.89810 1 0 -10 -8 16 16 0 -2 ; 0 20 24 136 320 1232 3598	
10 - 308	3.23607 1.41421 0.00000 0.00000 -1.23607 -1.41421 -2.00000 1 0 -10 -8 16 16 0 0 ; 0 20 24 136 320 1232 3584	
10 - 309	3.25296 1.26565 0.45118 -0.30167 -1.20582 -1.53876 -1.92354 1 0 -10 -8 16 14 -4 -2 ; 0 20 24 136 330 1256 3738	
10 - 310	3.25535 1.19799 0.61803 -0.53454 -1.00000 -1.61803 -1.91880 1 0 -10 -8 16 14 -5 -4 ; 0 20 24 136 330 1262 3752	
10 - 311	3.26685 1.10922 0.61803 -0.22649 -1.36723 -1.61803 -1.78234 1 0 -10 -8 16 12 -7 -2 ; 0 20 24 136 340 1274 3878	
10 - 312	3.22255 1.54294 0.00000 -0.54294 -1.00000 -1.00000 -2.22255 1 0 -10 -8 15 20 6 0 ; 0 20 24 140 300 1256 3360	
10 - 313	3.25307 1.36046 0.31903 -0.51935 -1.00000 -1.27641 -2.13680 1 0 -10 -8 15 16 0 -2 ; 0 20 24 140 320 1292 3654	
10 - 314	3.26415 1.29421 0.36315 -0.52793 -0.75230 -1.64128 -2.00000 1 0 -10 -8 15 14 -1 -2 ; 0 20 24 140 330 1298 3794	

Number	Eigenvalues Characteristic polynomial coefficients and spectral moments	Graph
10 - 315	3.27550 1.23312 0.20477 0.00000 -1.14502 -1.65978 -1.90858 1 0 -10 -8 15 12 -3 0 ; 0 20 24 140 340 1310 3920	
10 - 316	3.28140 1.00000 0.77188 -0.51249 -1.00000 -1.54079 -2.00000 1 0 -10 -8 15 12 -6 -4 ; 0 20 24 140 340 1328 3948	
10 - 317	3.30065 1.11295 0.30239 0.00000 -1.14904 -1.56695 -2.00000 1 0 -10 -8 14 10 -4 0 ; 0 20 24 144 350 1376 4116	
10 - 318	3.31321 0.86926 0.61803 0.00000 -1.27265 -1.61803 -1.90982 1 0 -10 -8 14 8 -7 0 ; 0 20 24 144 360 1394 4256	
10 - 319	3.30104 1.23230 0.00000 0.00000 -1.00000 -1.35295 -2.18039 1 0 -10 -8 13 12 0 0 ; 0 20 24 148 340 1412 4032	
10 - 320	3.30104 1.23230 0.00000 0.00000 -1.00000 -1.35295 -2.18039 1 0 -10 -8 13 12 0 0 ; 0 20 24 148 340 1412 4032	
10 - 321	3.30977 1.18153 0.00000 0.00000 -0.72309 -1.76821 -2.00000 1 0 -10 -8 13 10 0 0 ; 0 20 24 148 350 1412 4172	
10 - 322	3.31444 1.04422 0.51828 -0.43689 -0.78151 -1.54519 -2.11334 1 0 -10 -8 13 10 -3 -2 ; 0 20 24 148 350 1430 4186	
10 - 323	3.31573 1.00000 0.57551 -0.35010 -1.00000 -1.39483 -2.14631 1 0 -10 -8 13 10 -4 -2 ; 0 20 24 148 350 1436 4186	
10 - 324	3.32341 1.00000 0.35793 0.00000 -1.00000 -1.68133 -2.00000 1 0 -10 -8 13 8 -4 0 ; 0 20 24 148 360 1436 4312	
10 - 325	3.33225 1.09479 0.00000 0.00000 -0.60020 -1.82684 -2.00000 1 0 -10 -8 12 8 0 0 ; 0 20 24 152 360 1472 4368	
10 - 326	3.33472 1.03563 0.19999 0.00000 -0.85767 -1.59218 -2.12050 1 0 -10 -8 12 8 -2 0 ; 0 20 24 152 360 1484 4368	
10 - 327	3.34827 1.00000 0.20359 0.00000 -1.00000 -1.30698 -2.24487 1 0 -10 -8 11 8 -2 0 ; 0 20 24 156 360 1544 4424	
10 - 328	3.35017 0.80194 0.67353 -0.55496 -0.64104 -1.38266 -2.24698 1 0 -10 -8 11 8 -3 -2 ; 0 20 24 156 360 1550 4438	
10 - 329	3.35621 0.92203 0.24747 0.00000 -0.74106 -1.65409 -2.13056 1 0 -10 -8 11 6 -2 0 ; 0 20 24 156 370 1544 4564	

Number	Eigenvalues / Characteristic polynomial coefficients and spectral moments	Graph
10 - 330	3.35921 1.03137 0.00000 0.00000 -0.74724 -1.34381 -2.29952 1 0 -10 -8 10 8 0 0 ; 0 20 24 160 360 1592 4480	
10 - 331	3.37692 0.77429 0.32936 0.00000 -0.63725 -1.70172 -2.14160 1 0 -10 -8 10 4 -2 0 ; 0 20 24 160 380 1604 4760	
10 - 332	3.37692 0.77429 0.32936 0.00000 -0.63725 -1.70172 -2.14160 1·0 -10 -8 10 4 -2 0 ; 0 20 24 160 380 1604 4760	
10 - 333	3.41421 0.58579 0.00000 0.00000 0.00000 -2.00000 -2.00000 1 0 -10 -8 8 0 0 0 ; 0 20 24 168 400 1712 5152	
10 - 334	3.40496 0.86806 0.00000 0.00000 -0.83140 -1.00000 -2.44162 1 0 -10 -8 7 6 0 0 ; 0 20 24 172 370 1772 4788	
10 - 335	3.46087 0.34933 0.00000 0.00000 0.00000 -1.33871 -2.47148 1 0 -10 -8 4 0 0 0 ; 0 20 24 184 400 1952 5376	
10 - 336	3.12489 2.00000 -0.36333 -1.00000 -1.00000 -1.00000 -1.76156 1 0 -10 -10 19 38 22 4 ; 0 20 30 124 310 1028 2982	
10 - 337	3.28191 1.58570 0.16940 -1.00000 -1.00000 -1.32580 -1.71120 1 0 -10 -10 16 26 7 -2 ; 0 20 30 136 370 1298 4074	
10 - 338	3.32341 1.41421 0.35793 -1.00000 -1.00000 -1.41421 -1.68133 1 0 -10 -10 15 22 2 -4 ; 0 20 30 140 390 1388 4438	
10 - 339	3.35706 1.37014 0.22296 -1.00000 -1.00000 -1.00000 -1.95017 1 0 -10 -10 13 20 4 -2 ; 0 20 30 148 400 1496 4704	
10 - 340	3.35706 1.37015 0.22296 -1.00000 -1.00000 -1.00000 -1.95017 1 0 -10 -10 13 20 4 -2 ; 0 20 30 148 400 1496 4704	
10 - 341	3.38996 1.21495 0.30331 -0.59231 -1.00000 -1.44394 -1.87197 1 0 -10 -10 12 16 1 -2 ; 0 20 30 152 420 1574 5054	
10 - 342	3.38896 1.33155 0.00000 -0.63868 -1.00000 -1.00000 -2.08183 1 0 -10 -10 11 18 6 0 ; 0 20 30 156 410 1604 4970	
10 - 343	3.41143 1.11724 0.35134 -0.55715 -1.00000 -1.37916 -1.94371 1 0 -10 -10 11 14 0 -2 ; 0 20 30 156 430 1640 5264	
10 - 344	3.41143 1.11725 0.35134 -0.55715 -1.00000 -1.37916 -1.94371 1 0 -10 -10 11 14 0 -2 ; 0 20 30 156 430 1640 5264	

Number	Eigenvalues / Characteristic polynomial coefficients and spectral moments	Graph
10 - 345	3.43210 1.00000 0.42019 -0.52077 -1.00000 -1.33152 -2.00000 1 0 -10 -10 10 12 -1 -2 ; 0 20 30 160 440 1706 5474	
10 - 346	3.44117 0.79348 0.61803 -0.39903 -1.00000 -1.61803 -1.83562 1 0 -10 -10 10 10 -3 -2 ; 0 20 30 160 450 1718 5614	
10 - 347	3.44949 1.00000 0.00000 0.00000 -1.00000 -1.44949 -2.00000 1 0 -10 -10 9 10 0 0 ; 0 20 30 164 450 1760 5670	
10 - 348	3.45925 0.81294 0.31468 0.00000 -1.25369 -1.39554 -1.93764 1 0 -10 -10 9 8 -3 0 ; 0 20 30 164 460 1778 5810	
10 - 349	3.47735 0.61803 0.44807 0.00000 -1.00000 -1.61803 -1.92542 1 0 -10 -10 8 6 -3 0 ; 0 20 30 168 470 1838 6020	
10 - 350	3.48767 0.69264 0.28777 0.00000 -1.00000 -1.37363 -2.09445 1 0 -10 -10 7 6 -2 0 ; 0 20 30 172 470 1892 6090	
11 - 351	3.22869 0.75866 0.00000 0.00000 0.00000 -0.75866 -3.22869 1 0 -11 0 6 0 0 0 ; 0 22 0 218 0 2266 0	
11 - 352	3.17634 0.90381 0.61803 0.00000 -0.24579 -1.61803 -2.83436 1 0 -11 -2 17 -4 -2 0 ; 0 22 6 174 130 1564 1764	
11 - 353	3.23607 1.00000 0.00000 0.00000 0.00000 -1.23607 -3.00000 1 0 -11 -2 12 0 0 0 ; 0 22 6 194 110 1882 1526	
11 - 354	3.17741 1.00000 0.67836 0.41421 -1.00000 -1.85577 -2.41421 1 0 -11 -4 23 0 -13 4 ; 0 22 12 150 220 1270 2716	
11 - 355	3.22732 1.00000 0.62907 0.00000 -0.40125 -2.00000 -2.45513 1 0 -11 -4 20 -2 -4 0 ; 0 22 12 162 230 1414 2982	
11 - 356	3.26648 0.80780 0.64395 0.00000 -0.23876 -1.94560 -2.53388 1 0 -11 -4 18 -4 -2 0 ; 0 22 12 170 240 1534 3192	
11 - 357	3.26234 1.10419 0.43077 0.00000 -0.71098 -1.30227 -2.78405 1 0 -11 -4 16 4 -4 0 ; 0 22 12 178 200 1678 2632	
11 - 358	3.30278 0.73205 0.61803 0.00000 -0.30278 -1.61803 -2.73205 1 0 -11 -4 15 -2 -2 0 ; 0 22 12 182 230 1732 3122	
11 - 359	3.26869 1.27274 0.00000 0.00000 -0.66960 -1.00000 -2.87183 1 0 -11 -4 14 8 0 0 ; 0 22 12 186 180 1786 2380	

Number	Eigenvalues Characteristic polynomial coefficients and spectral moments	Graph
11 - 360	3.32308 0.80377 0.48379 0.00000 -0.39385 -1.39033 -2.82646 1 0 -11 -4 13 0 -2 0; 0 22 12 190 220 1864 3024	
11 - 361	3.33473 0.90698 0.00000 0.00000 0.00000 -1.39252 -2.84919 1 0 -11 -4 12 0 0 0; 0 22 12 194 220 1918 3052	
11 - 362	3.41421 0.58579 0.00000 0.00000 0.00000 -1.00000 -3.00000 1 0 -11 -4 6 0 0 0; 0 22 12 218 220 2314 3220	
11 - 363	3.17812 1.48803 0.61803 -0.14803 -1.27179 -1.61803 -2.24633 1 0 -11 -6 25 14 -12 -2; 0 22 18 142 260 1192 2968	
11 - 364	3.23607 1.00000 1.00000 0.00000 -1.23607 -2.00000 -2.00000 1 0 -11 -6 24 8 -16 0; 0 22 18 146 290 1282 3458	
11 - 365	3.24015 1.43518 0.26350 0.00000 -0.74350 -2.00000 -2.19533 1 0 -11 -6 22 10 -4 0; 0 22 18 154 280 1342 3388	
11 - 366	3.25083 1.26683 0.68473 -0.18075 -1.00000 -1.66517 -2.35647 1 0 -11 -6 22 10 -10 -2; 0 22 18 154 280 1378 3402	
11 - 367	3.26162 1.20334 0.65233 0.00000 -1.06579 -1.75650 -2.29500 1 0 -11 -6 22 8 -11 0; 0 22 18 154 290 1384 3542	
11 - 368	3.27366 1.08350 0.67476 0.17713 -1.17363 -1.79092 -2.24449 1 0 -11 -6 22 6 -13 2; 0 22 18 154 300 1396 3682	
11 - 369	3.24680 1.52045 0.00000 0.00000 -0.75236 -1.49714 -2.51775 1 0 -11 -6 20 14 0 0; 0 22 18 162 260 1450 3164	
11 - 370	3.29012 1.14383 0.70225 -0.22879 -0.82287 -1.65343 -2.43111 1 0 -11 -6 20 8 -8 -2; 0 22 18 162 290 1498 3640	
11 - 371	3.30150 1.00000 0.75988 0.00000 -1.00000 -1.65970 -2.40168 1 0 -11 -6 20 6 -10 0; 0 22 18 162 300 1510 3780	
11 - 372	3.30473 1.16001 0.53969 0.00000 -1.12548 -1.36854 -2.51042 1 0 -11 -6 19 8 -8 0; 0 22 18 166 290 1564 3668	
11 - 373	3.30474 1.16001 0.53969 0.00000 -1.12548 -1.36854 -2.51042 1 0 -11 -6 19 8 -8 0; 0 22 18 166 290 1564 3668	
11 - 374	3.31390 1.06503 0.61803 0.00000 -0.92290 -1.61803 -2.45603 1 0 -11 -6 19 6 -8 0; 0 22 18 166 300 1564 3822	

Number	Eigenvalues Characteristic polynomial coefficients and spectral moments	Graph
11 - 375	3.31390 1.06503 0.61803 0.00000 -0.92290 -1.61803 -2.45602 1 0 -11 -6 19 6 -8 0; 0 22 18 166 300 1564 3822	
11 - 376	3.33220 0.77366 0.70988 0.27958 -0.89729 -1.87651 -2.32151 1 0 -11 -6 19 2 -9 2; 0 22 18 166 320 1570 4116	
11 - 377	3.33807 1.12925 0.36714 0.00000 -0.72674 -1.56269 -2.54503 1 0 -11 -6 17 6 -4 0; 0 22 18 174 300 1672 3906	
11 - 378	3.35386 1.00000 0.41421 0.00000 -0.47645 -1.87740 -2.41421 1 0 -11 -6 17 2 -3 0; 0 22 18 174 320 1666 4214	
11 - 379	3.34386 1.15896 0.32750 0.00000 -1.00000 -1.19667 -2.63365 1 0 -11 -6 16 8 -4 0; 0 22 18 178 290 1738 3794	
11 - 380	3.36898 0.87315 0.55525 0.00000 -0.56701 -1.72161 -2.50876 1 0 -11 -6 16 2 -4 0; 0 22 18 178 320 1738 4256	
11 - 381	3.36007 1.00000 0.65576 -0.33910 -1.00000 -1.00000 -2.67673 1 0 -11 -6 15 8 -5 -2; 0 22 18 182 290 1810 3850	
11 - 382	3.37771 1.07591 0.22845 0.00000 -0.67677 -1.33105 -2.67425 1 0 -11 -6 14 6 -2 0; 0 22 18 186 300 1858 4032	
11 - 383	3.40666 0.85320 0.36825 0.00000 -0.46706 -1.50791 -2.65315 1 0 -11 -6 13 2 -2 0; 0 22 18 190 320 1924 4382	
11 - 384	3.39434 1.12631 0.00000 0.00000 -0.75724 -1.00000 -2.76342 1 0 -11 -6 12 8 0 0; 0 22 18 194 290 1978 3962	
11 - 385	3.42788 1.00000 0.00000 0.00000 -0.62434 -1.00000 -2.80354 1 0 -11 -6 10 6 0 0; 0 22 18 202 300 2110 4200	
11 - 386	3.43730 0.82191 0.32096 0.00000 -0.79069 -1.00000 -2.78949 1 0 -11 -6 10 4 -2 0; 0 22 18 202 310 2122 4354	
11 - 387	3.46153 0.68240 0.00000 0.00000 0.00000 -1.37699 -2.76695 1 0 -11 -6 9 0 0 0; 0 22 18 206 330 2176 4704	
11 - 388	3.17741 1.73205 0.67836 -1.00000 -1.00000 -1.73205 -1.85577 1 0 -11 -8 27 28 -9 -12; 0 22 24 134 300 1126 3192	
11 - 389	3.26473 1.53781 0.64907 -0.70130 -1.00000 -1.75031 -2.00000 1 0 -11 -8 24 22 -8 -8; 0 22 24 146 330 1318 3794	

Number	Eigenvalues Characteristic polynomial coefficients and spectral moments	Graph
11 - 390	3.29588 1.24698 0.93618 -0.44504 -1.47887 -1.75319 -1.80194 1 0 -11 -8 24 18 -15 -8 ; 0 22 24 146 350 1360 4102	
11 - 391	3.30104 1.23230 1.00000 -1.00000 -1.00000 -1.35295 -2.18039 1 0 -11 -8 23 20 -13 -12 ; 0 22 24 150 340 1414 4032	
11 - 392	3.25542 1.73501 0.20002 -0.60043 -1.00000 -1.27189 -2.31813 1 0 -11 -8 22 26 4 -2 ; 0 22 24 154 310 1378 3556	
11 - 393	3.30219 1.56926 0.39357 -0.65671 -1.00000 -1.28604 -2.32226 1 0 -11 -8 21 22 -1 -4 ; 0 22 24 158 330 1474 3934	
11 - 394	3.32481 1.46701 0.38134 -0.27546 -1.14089 -1.55144 -2.20537 1 0 -11 -8 21 18 -4 -2 ; 0 22 24 158 350 1492 4228	
11 - 395	3.34242 1.26032 0.74513 -0.53593 -1.00000 -1.64910 -2.16284 1 0 -11 -8 21 16 -9 -6 ; 0 22 24 158 360 1522 4410	
11 - 396	3.34587 1.06297 1.00000 -0.66052 -1.00000 -1.54701 -2.20130 1 0 -11 -8 21 16 -11 -8 ; 0 22 24 158 360 1534 4424	
11 - 397	3.36147 1.16745 0.61803 0.00000 -1.52892 -1.61803 -2.00000 1 0 -11 -8 21 12 -12 0 ; 0 22 24 158 380 1540 4676	
11 - 398	3.34561 1.44343 0.11051 0.00000 -1.00000 -1.71738 -2.18216 1 0 -11 -8 20 16 -2 0 ; 0 22 24 162 360 1546 4424	
11 - 399	3.36499 1.18727 0.71610 -0.35511 -1.13844 -1.56517 -2.20964 1 0 -11 -8 20 14 -9 -4 ; 0 22 24 162 370 1588 4606	
11 - 400	3.38393 1.00000 0.74244 0.00000 -1.32791 -1.79846 -2.00000 1 0 -11 -8 20 10 -12 0 ; 0 22 24 162 390 1606 4886	
11 - 401	3.33939 1.51394 0.26707 -0.50975 -1.00000 -1.21084 -2.39981 1 0 -11 -8 19 20 1 -2 ; 0 22 24 166 340 1594 4186	
11 - 402	3.38548 1.15322 0.61803 -0.20786 -1.10946 -1.61803 -2.22138 1 0 -11 -8 19 12 -8 -2 ; 0 22 24 166 380 1648 4802	
11 - 403	3.39526 1.05768 0.61803 0.00000 -1.28369 -1.61803 -2.16925 1 0 -11 -8 19 10 -10 0 ; 0 22 24 166 390 1660 4942	
11 - 404	3.38388 1.31684 0.38026 -0.46208 -0.67842 -1.62871 -2.31176 1 0 -11 -8 18 14 -2 -2 ; 0 22 24 170 370 1678 4704	

Number	Eigenvalues Characteristic polynomial coefficients and spectral moments	Graph
11 - 405	3.39327 1.24698 0.43557 -0.44504 -0.60980 -1.80194 -2.21904 1 0 -11 -8 18 12 -3 -2 ; 0 22 24 170 380 1684 4858	
11 - 406	3.39495 1.23855 0.30433 0.00000 -1.11398 -1.52731 -2.29653 1 0 -11 -8 18 12 -5 0 ; 0 22 24 170 380 1696 4844	
11 - 407	3.40595 1.05723 0.65346 -0.24127 -0.90475 -1.76721 -2.20341 1 0 -11 -8 18 10 -7 -2 ; 0 22 24 170 390 1708 5012	
11 - 408	3.39890 1.26637 0.41421 -0.35449 -1.00000 -1.31078 -2.41421 1 0 -11 -8 17 14 -3 -2 ; 0 22 24 174 370 1750 4760	
11 - 409	3.42617 1.00000 0.53166 0.00000 -1.00000 -1.70829 -2.24954 1 0 -11 -8 17 8 -7 0 ; 0 22 24 174 400 1774 5208	
11 - 410	3.43629 0.76630 0.61803 0.27353 -1.24399 -1.61803 -2.23213 1 0 -11 -8 17 6 -10 2 ; 0 22 24 174 410 1792 5348	
11 - 411	3.40826 1.31765 0.00000 0.00000 -1.00000 -1.26867 -2.45724 1 0 -11 -8 16 14 0 0 ; 0 22 24 178 370 1798 4802	
11 - 412	3.41611 1.27422 0.00000 0.00000 -0.74131 -1.55041 -2.39860 1 0 -11 -8 16 12 0 0 ; 0 22 24 178 380 1798 4956	
11 - 413	3.41837 1.23722 0.14205 0.00000 -1.00000 -1.37309 -2.42455 1 0 -11 -8 16 12 -2 0 ; 0 22 24 178 380 1810 4956	
11 - 414	3.42890 1.08538 0.53076 -0.37870 -0.71344 -1.57784 -2.37505 1 0 -11 -8 16 10 -4 -2 ; 0 22 24 178 390 1822 5124	
11 - 415	3.43359 1.13189 0.18779 0.00000 -0.66128 -1.84047 -2.25153 1 0 -11 -8 16 8 -2 0 ; 0 22 24 178 400 1810 5264	
11 - 416	3.43468 1.10509 0.26208 0.00000 -0.74413 -1.77704 -2.28067 1 0 -11 -8 16 8 -3 0 ; 0 22 24 178 400 1816 5264	
11 - 417	3.43601 1.00000 0.71582 -0.65006 -1.00000 -1.00000 -2.50177 1 0 -11 -8 15 12 -5 -4 ; 0 22 24 182 380 1894 5040	
11 - 418	3.45010 0.85121 0.72819 -0.34476 -0.72532 -1.55628 -2.40313 1 0 -11 -8 15 8 -5 -2 ; 0 22 24 182 400 1894 5334	
11 - 419	3.45670 0.90214 0.49202 0.00000 -0.82729 -1.68288 -2.34069 1 0 -11 -8 15 6 -5 0 ; 0 22 24 182 410 1894 5474	

Number	Eigenvalues Characteristic polynomial coefficients and spectral moments	Graph
11 - 420	3.47290 0.92125 0.28279 0.00000 -0.51584 -1.87220 -2.28891 1 0 -11 -8 14 4 -2 0 ; 0 22 24 186 420 1942 5684	
11 - 421	3.48113 1.05911 0.00000 0.00000 -0.62149 -1.36966 -2.54909 1 0 -11 -8 12 8 0 0 ; 0 22 24 194 400 2062 5488	
11 - 422	3.50816 0.79290 0.31582 0.00000 -0.61029 -1.47152 -2.53507 1 0 -11 -8 11 4 -2 0 ; 0 22 24 198 420 2140 5852	
11 - 423	3.50466 1.00000 0.00000 0.00000 -0.86464 -1.00000 -2.64002 1 0 -11 -8 10 8 0 0 ; 0 22 24 202 400 2194 5600	
11 - 424	3.51952 0.74542 0.33119 0.00000 -0.65853 -1.35170 -2.58590 1 0 -11 -8 10 4 -2 0 ; 0 22 24 202 420 2206 5908	
11 - 425	3.53040 0.66552 0.00000 0.00000 0.00000 -1.71682 -2.47910 1 0 -11 -8 10 0 0 0 ; 0 22 24 202 440 2194 6216	
11 - 426	3.31364 1.83604 0.00000 -0.66364 -1.00000 -1.48604 -2.00000 1 0 -11 -10 22 34 12 0 ; 0 22 30 154 380 1438 4312	
11 - 427	3.35386 1.73205 0.00000 -0.47645 -1.00000 -1.73205 -1.87740 1 0 -11 -10 21 30 9 0 ; 0 22 30 158 400 1522 4690	
11 - 428	3.39403 1.59296 0.20119 -0.55858 -1.00000 -1.77268 -1.85692 1 0 -11 -10 20 26 4 -2 ; 0 22 30 162 420 1618 5082	
11 - 429	3.41421 1.41421 0.58579 -1.00000 -1.00000 -1.41421 -2.00000 1 0 -11 -10 20 24 -4 -8 ; 0 22 30 162 430 1666 5278	
11 - 430	3.44494 1.27878 0.61803 -0.57166 -1.25772 -1.61804 -1.89434 1 0 -11 -10 19 20 -6 -6 ; 0 22 30 166 450 1744 5642	
11 - 431	3.44667 1.21704 0.73314 -0.81139 -1.00000 -1.70427 -1.88119 1 0 -11 -10 19 20 -7 -8 ; 0 22 30 166 450 1750 5656	
11 - 432	3.41678 1.59461 0.00000 -0.51403 -1.00000 -1.29990 -2.19747 1 0 -11 -10 18 26 8 0 ; 0 22 30 170 420 1726 5208	
11 - 433	3.46357 1.24698 0.53787 -0.44504 -1.17662 -1.80194 -1.82482 1 0 -11 -10 18 18 -5 -4 ; 0 22 30 170 460 1804 5852	
11 - 434	3.44839 1.49175 0.00000 -0.45245 -0.75647 -1.59600 -2.13522 1 0 -11 -10 17 22 6 0 ; 0 22 30 174 440 1804 5586	

Number	Eigenvalues Characteristic polynomial coefficients and spectral moments	Graph
11 - 435	3.49265 1.04720 0.61803 -0.19898 -1.46484 -1.61803 -1.87602 1 0 -11 -10 17 14 -8 -2 ; 0 22 30 174 480 1888 6216	
11 - 436	3.47359 1.37157 0.25349 -0.57554 -1.00000 -1.28652 -2.23659 1 0 -11 -10 16 20 2 -2 ; 0 22 30 178 450 1894 5824	
11 - 437	3.48929 1.28917 0.00000 0.00000 -1.00000 -1.77846 -2.00000 1 0 -11 -10 16 16 0 0 ; 0 22 30 178 470 1906 6118	
11 - 438	3.48929 1.28917 0.00000 0.00000 -1.00000 -1.77846 -2.00000 1 0 -11 -10 16 16 0 0 ; 0 22 30 178 470 1906 6118	
11 - 439	3.48929 1.28917 0.00000 0.00000 -1.00000 -1.77846 -2.00000 1 0 -11 -10 16 16 0 0 ; 0 22 30 178 470 1906 6118	
11 - 440	3.49086 1.24698 0.34338 -0.44504 -0.83424 -1.80194 -2.00000 1 0 -11 -10 16 16 -1 -2 ; 0 22 30 178 470 1912 6132	
11 - 441	3.51012 1.18677 0.12720 0.00000 -1.14623 -1.54117 -2.13668 1 0 -11 -10 15 14 -2 0 ; 0 22 30 182 480 1984 6342	
11 - 442	3.51162 1.12809 0.43078 -0.34548 -1.00000 -1.58575 -2.13926 1 0 -11 -10 15 14 -3 -2 ; 0 22 30 182 480 1990 6356	
11 - 443	3.51257 1.10240 0.46718 -0.29993 -1.23337 -1.37444 -2.17441 1 0 -11 -10 15 14 -4 -2 ; 0 22 30 182 480 1996 6356	
11 - 444	3.52015 1.00000 0.55754 -0.28408 -1.00000 -1.79360 -2.00000 1 0 -11 -10 15 12 -5 -2 ; 0 22 30 182 490 2002 6510	
11 - 445	3.52892 0.83255 0.61803 0.00000 -1.36147 -1.61803 -2.00000 1 0 -11 -10 15 10 -8 0 ; 0 22 30 182 500 2020 6650	
11 - 446	3.53663 0.61803 0.61803 0.30677 -1.61803 -1.61803 -1.84341 1 0 -11 -10 15 8 -10 2 ; 0 22 30 182 510 2032 6790	
11 - 447	3.51382 1.21087 0.32348 -0.51040 -1.00000 -1.23795 -2.29982 1 0 -11 -10 14 16 0 -2 ; 0 22 30 186 470 2038 6272	
11 - 448	3.52980 1.03513 0.47377 -0.38361 -0.84300 -1.66040 -2.15170 1 0 -11 -10 14 12 -3 -2 ; 0 22 30 186 490 2056 6580	
11 - 449	3.53663 1.00000 0.30678 0.00000 -1.00000 -1.84341 -2.00000 1 0 -11 -10 14 10 -4 0 ; 0 22 30 186 500 2062 6720	

Number	Eigenvalues Characteristic polynomial coefficients and spectral moments	Graph
11 - 450	3.53663 1.00000 0.30677 0.00000 -1.00000 -1.84340 -2.00000 1 0 -11 -10 14 10 -4 0 ;　　0 22 30 186 500 2062 6720	
11 - 451	3.53292 1.11583 0.37491 -0.47705 -1.00000 -1.21834 -2.32828 1 0 -11 -10 13 14 -1 -2 ;　　0 22 30 190 480 2110 6496	
11 - 452	3.53793 1.12751 0.00000 0.00000 -0.85421 -1.57442 -2.23681 1 0 -11 -10 13 12 0 0 ;　　0 22 30 190 490 2104 6636	
11 - 453	3.54695 1.00000 0.24185 0.00000 -1.00000 -1.59180 -2.19700 1 0 -11 -10 13 10 -3 0 ;　　0 22 30 190 500 2122 6790	
11 - 454	3.55539 1.05298 0.00000 0.00000 -0.72836 -1.62963 -2.25038 1 0 -11 -10 12 10 0 0 ;　　0 22 30 194 500 2170 6860	
11 - 455	3.55849 0.87201 0.56118 -0.42944 -0.78788 -1.47865 -2.29571 1 0 -11 -10 12 10 -3 -2 ;　　0 22 30 194 500 2188 6874	
11 - 456	3.55935 0.76853 0.67739 -0.34692 -1.00000 -1.34472 -2.31363 1 0 -11 -10 12 10 -4 -2 ;　　0 22 30 194 500 2194 6874	
11 - 457	3.56155 1.00000 0.00000 0.00000 -0.56155 -2.00000 -2.00000 1 0 -11 -10 12 8 0 0 ;　　0 22 30 194 510 2170 7014	
11 - 458	3.56583 0.77511 0.48942 0.00000 -1.16109 -1.40760 -2.26166 1 0 -11 -10 12 8 -5 0 ;　　0 22 30 194 510 2200 7014	
11 - 459	3.57405 0.90530 0.20530 0.00000 -0.85954 -1.51890 -2.30620 1 0 -11 -10 11 8 -2 0 ;　　0 22 30 198 510 2248 7084	
11 - 460	3.59630 0.67407 0.34402 0.00000 -0.60194 -1.80419 -2.20827 1 0 -11 -10 10 4 -2 0 ;　　0 22 30 202 530 2314 7462	
11 - 461	3.59930 0.85386 0.00000 0.00000 -0.53157 -1.54606 -2.37553 1 0 -11 -10 9 6 0 0 ;　　0 22 30 206 520 2368 7378	
11 - 462	3.70156 0.00000 0.00000 0.00000 0.00000 -1.00000 -2.70156 1 0 -11 -10 0 0 0 0 ;　　0 22 30 242 550 2962 8470	
11 - 463	3.35386 2.00000 -0.47645 -1.00000 -1.00000 -1.00000 -1.87740 1 0 -11 -12 21 46 29 6 ;　　0 22 36 158 430 1534 4816	
11 - 464	3.42194 1.84662 -0.32584 -1.00000 -1.00000 -1.00000 -1.94272 1 0 -11 -12 19 40 23 4 ;　　0 22 36 166 460 1702 5460	

Number	Eigenvalues Characteristic polynomial coefficients and spectral moments	Graph
11 - 465	3.47193 1.67912 0.00000 -1.00000 -1.00000 -1.29122 -1.85984 1 0 -11 -12 18 34 14 0; 0 22 36 170 490 1822 6034	
11 - 466	3.50535 1.55651 0.14599 -1.00000 -1.00000 -1.35557 -1.85228 1 0 -11 -12 17 30 9 -2; 0 22 36 174 510 1918 6440	
11 - 467	3.55336 1.38804 0.19638 -0.69991 -1.00000 -1.65150 -1.78636 1 0 -11 -12 15 24 5 -2; 0 22 36 182 540 2074 7070	
11 - 468	3.55568 1.34708 0.33197 -1.00000 -1.00000 -1.30069 -1.93405 1 0 -11 -12 15 24 3 -4; 0 22 36 182 540 2086 7084	
11 - 469	3.57189 1.32266 0.21598 -0.67325 -1.00000 -1.51268 -1.92460 1 0 -11 -12 14 22 4 -2; 0 22 36 186 550 2146 7308	
11 - 470	3.58065 1.24698 0.27552 -0.44504 -1.31808 -1.53809 -1.80194 1 0 -11 -12 14 20 1 -2; 0 22 36 186 560 2164 7462	
11 - 471	3.58207 1.31352 0.20259 -1.00000 -1.00000 -1.00000 -2.09819 1 0 -11 -12 13 22 5 -2; 0 22 36 190 550 2206 7392	
11 - 472	3.59959 1.11345 0.46863 -0.71915 -1.00000 -1.54168 -1.92084 1 0 -11 -12 13 18 -1 -4; 0 22 36 190 570 2242 7714	
11 - 473	3.61467 1.09992 0.33093 -0.48072 -1.00000 -1.66033 -1.90447 1 0 -11 -12 12 16 0 -2; 0 22 36 194 580 2302 7938	
11 - 474	3.61467 1.09992 0.33093 -0.48072 -1.00000 -1.66034 -1.90447 1 0 -11 -12 12 16 0 -2; 0 22 36 194 580 2302 7938	
11 - 475	3.62421 1.09575 0.30725 -0.58062 -1.00000 -1.34025 -2.10633 1 0 -11 -12 11 16 1 -2; 0 22 36 198 580 2362 8022	
11 - 476	3.63920 0.77360 0.61803 -0.31192 -1.19498 -1.61803 -1.90591 1 0 -11 -12 11 12 -4 -2; 0 22 36 198 600 2392 8330	
11 - 477	3.65338 0.84960 0.25874 0.00000 -1.14153 -1.74644 -1.87375 1 0 -11 -12 10 10 -3 0; 0 22 36 202 610 2452 8554	
11 - 478	3.64989 1.00000 0.33064 -0.73953 -1.00000 -1.00000 -2.24100 1 0 -11 -12 9 14 1 -2; 0 22 36 206 590 2494 8344	
11 - 479	3.66908 0.61803 0.47602 0.00000 -1.14510 -1.61803 -2.00000 1 0 -11 -12 9 8 -4 0; 0 22 36 206 620 2524 8792	

Number	Eigenvalues Characteristic polynomial coefficients and spectral moments	Graph
11 - 480	3.67059 0.90139 0.00000 0.00000 -1.00000 -1.37689 -2.19509 1 0 -11 -12 8 10 0 0; 0 22 36 210 610 2566 8722	
11 - 481	3.70090 0.57300 0.31439 0.00000 -1.00000 -1.32605 -2.26224 1 0 -11 -12 6 6 -2 0; 0 22 36 218 630 2710 9198	
11 - 482	3.69400 1.25200 0.00000 -1.00000 -1.00000 -1.00000 -1.94600 1 0 -11 -14 9 22 9 0; 0 22 42 206 660 2602 9282	
11 - 483	3.71565 1.13709 0.00000 -0.51704 -1.00000 -1.48048 -1.85522 1 0 -11 -14 8 18 6 0; 0 22 42 210 680 2686 9688	
11 - 484	3.73205 1.00000 0.26795 -1.00000 -1.00000 -1.00000 -2.00000 1 0 -11 -14 7 16 3 -2; 0 22 42 214 690 2770 9954	
11 - 485	3.73878 1.06877 0.00000 -0.71890 -1.00000 -1.00000 -2.08865 1 0 -11 -14 6 16 6 0; 0 22 42 218 690 2818 10038	
11 - 486	3.75239 0.77780 0.42212 -0.58640 -1.00000 -1.43004 -1.93586 1 0 -11 -14 6 12 0 -2; 0 22 42 218 710 2854 10360	
11 - 487	3.77115 0.61803 0.24297 0.00000 -1.20959 -1.61803 -1.80452 1 0 -11 -14 5 8 -2 0; 0 22 42 222 730 2932 10752	
11 - 488	3.77846 0.71083 0.00000 0.00000 -1.00000 -1.48929 -2.00000 1 0 -11 -14 4 8 0 0; 0 22 42 226 730 2986 10850	
12 - 489	3.46410 0.00000 0.00000 0.00000 0.00000 0.00000 -3.46410 1 0 -12 0 0 0 0 0; 0 24 0 288 0 3456 0	
12 - 490	3.48715 0.87045 0.00000 0.00000 0.00000 -1.28782 -3.06978 1 0 -12 -4 12 0 0 0; 0 24 12 240 240 2640 3696	
12 - 491	3.47419 1.11268 0.00000 0.00000 0.00000 -2.00000 -2.58687 1 0 -12 -6 20 0 0 0; 0 24 18 208 360 2124 5208	
12 - 492	3.50322 0.73205 0.70177 0.00000 -0.46848 -1.73651 -2.73205 1 0 -12 -6 18 0 -4 0; 0 24 18 216 360 2292 5292	
12 - 493	3.56959 0.83639 0.30890 0.00000 -0.72542 -1.00000 -2.98946 1 0 -12 -6 11 4 -2 0; 0 24 18 244 340 2784 5250	
12 - 494	3.59101 0.70745 0.00000 0.00000 0.00000 -1.32286 -2.97560 1 0 -12 -6 10 0 0 0; 0 24 18 248 360 2844 5628	

Number	Eigenvalues Characteristic polynomial coefficients and spectral moments	Graph
12 - 495	3.48803 1.16397 0.63014 0.00000 -1.05132 -2.00000 -2.23081 1 0 -12 -8 24 10 -12 0 ; 0 24 24 192 430 1992 5880	
12 - 496	3.50790 0.75564 0.61803 0.61803 -1.61803 -1.61803 -2.26354 1 0 -12 -8 24 6 -19 6 ; 0 24 24 192 450 2034 6174	
12 - 497	3.48310 1.41421 0.20013 0.00000 -1.11881 -1.41421 -2.56442 1 0 -12 -8 22 16 -4 0 ; 0 24 24 200 400 2088 5488	
12 - 498	3.50367 1.19137 0.61803 -0.19137 -1.00000 -1.61803 -2.50367 1 0 -12 -8 22 12 -9 -2 ; 0 24 24 200 420 2118 5838	
12 - 499	3.52624 1.32069 0.12345 0.00000 -1.00000 -1.30537 -2.66500 1 0 -12 -8 19 14 -2 0 ; 0 24 24 212 410 2292 5824	
12 - 500	3.55220 1.00000 0.56857 0.00000 -1.00000 -1.52735 -2.59342 1 0 -12 -8 19 8 -8 0 ; 0 24 24 212 440 2328 6328	
12 - 501	3.55220 1.00000 0.56857 0.00000 -1.00000 -1.52735 -2.59342 1 0 -12 -8 19 8 -8 0 ; 0 24 24 212 440 2328 6328	
12 - 502	3.55872 0.85455 0.69955 0.00000 -0.87072 -1.69674 -2.54535 1 0 -12 -8 19 6 -8 0 ; 0 24 24 212 450 2328 6496	
12 - 503	3.55872 0.85455 0.69955 0.00000 -0.87072 -1.69674 -2.54535 1 0 -12 -8 19 6 -8 0 ; 0 24 24 212 450 2328 6496	
12 - 504	3.56885 0.80194 0.63555 0.00000 -0.55496 -2.20440 -2.24698 1 0 -12 -8 19 2 -5 0 ; 0 24 24 212 470 2310 6832	
12 - 505	3.55686 1.22242 0.14083 0.00000 -1.00000 -1.20145 -2.71866 1 0 -12 -8 17 12 -2 0 ; 0 24 24 220 420 2436 6104	
12 - 506	3.59516 0.86183 0.46899 0.00000 -0.62979 -1.65468 -2.64150 1 0 -12 -8 16 4 -4 0 ; 0 24 24 224 460 2520 6832	
12 - 507	3.59222 1.06205 0.19072 0.00000 -0.72901 -1.37624 -2.73974 1 0 -12 -8 15 8 -2 0 ; 0 24 24 228 440 2580 6552	
12 - 508	3.61507 0.88891 0.28351 0.00000 -0.52257 -1.54385 -2.72106 1 0 -12 -8 14 4 -2 0 ; 0 24 24 232 460 2652 6944	
12 - 509	3.62512 0.81305 0.00000 0.00000 0.00000 -1.80148 -2.63669 1 0 -12 -8 14 0 0 0 ; 0 24 24 232 480 2640 7280	

Number	Eigenvalues Characteristic polynomial coefficients and spectral moments	Graph
12 - 510	3.61393　1.00000　0.19689　0.00000　-1.00000　-1.00000　-2.81082 1 0 -12 -8　13　8 -2　0 ;　　0　24　24　236　440　2724　6664	
12 - 511	3.64575　0.73205　0.00000　0.00000　0.00000　-1.64575　-2.73205 1 0 -12 -8　12　0　0　0 ;　　0　24　24　240　480　2784　7392	
12 - 512	3.65946　0.88249　0.00000　0.00000　-0.64029　-1.00000　-2.90166 1 0 -12 -8　9　6　0　0 ;　　0　24　24　252　450　3000　7056	
12 - 513	3.68534　0.55299　0.00000　0.00000　0.00000　-1.36726　-2.87108 1 0 -12 -8　8　0　0　0 ;　　0　24　24　256　480　3072　7616	
12 - 514	3.48316　1.54736　0.61803　-0.85147　-1.00000　-1.61803　-2.17905 1 0 -12 -10　26　28 -7 -10 ;　　0　24　30　184　460　1926　5978	
12 - 515	3.50323　1.41421　0.70177　-0.46848　-1.41421　-1.73651　-2.00000 1 0 -12 -10　26　24 -12 -8 ;　　0　24　30　184　480　1956　6300	
12 - 516	3.51414　1.57199　0.00000　0.00000　-1.08613　-2.00000　-2.00000 1 0 -12 -10　24　24　0　0 ;　　0　24　30　192　480　2028　6384	
12 - 517	3.55147　1.29678　0.65846　-0.46626　-1.12680　-1.70532　-2.20832 1 0 -12 -10　23　20 -9 -6 ;　　0　24　30　196　500　2154　6832	
12 - 518	3.55391　1.20883　0.80194　-0.55496　-1.16683　-1.59592　-2.24698 1 0 -12 -10　23　20 -11 -8 ;　　0　24　30　196　500　2166　6846	
12 - 519	3.56583　1.42134　0.28353　-0.46454　-0.73642　-1.76593　-2.30382 1 0 -12 -10　21　20　0 -2 ;　　0　24　30　204　500　2244　6944	
12 - 520	3.59509　1.05584　0.69774　-0.15803　-1.25448　-1.71476　-2.22140 1 0 -12 -10　21　14 -11 -2 ;　　0　24　30　204　530　2310　7448	
12 - 521	3.58903　1.26018　0.52485　-0.46570　-1.00000　-1.50648　-2.40188 1 0 -12 -10　20　18 -5 -4 ;　　0　24　30　208　510　2346　7196	
12 - 522	3.60148　1.20846　0.31659　0.00000　-1.07475　-1.79657　-2.25521 1 0 -12 -10　20　14 -6　0 ;　　0　24　30　208　530　2352　7504	
12 - 523	3.61186　1.00000　0.61803　0.00000　-1.34133　-1.61803　-2.27053 1 0 -12 -10　20　12 -11　0 ;　　0　24　30　208　540　2382　7672	
12 - 524	3.61908　0.88138　0.61803　0.19166　-1.47670　-1.61803　-2.21541 1 0 -12 -10　20　10 -13　2 ;　　0　24　30　208　550　2394　7826	

Number	Eigenvalues Characteristic polynomial coefficients and spectral moments	Graph
12 - 525	3.56155 1.56155 0.00000 -0.56155 -1.00000 -1.00000 -2.56155 1 0 -12 -10 19 26 8 0 ; 0 24 30 212 470 2340 6566	
12 - 526	3.62071 1.00000 0.65386 -0.20818 -1.00000 -1.75764 -2.30875 1 0 -12 -10 19 12 -8 -2 ; 0 24 30 212 540 2436 7756	
12 - 527	3.61372 1.25841 0.33785 -0.50373 -0.66256 -1.58890 -2.45479 1 0 -12 -10 18 16 -1 -2 ; 0 24 30 216 520 2466 7490	
12 - 528	3.61409 1.26764 0.11191 0.00000 -1.16580 -1.35104 -2.47679 1 0 -12 -10 18 16 -2 0 ; 0 24 30 216 520 2472 7476	
12 - 529	3.63363 1.09187 0.28488 0.00000 -0.80085 -2.00000 -2.20953 1 0 -12 -10 18 10 -4 0 ; 0 24 30 216 550 2484 7980	
12 - 530	3.64261 0.82396 0.65128 0.00000 -0.93932 -2.00000 -2.17853 1 0 -12 -10 18 8 -8 0 ; 0 24 30 216 560 2508 8148	
12 - 531	3.65109 0.80194 0.72611 -0.37720 -0.55496 -2.00000 -2.24698 1 0 -12 -10 17 8 -5 -2 ; 0 24 30 220 560 2562 8232	
12 - 532	3.64854 1.06477 0.45777 -0.41028 -0.70646 -1.54828 -2.50605 1 0 -12 -10 16 12 -3 -2 ; 0 24 30 224 540 2622 7966	
12 - 533	3.64889 1.08704 0.26164 0.00000 -1.18761 -1.28566 -2.52431 1 0 -12 -10 16 12 -4 0 ; 0 24 30 224 540 2628 7952	
12 - 534	3.66247 0.78191 0.61803 0.00000 -1.00000 -1.61803 -2.44438 1 0 -12 -10 16 8 -7 0 ; 0 24 30 224 560 2646 8288	
12 - 535	3.67033 0.93164 0.34762 0.00000 -0.82653 -1.63906 -2.48399 1 0 -12 -10 15 8 -4 0 ; 0 24 30 228 560 2700 8358	
12 - 536	3.68340 1.00000 0.16876 0.00000 -1.00000 -1.22448 -2.62768 1 0 -12 -10 13 10 -2 0 ; 0 24 30 236 550 2832 8330	
12 - 537	3.68876 0.94074 0.19820 0.00000 -0.75918 -1.47929 -2.58923 1 0 -12 -10 13 8 -2 0 ; 0 24 30 236 560 2832 8498	
12 - 538	3.70891 0.74346 0.31205 0.00000 -0.54779 -1.65899 -2.55764 1 0 -12 -10 12 4 -2 0 ; 0 24 30 240 580 2904 8904	
12 - 539	3.55503 1.74772 0.11890 -1.00000 -1.00000 -1.24217 -2.17948 1 0 -12 -12 23 38 12 -2 ; 0 24 36 196 530 2160 6986	

Number	Eigenvalues Characteristic polynomial coefficients and spectral moments	Graph
12 - 540	3.62544 1.33374 0.61803 -0.58653 -1.53486 -1.61803 -1.83779 1 0 -12 -12 22 26 -7 -8 ; 0 24 36 200 590 2346 8120	
12 - 541	3.60344 1.60547 0.15085 -0.72487 -1.00000 -1.44114 -2.19376 1 0 -12 -12 21 32 8 -2 ; 0 24 36 204 560 2328 7658	
12 - 542	3.64575 1.00000 1.00000 -1.00000 -1.00000 -1.64575 -2.00000 1 0 -12 -12 21 24 -10 -12 ; 0 24 36 204 600 2436 8400	
12 - 543	3.64575 1.00000 1.00000 -1.00000 -1.00000 -1.64575 -2.00000 1 0 -12 -12 21 24 -10 -12 ; 0 24 36 204 600 2436 8400	
12 - 544	3.63001 1.50738 0.18539 -0.57455 -1.00000 -1.58973 -2.15850 1 0 -12 -12 20 28 5 -2 ; 0 24 36 208 580 2418 8078	
12 - 545	3.64405 1.34011 0.55472 -1.00000 -1.00000 -1.34785 -2.19103 1 0 -12 -12 20 26 -3 -8 ; 0 24 36 208 590 2466 8288	
12 - 546	3.64575 1.41421 0.00000 0.00000 -1.41421 -1.64575 -2.00000 1 0 -12 -12 20 24 0 0 ; 0 24 36 208 600 2448 8400	
12 - 547	3.65973 1.14613 0.73566 -0.62643 -1.22276 -1.69234 -2.00000 1 0 -12 -12 20 22 -8 -8 ; 0 24 36 208 610 2496 8624	
12 - 548	3.67579 1.00000 0.84457 -0.71284 -1.00000 -1.80752 -2.00000 1 0 -12 -12 19 20 -8 -8 ; 0 24 36 212 620 2568 8876	
12 - 549	3.68333 1.14192 0.61803 -0.65830 -1.00000 -1.61803 -2.16696 1 0 -12 -12 18 20 -5 -6 ; 0 24 36 216 620 2622 8946	
12 - 550	3.69639 1.00000 0.61803 -0.17819 -1.51820 -1.61804 -2.00000 1 0 -12 -12 18 16 -9 -2 ; 0 24 36 216 640 2646 9254	
12 - 551	3.67846 1.38605 0.00000 -0.50951 -0.59757 -1.75463 -2.20280 1 0 -12 -12 17 22 6 0 ; 0 24 36 220 610 2628 8820	
12 - 552	3.68819 1.27879 0.27128 -0.50202 -0.82714 -1.71831 -2.19080 1 0 -12 -12 17 20 1 -2 ; 0 24 36 220 620 2658 9002	
12 - 553	3.68894 1.26501 0.29089 -0.41189 -1.00000 -1.60701 -2.22593 1 0 -12 -12 17 20 0 -2 ; 0 24 36 220 620 2664 9002	
12 - 554	3.69844 1.09908 0.55962 -0.44203 -1.18690 -1.51304 -2.21517 1 0 -12 -12 17 18 -5 -4 ; 0 24 36 220 630 2694 9184	

Number	Eigenvalues Characteristic polynomial coefficients and spectral moments	Graph
12 - 555	3.70928 1.19394 0.00000 0.00000 -0.90321 -2.00000 -2.00000 1 0 -12 -12 16 16 0 0 ; 0 24 36 224 640 2736 9408	
12 - 556	3.71070 1.16117 0.11336 0.00000 -1.10776 -1.68872 -2.18875 1 0 -12 -12 16 16 -2 0 ; 0 24 36 224 640 2748 9408	
12 - 557	3.71982 0.91515 0.61803 -0.21758 -1.24023 -1.61803 -2.17715 1 0 -12 -12 16 14 -7 -2 ; 0 24 36 224 650 2778 9590	
12 - 558	3.71596 1.11219 0.48525 -0.65392 -1.00000 -1.28399 -2.37550 1 0 -12 -12 15 18 -2 -4 ; 0 24 36 228 630 2820 9352	
12 - 559	3.72627 1.07488 0.18426 0.00000 -1.10918 -1.63579 -2.24043 1 0 -12 -12 15 14 -3 0 ; 0 24 36 228 650 2826 9660	
12 - 560	3.74775 0.81173 0.47272 0.00000 -1.08886 -1.73559 -2.20774 1 0 -12 -12 14 10 -6 0 ; 0 24 36 232 670 2916 10080	
12 - 561	3.72377 1.27906 0.00000 -0.50416 -1.00000 -1.00000 -2.49867 1 0 -12 -12 13 20 6 0 ; 0 24 36 236 620 2916 9324	
12 - 562	3.74811 1.04572 0.00000 0.00000 -0.77702 -1.70266 -2.31415 1 0 -12 -12 13 12 0 0 ; 0 24 36 236 660 2952 9996	
12 - 563	3.74943 1.00000 0.14692 0.00000 -1.00000 -1.54233 -2.35402 1 0 -12 -12 13 12 -2 0 ; 0 24 36 236 660 2964 9996	
12 - 564	3.75534 0.80194 0.57376 -0.51037 -0.55496 -1.81872 -2.24698 1 0 -12 -12 13 10 -3 -2 ; 0 24 36 236 670 2970 10178	
12 - 565	3.75962 0.86402 0.51409 -0.36959 -1.00000 -1.32798 -2.44015 1 0 -12 -12 12 12 -3 -2 ; 0 24 36 240 660 3042 10094	
12 - 566	3.76445 0.73540 0.61803 -0.44742 -0.68084 -1.61803 -2.37158 1 0 -12 -12 12 10 -3 -2 ; 0 24 36 240 670 3042 10262	
12 - 567	3.76826 0.85855 0.20320 0.00000 -0.73296 -1.83451 -2.26255 1 0 -12 -12 12 8 -2 0 ; 0 24 36 240 680 3036 10416	
12 - 568	3.76195 1.00000 0.35510 -0.59321 -1.00000 -1.00000 -2.52383 1 0 -12 -12 11 14 0 -2 ; 0 24 36 244 650 3096 10010	
12 - 569	3.78483 0.87602 0.00000 0.00000 -0.61623 -1.60488 -2.43973 1 0 -12 -12 10 8 0 0 ; 0 24 36 248 680 3168 10584	

Number	Eigenvalues Characteristic polynomial coefficients and spectral moments	Graph
12 - 570	3.78667 0.74645 0.31284 0.00000 -1.00000 -1.37054 -2.47542 1 0 -12 -12 10 8 -3 0; 0 24 36 248 680 3186 10584	
12 - 571	3.80225 0.82725 0.00000 0.00000 -0.76042 -1.30390 -2.56518 1 0 -12 -12 8 8 0 0; 0 24 36 256 680 3312 10752	
12 - 572	3.81528 0.73889 0.00000 0.00000 -0.59937 -1.37814 -2.57667 1 0 -12 -12 7 6 0 0; 0 24 36 260 690 3384 11004	
12 - 573	3.83668 0.36690 0.00000 0.00000 0.00000 -1.70792 -2.49566 1 0 -12 -12 6 0 0 0; 0 24 36 264 720 3456 11592	
12 - 574	3.65186 1.75665 -0.40272 -0.51119 -1.00000 -1.59158 -1.90301 1 0 -12 -14 20 42 23 4; 0 24 42 208 630 2466 8596	
12 - 575	3.71609 1.46826 0.25139 -1.00000 -1.00000 -1.53135 -1.90439 1 0 -12 -14 18 32 7 -4; 0 24 42 216 680 2706 9688	
12 - 576	3.73494 1.23838 0.61803 -1.00000 -1.26726 -1.61803 -1.70606 1 0 -12 -14 18 28 -3 -10; 0 24 42 216 700 2766 10066	
12 - 577	3.72170 1.51266 0.00000 -0.69020 -1.00000 -1.54416 -2.00000 1 0 -12 -14 17 32 12 0; 0 24 42 220 680 2748 9758	
12 - 578	3.75301 1.33460 0.20366 -0.50118 -1.26364 -1.64836 -1.87810 1 0 -12 -14 16 26 4 -2; 0 24 42 224 710 2868 10374	
12 - 579	3.76372 1.12525 0.61803 -1.00000 -1.00000 -1.61803 -1.88897 1 0 -12 -14 16 24 -3 -8; 0 24 42 224 720 2910 10584	
12 - 580	3.75942 1.37107 0.00000 -0.46632 -1.00000 -1.66417 -2.00000 1 0 -12 -14 15 26 8 0; 0 24 42 228 710 2916 10458	
12 - 581	3.77586 1.16189 0.42093 -0.54776 -1.25030 -1.69836 -1.86226 1 0 -12 -14 15 22 -1 -4; 0 24 42 228 730 2970 10822	
12 - 582	3.79050 1.21797 0.21700 -0.66090 -1.00000 -1.38749 -2.17708 1 0 -12 -14 13 22 4 -2; 0 24 42 236 730 3084 11004	
12 - 583	3.80389 1.00000 0.50775 -0.60772 -1.00000 -1.70391 -2.00000 1 0 -12 -14 13 18 -2 -4; 0 24 42 236 750 3120 11354	
12 - 584	3.81052 1.08379 0.29212 -0.50169 -1.00000 -1.54262 -2.14212 1 0 -12 -14 12 18 1 -2; 0 24 42 240 750 3174 11438	

Number	Eigenvalues Characteristic polynomial coefficients and spectral moments	Graph
12 - 585	3.81052 1.08379 0.29212 -0.50169 -1.00000 -1.54262 -2.14212 1 0 -12 -14 12 18 1 -2 ; 0 24 42 240 750 3174 11438	
12 - 586	3.81536 1.06069 0.00000 0.00000 -1.13622 -1.73983 -2.00000 1 0 -12 -14 12 16 0 0 ; 0 24 42 240 760 3180 11592	
12 - 587	3.81536 1.06069 0.00000 0.00000 -1.13622 -1.73983 -2.00000 1 0 -12 -14 12 16 0 0 ; 0 24 42 240 760 3180 11592	
12 - 588	3.82843 0.61803 0.61803 0.00000 -1.61803 -1.61803 -1.82843 1 0 -12 -14 12 12 -7 0 ; 0 24 42 240 780 3222 11928	
12 - 589	3.82843 0.61803 0.61803 0.00000 -1.61803 -1.61803 -1.82843 1 0 -12 -14 12 12 -7 0 ; 0 24 42 240 780 3222 11928	
12 - 590	3.80552 1.24849 0.00000 -0.72206 -1.00000 -1.00000 -2.33194 1 0 -12 -14 11 22 8 0 ; 0 24 42 244 730 3204 11186	
12 - 591	3.83046 0.85046 0.48469 -0.33003 -1.19958 -1.49305 -2.14295 1 0 -12 -14 11 14 -3 -2 ; 0 24 42 244 770 3270 11872	
12 - 592	3.84782 0.72507 0.35548 0.00000 -1.10746 -1.82091 -2.00000 1 0 -12 -14 10 10 -4 0 ; 0 24 42 248 790 3348 12292	
12 - 593	3.83307 1.13201 0.00000 -0.57969 -1.00000 -1.00000 -2.38539 1 0 -12 -14 9 18 6 0 ; 0 24 42 252 750 3360 11718	
12 - 594	3.85494 0.79755 0.18023 0.00000 -1.00000 -1.66536 -2.16736 1 0 -12 -14 9 10 -2 0 ; 0 24 42 252 790 3408 12390	
12 - 595	3.86620 0.78924 0.41421 -0.65544 -1.00000 -1.00000 -2.41421 1 0 -12 -14 7 12 0 -2 ; 0 24 42 260 780 3540 12432	
12 - 596	3.87530 0.66759 0.22922 0.00000 -1.00000 -1.45627 -2.31584 1 0 -12 -14 7 8 -2 0 ; 0 24 42 260 800 3552 12754	
12 - 597	3.90952 0.60930 0.00000 0.00000 -1.00000 -1.00000 -2.51882 1 0 -12 -14 3 6 0 0 ; 0 24 42 276 810 3828 13314	
12 - 598	3.64575 2.00000 -1.00000 -1.00000 -1.00000 -1.00000 -1.64575 1 0 -12 -16 21 60 46 12 ; 0 24 48 204 660 2436 8652	
12 - 599	3.77620 1.65615 -0.49528 -1.00000 -1.00000 -1.00000 -1.93706 1 0 -12 -16 15 42 28 6 ; 0 24 48 228 750 2976 10878	

Number	Eigenvalues Characteristic polynomial coefficients and spectral moments	Graph
12 - 600	3.82843 1.41421 0.00000 -1.00000 -1.00000 -1.41421 -1.82843 1 0 -12 -16 13 32 14 0 ; 0 24 48 236 800 3204 11984	
12 - 601	3.86267 1.24320 0.16125 -1.00000 -1.00000 -1.34097 -1.92615 1 0 -12 -16 11 26 8 -2 ; 0 24 48 244 830 3384 12726	
12 - 602	3.87439 1.22867 0.00000 -0.63597 -1.00000 -1.55369 -1.91339 1 0 -12 -16 10 24 9 0 ; 0 24 48 248 840 3450 12992	
12 - 603	3.89379 1.07008 0.22743 -0.69757 -1.00000 -1.58569 -1.90804 1 0 -12 -16 9 20 4 -2 ; 0 24 48 252 860 3552 13454	
12 - 604	3.89511 1.00000 0.39730 -1.00000 -1.00000 -1.29240 -2.00000 1 0 -12 -16 9 20 2 -4 ; 0 24 48 252 860 3564 13468	
12 - 605	3.90628 1.00000 0.25600 -0.67021 -1.00000 -1.49207 -2.00000 1 0 -12 -16 8 18 3 -2 ; 0 24 48 256 870 3630 13734	
12 - 606	3.91181 0.89050 0.34995 -0.43820 -1.32648 -1.47835 -1.90923 1 0 -12 -16 8 16 0 -2 ; 0 24 48 256 880 3648 13902	
12 - 607	3.91231 1.06224 0.00000 -0.50328 -1.00000 -1.35656 -2.11471 1 0 -12 -16 7 18 6 0 ; 0 24 48 260 870 3684 13832	
12 - 608	3.92347 0.81289 0.38245 -0.50178 -1.00000 -1.75339 -1.86364 1 0 -12 -16 7 14 0 -2 ; 0 24 48 260 890 3720 14182	
12 - 609	3.93061 0.82439 0.34789 -0.60848 -1.00000 -1.37707 -2.11734 1 0 -12 -16 6 14 1 -2 ; 0 24 48 264 890 3786 14294	
12 - 610	3.94600 0.74800 0.00000 0.00000 -1.00000 -1.69399 -2.00000 1 0 -12 -16 5 10 0 0 ; 0 24 48 268 910 3864 14728	
12 - 611	3.95805 0.53676 0.25757 0.00000 -1.23943 -1.38720 -2.12576 1 0 -12 -16 4 8 -2 0 ; 0 24 48 272 920 3948 15008	
12 - 612	4.05480 1.16104 -0.49324 -1.00000 -1.00000 -1.00000 -1.72259 1 0 -12 -20 1 24 18 4 ; 0 24 60 284 1080 4476 17976	
12 - 613	4.10043 0.61803 0.33888 -1.00000 -1.00000 -1.43931 -1.61803 1 0 -12 -20 -2 12 3 -2 ; 0 24 60 296 1140 4782 19446	
12 - 614	4.10548 0.77654 0.00000 -1.00000 -1.00000 -1.00000 -1.88202 1 0 -12 -20 -3 12 6 0 ; 0 24 60 300 1140 4836 19572	

Number	Eigenvalues Characteristic polynomial coefficients and spectral moments	Graph
13 - 615	3.76437 0.48980 0.00000 0.00000 0.00000 -1.00000 -3.25417 1 0 -13 -6 6 0 0 0; 0 26 18 314 390 4034 6846	
13 - 616	3.77846 0.71083 0.00000 0.00000 0.00000 -1.48929 -3.00000 1 0 -13 -8 12 0 0 0; 0 26 24 290 520 3650 8792	
13 - 617	3.84822 0.32733 0.00000 0.00000 0.00000 -1.00000 -3.17554 1 0 -13 -8 4 0 0 0; 0 26 24 322 520 4274 9240	
13 - 618	3.76195 1.00000 0.35510 0.00000 -0.59321 -2.00000 -2.52383 1 0 -13 -10 20 6 -4 0; 0 26 30 258 620 3158 9884	
13 - 619	3.75938 1.19116 0.12452 0.00000 -1.00000 -1.28598 -2.78909 1 0 -13 -10 18 14 -2 0; 0 26 30 266 580 3302 9296	
13 - 620	3.80691 0.90996 0.22854 0.00000 -0.57281 -1.57833 -2.79427 1 0 -13 -10 15 6 -2 0; 0 26 30 278 620 3536 10234	
13 - 621	3.82923 0.76098 0.30059 0.00000 -0.52842 -1.52044 -2.84194 1 0 -13 -10 13 4 -2 0; 0 26 30 286 630 3692 10556	
13 - 622	3.75937 1.24698 0.63306 -0.44504 -1.09951 -1.80194 -2.29292 1 0 -13 -12 24 22 -9 -6; 0 26 36 242 670 3008 10220	
13 - 623	3.77846 1.00000 0.71083 0.00000 -1.48929 -2.00000 -2.00000 1 0 -13 -12 24 16 -16 0; 0 26 36 242 700 3050 10724	
13 - 624	3.78447 1.27197 0.18533 0.00000 -1.00000 -2.00000 -2.24177 1 0 -13 -12 22 18 -4 0; 0 26 36 250 690 3134 10710	
13 - 625	3.80270 1.00000 0.71897 -0.34152 -1.00000 -1.79930 -2.38085 1 0 -13 -12 21 16 -9 -4; 0 26 36 254 700 3242 11004	
13 - 626	3.81446 0.84473 0.61803 0.17705 -1.50179 -1.61803 -2.33444 1 0 -13 -12 21 12 -14 2; 0 26 36 254 720 3272 11326	
13 - 627	3.81718 1.16511 0.20719 0.00000 -1.08992 -1.58204 -2.51753 1 0 -13 -12 19 16 -4 0; 0 26 36 262 700 3368 11144	
13 - 628	3.82874 0.91401 0.63421 -0.23128 -0.86111 -1.88807 -2.39651 1 0 -13 -12 19 12 -7 -2; 0 26 36 262 720 3386 11522	
13 - 629	3.85111 0.76112 0.61803 0.00000 -1.07620 -1.61803 -2.53603 1 0 -13 -12 17 10 -8 0; 0 26 36 270 730 3548 11858	

Number	Eigenvalues Characteristic polynomial coefficients and spectral moments	Graph
13 - 630	3.85111 0.76112 0.61803 0.00000 -1.07620 -1.61803 -2.53603 1 0 -13 -12 17 10 -8 0; 0 26 36 270 730 3548 11858	
13 - 631	3.84153 1.16400 0.00000 0.00000 -1.00000 -1.34485 -2.66067 1 0 -13 -12 16 16 0 0; 0 26 36 274 700 3578 11396	
13 - 632	3.86067 0.94698 0.19013 0.00000 -0.60640 -1.91958 -2.47180 1 0 -13 -12 16 8 -2 0; 0 26 36 274 740 3590 12124	
13 - 633	3.88607 0.74200 0.29596 0.00000 -0.48371 -1.93022 -2.51010 1 0 -13 -12 14 4 -2 0; 0 26 36 282 760 3746 12656	
13 - 634	3.88982 0.89919 0.17180 0.00000 -1.00000 -1.20994 -2.75087 1 0 -13 -12 12 10 -2 0; 0 26 36 290 730 3902 12278	
13 - 635	3.89511 0.73205 0.39730 0.00000 -1.00000 -1.29240 -2.73205 1 0 -13 -12 12 8 -4 0; 0 26 36 290 740 3914 12460	
13 - 636	3.91038 0.61803 0.34232 0.00000 -0.55355 -1.61803 -2.69914 1 0 -13 -12 11 4 -2 0; 0 26 36 294 760 3980 12908	
13 - 637	3.90915 0.85800 0.00000 0.00000 -0.63856 -1.33900 -2.78958 1 0 -13 -12 10 8 0 0; 0 26 36 298 740 4046 12628	
13 - 638	3.95601 0.36057 0.00000 0.00000 0.00000 -1.48612 -2.83046 1 0 -13 -12 6 0 0 0; 0 26 36 314 780 4358 13692	
13 - 639	3.82557 1.27091 0.61803 -0.52280 -1.57368 -1.61804 -2.00000 1 0 -13 -14 23 28 -8 -8; 0 26 42 246 770 3236 11816	
13 - 640	3.81939 1.50720 0.24489 -1.00000 -1.00000 -1.19289 -2.37859 1 0 -13 -14 21 34 7 -4; 0 26 42 254 740 3302 11438	
13 - 641	3.83063 1.44499 0.19199 -0.44698 -1.17328 -1.58907 -2.25829 1 0 -13 -14 21 30 4 -2; 0 26 42 254 760 3320 11788	
13 - 642	3.84805 1.15629 0.71615 -0.85914 -1.00000 -1.65729 -2.20406 1 0 -13 -14 21 26 -7 -10; 0 26 42 254 780 3386 12208	
13 - 643	3.85577 1.32164 0.41421 -1.00000 -1.00000 -1.17741 -2.41421 1 0 -13 -14 19 28 1 -6; 0 26 42 262 770 3494 12194	
13 - 644	3.87735 1.24698 0.26106 -0.44504 -0.87165 -1.80194 -2.26676 1 0 -13 -14 18 22 1 -2; 0 26 42 266 800 3572 12810	

Number	Eigenvalues Characteristic polynomial coefficients and spectral moments	Graph
13 - 645	3.88400 1.15580 0.35735 -0.28558 -1.10321 -1.76035 -2.24801 1 0 -13 -14 18 20 -3 -2 ; 0 26 42 266 810 3596 12992	
13 - 646	3.88541 1.08357 0.52938 -0.41740 -1.12856 -1.66742 -2.28497 1 0 -13 -14 18 20 -5 -4 ; 0 26 42 266 810 3608 13006	
13 - 647	3.90864 0.78394 0.61803 0.00000 -1.46522 -1.61803 -2.22736 1 0 -13 -14 17 14 -10 0 ; 0 26 42 270 840 3716 13622	
13 - 648	3.89122 1.28543 0.00000 -0.47877 -0.62762 -1.64820 -2.42207 1 0 -13 -14 16 22 6 0 ; 0 26 42 274 800 3698 12992	
13 - 649	3.89910 1.16077 0.29383 -0.40804 -1.00000 -1.51857 -2.42709 1 0 -13 -14 16 20 0 -2 ; 0 26 42 274 810 3734 13188	
13 - 650	3.91544 0.74654 0.70877 -0.21797 -1.08624 -1.79468 -2.27186 1 0 -13 -14 16 14 -7 -2 ; 0 26 42 274 840 3776 13734	
13 - 651	3.91729 1.00000 0.41421 -0.31955 -1.00000 -1.59774 -2.41421 1 0 -13 -14 15 16 -3 -2 ; 0 26 42 278 830 3830 13650	
13 - 652	3.92140 0.94621 0.44621 -0.35353 -0.79941 -1.84830 -2.31257 1 0 -13 -14 15 14 -3 -2 ; 0 26 42 278 840 3830 13832	
13 - 653	3.92780 0.68486 0.61803 0.00000 -1.26889 -1.61803 -2.34377 1 0 -13 -14 15 12 -8 0 ; 0 26 42 278 850 3860 14000	
13 - 654	3.93552 1.02005 0.00000 0.00000 -0.88480 -1.58667 -2.48410 1 0 -13 -14 13 14 0 0 ; 0 26 42 286 840 3968 14014	
13 - 655	3.94154 0.87163 0.28051 0.00000 -1.11269 -1.50903 -2.47196 1 0 -13 -14 13 12 -4 0 ; 0 26 42 286 850 3992 14196	
13 - 656	3.94600 0.74800 0.41421 0.00000 -1.00000 -1.69400 -2.41421 1 0 -13 -14 13 10 -5 0 ; 0 26 42 286 860 3998 14378	
13 - 657	3.94338 1.00000 0.00000 0.00000 -1.00000 -1.39091 -2.55247 1 0 -13 -14 12 14 0 0 ; 0 26 42 290 840 4046 14112	
13 - 658	3.94698 1.00000 0.30923 -0.62207 -1.00000 -1.00000 -2.63415 1 0 -13 -14 11 16 1 -2 ; 0 26 42 294 830 4118 14042	
13 - 659	3.97038 0.81971 0.00000 0.00000 -0.57717 -1.68417 -2.52875 1 0 -13 -14 10 8 0 0 ; 0 26 42 298 870 4202 14854	

Number	Eigenvalues Characteristic polynomial coefficients and spectral moments	Graph
13 - 660	3.99636 0.69169 0.00000 0.00000 -0.55249 -1.47870 -2.65685 1 0 -13 -14 7 6 0 0; 0 26 42 310 880 4436 15330	
13 - 661	4.00000 0.73205 0.00000 0.00000 -1.00000 -1.00000 -2.73205 1 0 -13 -14 6 8 0 0; 0 26 42 314 870 4514 15246	
13 - 662	3.87806 1.58343 0.00000 -0.77037 -1.00000 -1.69113 -2.00000 1 0 -13 -16 20 40 16 0; 0 26 48 258 840 3506 13048	
13 - 663	3.92107 1.19018 0.61803 -0.77870 -1.50919 -1.61803 -1.82337 1 0 -13 -16 19 30 -4 -10; 0 26 48 262 890 3704 14140	
13 - 664	3.91083 1.46600 0.00000 -0.58360 -1.00000 -1.79323 -2.00000 1 0 -13 -16 18 34 12 0; 0 26 48 266 870 3686 13818	
13 - 665	3.94523 1.08564 0.61803 -0.70367 -1.32719 -1.61803 -2.00000 1 0 -13 -16 17 26 -4 -8; 0 26 48 270 910 3860 14714	
13 - 666	3.93295 1.38731 0.13672 -1.00000 -1.00000 -1.17478 -2.28220 1 0 -13 -16 16 32 10 -2; 0 26 48 274 880 3854 14238	
13 - 667	3.95712 1.00000 0.67072 -0.84564 -1.00000 -1.78220 -2.00000 1 0 -13 -16 16 24 -4 -8; 0 26 48 274 920 3938 15008	
13 - 668	3.95573 1.25175 0.19203 -0.59324 -1.00000 -1.62679 -2.17947 1 0 -13 -16 15 26 5 -2; 0 26 48 278 910 3962 14896	
13 - 669	3.96442 1.00000 0.65033 -1.00000 -1.00000 -1.40287 -2.21188 1 0 -13 -16 15 24 -3 -8; 0 26 48 278 920 4010 15120	
13 - 670	3.96182 1.28156 0.00000 -0.45838 -1.00000 -1.51279 -2.27221 1 0 -13 -16 14 26 8 0; 0 26 48 282 910 4022 14994	
13 - 671	3.97333 1.12664 0.26679 -0.40731 -1.17737 -1.60124 -2.18085 1 0 -13 -16 14 22 1 -2; 0 26 48 282 930 4064 15372	
13 - 672	3.98316 1.00000 0.19947 0.00000 -1.46866 -1.71398 -2.00000 1 0 -13 -16 14 18 -4 0; 0 26 48 282 950 4094 15722	
13 - 673	3.98316 1.00000 0.19947 0.00000 -1.46865 -1.71397 -2.00000 1 0 -13 -16 14 18 -4 0; 0 26 48 282 950 4094 15722	
13 - 674	3.98526 1.05777 0.30075 -0.38774 -1.18461 -1.53722 -2.23422 1 0 -13 -16 13 20 0 -2; 0 26 48 286 940 4148 15666	

Number	Eigenvalues Characteristic polynomial coefficients and spectral moments	Graph
13 - 675	3.99884 1.06976 0.24292 -0.62067 -1.00000 -1.29339 -2.39746 1 0 -13 -16 11 20 3 -2 ; 0 26 48 294 940 4286 15890	
13 - 676	4.01261 0.77234 0.50323 -0.33282 -1.00000 -1.73685 -2.21851 1 0 -13 -16 11 14 -3 -2 ; 0 26 48 294 970 4322 16436	
13 - 677	4.01941 0.79288 0.45560 -0.37982 -1.00000 -1.55334 -2.33473 1 0 -13 -16 10 14 -2 -2 ; 0 26 48 298 970 4394 16548	
13 - 678	4.02278 0.82548 0.15259 0.00000 -1.00000 -1.76676 -2.23409 1 0 -13 -16 10 12 -2 0 ; 0 26 48 298 980 4394 16716	
13 - 679	4.03616 0.84326 0.00000 0.00000 -1.00000 -1.45319 -2.42623 1 0 -13 -16 8 12 0 0 ; 0 26 48 306 980 4538 16940	
13 - 680	4.04313 0.74315 0.00000 0.00000 -0.61239 -2.00000 -2.17389 1 0 -13 -16 8 8 0 0 ; 0 26 48 306 1000 4538 17304	
13 - 681	3.91729 1.73205 -0.31955 -1.00000 -1.00000 -1.59774 -1.73205 1 0 -13 -18 19 52 33 6 ; 0 26 54 262 910 3686 14126	
13 - 682	3.96942 1.59698 -0.44648 -1.00000 -1.00000 -1.00000 -2.11992 1 0 -13 -18 15 44 29 6 ; 0 26 54 278 950 4022 15358	
13 - 683	3.98064 1.52464 -0.35169 -0.59247 -1.00000 -1.69536 -1.86576 1 0 -13 -18 15 40 23 4 ; 0 26 54 278 970 4058 15736	
13 - 684	4.00592 1.35409 0.12243 -1.00000 -1.00000 -1.59913 -1.88331 1 0 -13 -18 14 34 12 -2 ; 0 26 54 282 1000 4202 16450	
13 - 685	4.02334 1.23540 0.26089 -1.00000 -1.00000 -1.64892 -1.87070 1 0 -13 -18 13 30 7 -4 ; 0 26 54 286 1020 4310 16954	
13 - 686	4.03294 1.24698 0.00000 -0.44504 -1.25561 -1.77733 -1.80194 1 0 -13 -18 12 28 9 0 ; 0 26 54 290 1030 4376 17234	
13 - 687	4.04417 1.19965 0.00000 -0.41594 -1.28342 -1.54446 -2.00000 1 0 -13 -18 11 26 8 0 ; 0 26 54 294 1040 4460 17542	
13 - 688	4.05137 1.00000 0.48270 -1.00000 -1.00000 -1.53407 -2.00000 1 0 -13 -18 11 24 1 -6 ; 0 26 54 294 1050 4502 17766	
13 - 689	4.05615 0.85631 0.61803 -0.67339 -1.37953 -1.61803 -1.85954 1 0 -13 -18 11 22 -2 -6 ; 0 26 54 294 1060 4520 17948	

Number	Eigenvalues Characteristic polynomial coefficients and spectral moments	Graph
13 - 690	4.06674 1.06158 0.20565 -0.71022 -1.00000 -1.47857 -2.14518 1 0 -13 -18 9 22 5 -2 ; 0 26 54 302 1060 4634 18172	
13 - 691	4.07642 1.05685 0.00000 -0.41291 -1.00000 -1.56456 -2.15580 1 0 -13 -18 8 20 6 0 ; 0 26 54 306 1070 4706 18466	
13 - 692	4.08617 0.80510 0.39151 -0.38312 -1.17513 -1.72454 -2.00000 1 0 -13 -18 8 16 -1 -2 ; 0 26 54 306 1090 4748 18844	
13 - 693	4.08387 1.00000 0.21320 -1.00000 -1.00000 -1.00000 -2.29707 1 0 -13 -18 7 20 5 -2 ; 0 26 54 310 1070 4790 18606	
13 - 694	4.09335 1.00000 0.00000 -0.49053 -1.00000 -1.29457 -2.30825 1 0 -13 -18 6 18 6 0 ; 0 26 54 314 1080 4862 18900	
13 - 695	4.10867 0.75446 0.35680 -0.59660 -1.00000 -1.31056 -2.31275 1 0 -13 -18 5 14 1 -2 ; 0 26 54 318 1100 4970 19404	
13 - 696	4.11207 0.75884 0.00000 0.00000 -1.22646 -1.39218 -2.25227 1 0 -13 -18 5 12 0 0 ; 0 26 54 318 1110 4976 19572	
13 - 697	4.11602 0.61803 0.19627 0.00000 -1.13930 -1.61803 -2.17299 1 0 -13 -18 5 10 -2 0 ; 0 26 54 318 1120 4988 19754	
13 - 698	4.12564 0.49380 0.26460 0.00000 -1.00000 -1.69468 -2.18935 1 0 -13 -18 4 8 -2 0 ; 0 26 54 322 1130 5066 20062	
13 - 699	4.13769 0.59969 0.00000 0.00000 -1.00000 -1.35107 -2.38631 1 0 -13 -18 2 8 0 0 ; 0 26 54 330 1130 5210 20314	
13 - 700	4.13264 1.14044 0.00000 -1.00000 -1.00000 -1.27307 -2.00000 1 0 -13 -20 6 26 12 0 ; 0 26 60 314 1170 5054 20454	
13 - 701	4.14650 1.04998 0.00000 -0.66709 -1.00000 -1.64082 -1.88858 1 0 -13 -20 5 22 9 0 ; 0 26 60 318 1190 5150 20958	
13 - 702	4.16100 0.87414 0.24204 -0.72063 -1.00000 -1.67934 -1.87720 1 0 -13 -20 4 18 4 -2 ; 0 26 60 322 1210 5258 21476	
13 - 703	4.17466 0.61803 0.43895 -0.45367 -1.27907 -1.61803 -1.88087 1 0 -13 -20 3 14 0 -2 ; 0 26 60 326 1230 5360 21980	
13 - 704	4.19079 0.83618 0.00000 -0.75294 -1.00000 -1.00000 -2.27402 1 0 -13 -20 0 14 6 0 ; 0 26 60 338 1230 5558 22386	

Number	Eigenvalues Characteristic polynomial coefficients and spectral moments	Graph
13 - 705	4.21617 0.44578 0.00000 0.00000 -1.00000 -1.43083 -2.23112 1 0 -13 -20 -2 6 0 0 ; 0 26 60 346 1270 5750 23394	
13 - 706	4.14134 1.48486 -1.00000 -1.00000 -1.00000 -1.00000 -1.62620 1 0 -13 -22 7 44 37 10 ; 0 26 66 310 1210 5078 20874	
13 - 707	4.21356 1.14858 -0.42714 -1.00000 -1.00000 -1.00000 -1.93500 1 0 -13 -22 1 26 19 4 ; 0 26 66 334 1300 5654 23478	
13 - 708	4.24447 0.76356 0.23779 -1.00000 -1.00000 -1.42640 -1.81942 1 0 -13 -22 -1 16 5 -2 ; 0 26 66 342 1350 5894 24738	
13 - 709	4.25327 0.68183 0.27820 -1.00000 -1.00000 -1.28669 -1.92661 1 0 -13 -22 -2 14 4 -2 ; 0 26 66 346 1360 5978 25074	
14 - 710	4.00000 1.00000 0.00000 0.00000 -1.00000 -1.00000 -3.00000 1 0 -14 -12 13 12 0 0 ; 0 28 36 340 780 4828 14196	
14 - 711	4.07245 0.35461 0.00000 0.00000 0.00000 -1.35041 -3.07665 1 0 -14 -12 6 0 0 0 ; 0 28 36 368 840 5416 15960	
14 - 712	4.00000 0.80194 0.80194 -0.55496 -0.55496 -2.24698 -2.24698 1 0 -14 -14 21 14 -7 -4 ; 0 28 42 308 910 4354 15806	
14 - 713	4.06038 0.82862 0.19776 0.00000 -0.63967 -1.72848 -2.71861 1 0 -14 -14 14 8 -2 0 ; 0 28 42 336 940 4912 17052	
14 - 714	4.11232 0.52677 0.31277 0.00000 -1.00000 -1.00000 -2.95186 1 0 -14 -14 7 6 -2 0 ; 0 28 42 364 950 5500 17934	
14 - 715	4.05849 0.93225 0.61803 -0.14835 -1.56369 -1.61803 -2.27870 1 0 -14 -16 20 20 -11 -2 ; 0 28 48 312 1020 4642 17766	
14 - 716	4.06169 1.15609 0.39806 -0.54396 -1.00000 -1.57673 -2.49515 1 0 -14 -16 18 24 -1 -4 ; 0 28 48 320 1000 4750 17612	
14 - 717	4.07325 1.05987 0.19256 0.00000 -1.06358 -2.00000 -2.26209 1 0 -14 -16 18 18 -4 0 ; 0 28 48 320 1030 4768 18172	
14 - 718	4.09708 1.00000 0.11438 0.00000 -1.00000 -1.69791 -2.51355 1 0 -14 -16 15 16 -2 0 ; 0 28 48 332 1040 5008 18704	
14 - 719	4.10446 0.80194 0.50689 -0.44342 -0.55496 -2.16792 -2.24698 1 0 -14 -16 15 12 -3 -2 ; 0 28 48 332 1060 5014 19110	

Number	Eigenvalues Characteristic polynomial coefficients and spectral moments	Graph
14 - 720	4.11342 0.96312 0.00000 0.00000 -0.81475 -1.67986 -2.58194 1 0 -14 -16 13 14 0 0 ; 0 28 48 340 1050 5164 19124	
14 - 721	4.11834 0.81008 0.28404 0.00000 -1.00000 -1.64254 -2.56991 1 0 -14 -16 13 12 -4 0 ; 0 28 48 340 1060 5188 19320	
14 - 722	4.14327 0.77245 0.00000 0.00000 -0.54603 -1.74247 -2.62722 1 0 -14 -16 10 8 0 0 ; 0 28 48 352 1080 5416 20048	
14 - 723	4.14743 0.73205 0.18144 0.00000 -1.00000 -1.32887 -2.73205 1 0 -14 -16 9 10 -2 0 ; 0 28 48 356 1070 5512 19964	
14 - 724	4.15633 0.73205 0.00000 0.00000 -0.63090 -1.52543 -2.73205 1 0 -14 -16 8 8 0 0 ; 0 28 48 360 1080 5584 20272	
14 - 725	4.16867 0.37694 0.00000 0.00000 0.00000 -2.00000 -2.54561 1 0 -14 -16 8 0 0 0 ; 0 28 48 360 1120 5584 21056	
14 - 726	4.21819 0.00000 0.00000 0.00000 0.00000 -1.29966 -2.91852 1 0 -14 -16 0 0 0 0 ; 0 28 48 392 1120 6256 21952	
14 - 727	4.11391 1.26860 0.18295 -0.47647 -1.16295 -1.69234 -2.23370 1 0 -14 -18 17 30 5 -2 ; 0 28 54 324 1110 5002 19628	
14 - 728	4.12488 1.00000 0.63667 -0.76156 -1.00000 -2.00000 -2.00000 1 0 -14 -18 17 26 -4 -8 ; 0 28 54 324 1130 5056 20062	
14 - 729	4.11718 1.36817 0.00000 -1.00000 -1.00000 -1.00000 -2.48535 1 0 -14 -18 15 34 14 0 ; 0 28 54 332 1090 5116 19474	
14 - 730	4.12754 1.23265 0.27057 -1.00000 -1.00000 -1.19101 -2.43974 1 0 -14 -18 15 30 6 -4 ; 0 28 54 332 1110 5164 19894	
14 - 731	4.13475 1.17838 0.20468 -0.51185 -1.00000 -1.69671 -2.30924 1 0 -14 -18 15 26 4 -2 ; 0 28 54 332 1130 5176 20272	
14 - 732	4.14516 0.80194 0.72872 -0.55496 -1.15678 -1.71710 -2.24698 1 0 -14 -18 15 22 -5 -6 ; 0 28 54 332 1150 5230 20692	
14 - 733	4.15856 0.73458 0.61803 -0.20135 -1.45118 -1.61803 -2.24061 1 0 -14 -18 14 18 -7 -2 ; 0 28 54 336 1170 5326 21182	
14 - 734	4.15856 0.73458 0.61803 -0.20135 -1.45118 -1.61803 -2.24061 1 0 -14 -18 14 18 -7 -2 ; 0 28 54 336 1170 5326 21182	

Number	Eigenvalues Characteristic polynomial coefficients and spectral moments	Graph
14 - 735	4.17190 0.94867 0.00000 0.00000 -0.91752 -2.00000 -2.20305 1 0 -14 -18 12 16 0 0; 0 28 54 344 1180 5452 21616	
14 - 736	4.18527 0.83472 0.38059 -0.39985 -1.00000 -1.51092 -2.48981 1 0 -14 -18 10 16 -1 -2; 0 28 54 352 1180 5626 21882	
14 - 737	4.19409 0.85986 0.00000 0.00000 -1.00000 -1.55102 -2.50293 1 0 -14 -18 9 14 0 0; 0 28 54 356 1190 5704 22190	
14 - 738	4.20403 0.73750 0.15677 0.00000 -1.15371 -1.40365 -2.54094 1 0 -14 -18 8 12 -2 0; 0 28 54 360 1200 5800 22512	
14 - 739	4.22822 0.51745 0.25374 0.00000 -1.00000 -1.37027 -2.62914 1 0 -14 -18 5 8 -2 0; 0 28 54 372 1220 6052 23282	
14 - 740	4.15632 1.41421 0.00000 -0.63090 -1.41421 -1.52543 -2.00000 1 0 -14 -20 16 40 16 0; 0 28 60 328 1200 5248 21280	
14 - 741	4.17356 1.09134 0.61803 -1.00000 -1.45839 -1.61803 -1.80651 1 0 -14 -20 16 34 -1 -12; 0 28 60 328 1230 5350 21952	
14 - 742	4.18071 1.30574 0.00000 -0.51221 -1.23086 -1.74338 -2.00000 1 0 -14 -20 14 34 12 0; 0 28 60 336 1230 5440 22148	
14 - 743	4.20147 1.00000 0.54510 -1.00000 -1.00000 -1.74657 -2.00000 1 0 -14 -20 13 28 0 -8; 0 28 60 340 1260 5596 22932	
14 - 744	4.20105 1.20008 0.14705 -0.77967 -1.00000 -1.58371 -2.18481 1 0 -14 -20 12 30 9 -2; 0 28 60 344 1250 5626 22834	
14 - 745	4.21478 1.09702 0.19421 -0.52832 -1.25523 -1.53617 -2.18630 1 0 -14 -20 11 26 5 -2; 0 28 60 348 1270 5734 23366	
14 - 746	4.22926 0.78712 0.61803 -0.82097 -1.00000 -1.61803 -2.19541 1 0 -14 -20 10 22 -1 -6; 0 28 60 352 1290 5854 23926	
14 - 747	4.22892 1.09940 0.00000 -0.48461 -1.00000 -1.54393 -2.29979 1 0 -14 -20 9 24 8 0; 0 28 60 356 1280 5884 23828	
14 - 748	4.23725 0.91701 0.28827 -0.42006 -1.17145 -1.64453 -2.20648 1 0 -14 -20 9 20 1 -2; 0 28 60 356 1300 5926 24234	
14 - 749	4.24914 0.85363 0.00000 0.00000 -1.10278 -2.00000 -2.00000 1 0 -14 -20 8 16 0 0; 0 28 60 360 1320 6016 24752	

Number	Eigenvalues Characteristic polynomial coefficients and spectral moments	Graph
14 - 750	4.24399 1.04986 0.00000 -0.56882 -1.00000 -1.30356 -2.42147 1 0 -14 -20 7 22 8 0 ; 0 28 60 364 1290 6052 24304	
14 - 751	4.25526 0.77099 0.36210 -0.44358 -1.00000 -1.66430 -2.28047 1 0 -14 -20 7 16 0 -2 ; 0 28 60 364 1320 6100 24906	
14 - 752	4.26086 0.78358 0.33036 -0.51506 -1.00000 -1.47832 -2.38143 1 0 -14 -20 6 16 1 -2 ; 0 28 60 368 1320 6178 25046	
14 - 753	4.26720 0.66945 0.16144 0.00000 -1.10516 -1.74783 -2.24510 1 0 -14 -20 6 12 -2 0 ; 0 28 60 368 1340 6196 25424	
14 - 754	4.27668 0.89367 0.00000 -0.61415 -1.00000 -1.00000 -2.55620 1 0 -14 -20 3 16 6 0 ; 0 28 60 380 1320 6400 25452	
14 - 755	4.21210 1.51667 -0.32906 -1.00000 -1.00000 -1.51198 -1.88773 1 0 -14 -22 13 48 32 6 ; 0 28 66 340 1300 5656 23436	
14 - 756	4.25841 1.32258 -0.37776 -0.57864 -1.00000 -1.62458 -2.00000 1 0 -14 -22 9 36 22 4 ; 0 28 66 356 1360 6052 25242	
14 - 757	4.28059 1.08590 0.15920 -0.63680 -1.38355 -1.68709 -1.81825 1 0 -14 -22 8 28 8 -2 ; 0 28 66 360 1400 6220 26222	
14 - 758	4.28600 0.80982 0.61803 -1.00000 -1.24600 -1.61803 -1.84982 1 0 -14 -22 8 26 1 -8 ; 0 28 66 360 1410 6262 26460	
14 - 759	4.28165 1.16815 0.00000 -1.00000 -1.00000 -1.30517 -2.14463 1 0 -14 -22 7 30 14 0 ; 0 28 66 364 1390 6268 26166	
14 - 760	4.30339 1.00000 0.16639 -1.00000 -1.00000 -1.26933 -2.20045 1 0 -14 -22 5 24 8 -2 ; 0 28 66 372 1420 6472 27076	
14 - 761	4.31021 0.86638 0.25174 -0.51086 -1.24509 -1.67239 -2.00000 1 0 -14 -22 5 20 3 -2 ; 0 28 66 372 1440 6502 27468	
14 - 762	4.32303 0.89908 0.00000 -0.44241 -1.00000 -1.60318 -2.17652 1 0 -14 -22 3 18 6 0 ; 0 28 66 380 1450 6652 27958	
14 - 763	4.33530 0.61803 0.38931 -0.54322 -1.00000 -1.61803 -2.18140 1 0 -14 -22 2 14 1 -2 ; 0 28 66 384 1470 6766 28518	
14 - 764	4.37228 0.41421 0.00000 0.00000 -1.00000 -1.37228 -2.41421 1 0 -14 -22 -3 6 0 0 ; 0 28 66 404 1510 7192 30058	

Number	Eigenvalues Characteristic polynomial coefficients and spectral moments	Graph
14 - 765	4.31874 1.34406 -0.38740 -1.00000 -1.00000 -1.51993 -1.75547 1 0 -14 -24 6 40 29 6 ; 0 28 72 368 1480 6538 27958	
14 - 766	4.37228 1.00000 0.00000 -1.00000 -1.00000 -1.37228 -2.00000 1 0 -14 -24 1 24 12 0 ; 0 28 72 388 1560 7060 30408	
14 - 767	4.38762 0.61803 0.48646 -1.00000 -1.00000 -1.61803 -1.87408 1 0 -14 -24 0 18 3 -4 ; 0 28 72 392 1590 7198 31192	
14 - 768	4.40405 0.81763 0.00000 -1.00000 -1.00000 -1.00000 -2.22168 1 0 -14 -24 -3 16 8 0 ; 0 28 72 404 1600 7420 31864	
14 - 769	4.40692 0.75807 0.00000 -0.58192 -1.00000 -1.44043 -2.14263 1 0 -14 -24 -3 14 6 0 ; 0 28 72 404 1610 7432 32060	
14 - 770	4.42055 0.45243 0.00000 0.00000 -1.18448 -1.68850 -2.00000 1 0 -14 -24 -4 8 0 0 ; 0 28 72 408 1640 7552 32816	
14 - 771	4.44949 1.00000 -0.44949 -1.00000 -1.00000 -1.00000 -2.00000 1 0 -14 -26 -5 22 18 4 ; 0 28 78 412 1710 7828 34398	
14 - 772	4.46357 0.78803 0.00000 -1.00000 -1.00000 -1.33505 -1.91655 1 0 -14 -26 -6 16 9 0 ; 0 28 78 416 1740 7966 35196	
14 - 773	4.47415 0.50995 0.32755 -1.00000 -1.00000 -1.39974 -1.91191 1 0 -14 -26 -7 12 4 -2 ; 0 28 78 420 1760 8080 35784	
15 - 774	4.32584 0.61803 0.00000 0.00000 -0.48892 -1.61803 -2.83692 1 0 -15 -18 7 6 0 0 ; 0 30 54 422 1320 7092 26838	
15 - 775	4.37228 0.00000 0.00000 0.00000 0.00000 -1.37228 -3.00000 1 0 -15 -18 0 0 0 0 ; 0 30 54 450 1350 7722 28350	
15 - 776	4.31663 0.61803 0.61803 0.00000 -1.61803 -1.61803 -2.31662 1 0 -15 -20 15 18 -10 0 ; 0 30 60 390 1410 6660 27510	
15 - 777	4.32620 0.83819 0.41421 -0.29505 -1.00000 -1.86933 -2.41421 1 0 -15 -20 13 18 -3 -2 ; 0 30 60 398 1410 6798 27804	
15 - 778	4.32803 1.00000 0.34567 -1.00000 -1.00000 -1.00000 -2.67370 1 0 -15 -20 11 24 3 -4 ; 0 30 60 406 1380 6942 27468	
15 - 779	4.36633 0.67015 0.00000 0.00000 -0.54972 -2.00000 -2.48675 1 0 -15 -20 8 8 0 0 ; 0 30 60 418 1460 7230 29540	

Number	Eigenvalues Characteristic polynomial coefficients and spectral moments	Graph
15 - 780	4.38865 0.44666 0.27473 0.00000 -1.00000 -1.34139 -2.76865 1 0 -15 -20 4 8 -2 0; 0 30 60 434 1460 7602 30100	
15 - 781	4.41175 0.47167 0.00000 0.00000 -1.00000 -1.00000 -2.88341 1 0 -15 -20 0 6 0 0; 0 30 60 450 1470 7950 30870	
15 - 782	4.37228 1.00000 0.00000 0.00000 -1.37228 -2.00000 -2.00000 1 0 -15 -22 12 24 0 0; 0 30 66 402 1530 7122 30282	
15 - 783	4.37228 1.00000 0.00000 0.00000 -1.37228 -2.00000 -2.00000 1 0 -15 -22 12 24 0 0; 0 30 66 402 1530 7122 30282	
15 - 784	4.37852 1.09235 0.00000 -0.42123 -1.00000 -1.66556 -2.38408 1 0 -15 -22 10 26 8 0; 0 30 66 410 1520 7254 30380	
15 - 785	4.38568 0.92460 0.27520 -0.39148 -1.12392 -1.77421 -2.29586 1 0 -15 -22 10 22 1 -2; 0 30 66 410 1540 7296 30814	
15 - 786	4.39874 0.87998 0.26862 -0.48487 -1.00000 -1.63268 -2.42979 1 0 -15 -22 8 20 2 -2; 0 30 66 418 1550 7470 31332	
15 - 787	4.42235 0.73597 0.00000 0.00000 -1.14128 -1.49357 -2.52346 1 0 -15 -22 5 14 0 0; 0 30 66 430 1580 7752 32410	
15 - 788	4.42526 0.61803 0.16447 0.00000 -1.10674 -1.61803 -2.48298 1 0 -15 -22 5 12 -2 0; 0 30 66 430 1590 7764 32620	
15 - 789	4.45508 0.51943 0.00000 0.00000 -1.00000 -1.28577 -2.68873 1 0 -15 -22 0 8 0 0; 0 30 66 450 1610 8202 33810	
15 - 790	4.42065 1.16524 0.11719 -1.00000 -1.00000 -1.51241 -2.19066 1 0 -15 -24 9 34 13 -2; 0 30 72 414· 1630 7590 32732	
15 - 791	4.43173 0.79950 0.61803 -0.77551 -1.45573 -1.61803 -2.00000 1 0 -15 -24 9 28 0 -8; 0 30 72 414 1660 7668 33404	
15 - 792	4.43947 1.04518 0.15055 -0.79484 -1.00000 -1.62850 -2.21186 1 0 -15 -24 7 28 9 -2; 0 30 72 422 1660 7794 33698	
15 - 793	4.46850 0.73263 0.32075 -0.43316 -1.16629 -1.68527 -2.23717 1 0 -15 -24 4 18 1 -2; 0 30 72 434 1710 8112 35252	
15 - 794	4.47090 0.73929 0.41421 -1.00000 -1.00000 -1.21018 -2.41421 1 0 -15 -24 3 20 3 -4; 0 30 72 438 1700 8190 35224	

Number	Eigenvalues Characteristic polynomial coefficients and spectral moments	Graph
15 - 795	4.47719 0.85841 0.00000 -0.43415 -1.00000 -1.49313 -2.40832 1 0 -15 -24 2 18 6 0 ; 0 30 72 442 1710 8262 35574	
15 - 796	4.50927 0.47648 0.00000 0.00000 -1.00000 -1.49469 -2.49106 1 0 -15 -24 -2 8 0 0 ; 0 30 72 458 1760 8658 37296	
15 - 797	4.47512 1.24698 -0.37349 -0.44504 -1.44188 -1.65975 -1.80194 1 0 -15 -26 6 38 23 4 ; 0 30 78 426 1760 8100 35840	
15 - 798	4.49396 1.10992 0.00000 -1.00000 -1.00000 -1.60388 -2.00000 1 0 -15 -26 4 32 16 0 ; 0 30 78 434 1790 8322 36862	
15 - 799	4.50467 1.00000 0.13536 -1.00000 -1.00000 -1.64002 -2.00000 1 0 -15 -26 3 28 11 -2 ; 0 30 78 438 1810 8442 37478	
15 - 800	4.51141 0.61803 0.61803 -0.75889 -1.61803 -1.61803 -1.75252 1 0 -15 -26 3 24 2 -6 ; 0 30 78 438 1830 8496 37926	
15 - 801	4.51140 1.00000 0.00000 -0.75889 -1.00000 -1.75252 -2.00000 1 0 -15 -26 2 26 12 0 ; 0 30 78 442 1820 8526 37856	
15 - 802	4.53582 0.80894 0.00000 -0.39257 -1.29019 -1.47881 -2.18320 1 0 -15 -26 -1 18 6 0 ; 0 30 78 454 1860 8832 39242	
15 - 803	4.54067 0.75346 0.21175 -1.00000 -1.00000 -1.19429 -2.31159 1 0 -15 -26 -2 18 6 -2 ; 0 30 78 458 1860 8922 39438	
15 - 804	4.54265 0.77336 0.00000 -0.47245 -1.00000 -1.64200 -2.20156 1 0 -15 -26 -2 16 6 0 ; 0 30 78 458 1870 8922 39634	
15 - 805	4.56155 0.43845 0.00000 0.00000 -1.00000 -2.00000 -2.00000 1 0 -15 -26 -4 8 0 0 ; 0 30 78 466 1910 9138 40838	
15 - 806	4.60555 0.00000 0.00000 0.00000 -1.00000 -1.00000 -2.60555 1 0 -15 -26 -12 0 0 0 ; 0 30 78 498 1950 9858 43134	
15 - 807	4.51882 1.39070 -1.00000 -1.00000 -1.00000 -1.00000 -1.90952 1 0 -15 -28 3 48 43 12 ; 0 30 84 438 1860 8574 38388	
15 - 808	4.56021 1.12124 -0.27208 -1.00000 -1.00000 -1.52952 -1.87985 1 0 -15 -28 -1 32 23 4 ; 0 30 84 454 1940 9054 40908	
15 - 809	4.57107 1.11516 -0.55134 -1.00000 -1.00000 -1.00000 -2.13489 1 0 -15 -28 -3 30 25 6 ; 0 30 84 462 1950 9222 41496	

Number	Eigenvalues Characteristic polynomial coefficients and spectral moments	Graph
15 - 810	4.58982 0.77980 0.17769 -1.00000 -1.00000 -1.74054 -1.80677 1 0 -15 -28 -4 20 8 -2; 0 30 84 466 2000 9414 42798	
15 - 811	4.60981 0.58685 0.26262 -1.00000 -1.00000 -1.30953 -2.14974 1 0 -15 -28 -7 14 5 -2; 0 30 84 478 2030 9702 44016	
15 - 812	4.73065 0.76692 -0.57282 -1.00000 -1.00000 -1.00000 -1.92475 1 0 -15 -32 -17 12 15 4; 0 30 96 518 2340 11262 52920	
15 - 813	4.74146 0.49006 0.00000 -1.00000 -1.00000 -1.44695 -1.78457 1 0 -15 -32 -18 6 6 0; 0 30 96 522 2370 11406 53802	
16 - 814	4.60555 0.00000 0.00000 0.00000 0.00000 -2.00000 -2.60555 1 0 -16 -24 0 0 0 0; 0 32 72 512 1920 9920 43008	
16 - 815	4.61354 0.71334 0.00000 0.00000 -1.08363 -2.00000 -2.24325 1 0 -16 -26 4 16 0 0; 0 32 78 496 2000 9836 44072	
16 - 816	4.62253 0.61803 0.36081 -0.47879 -1.00000 -1.61803 -2.50455 1 0 -16 -26 2 16 1 -2; 0 32 78 504 2000 10022 44450	
16 - 817	4.66329 0.36300 0.00000 0.00000 -1.00000 -1.30022 -2.72607 1 0 -16 -26 -5 6 0 0; 0 32 78 532 2050 10700 46830	
16 - 818	4.65427 1.00000 0.00000 -1.00000 -1.00000 -1.25230 -2.40198 1 0 -16 -28 1 28 14 0; 0 32 84 508 2100 10364 46844	
16 - 819	4.66190 0.80194 0.21238 -0.55496 -1.22404 -1.65025 -2.24698 1 0 -16 -28 1 22 5 -2; 0 32 84 508 2130 10418 47530	
16 - 820	4.67412 0.78697 0.00000 -0.40284 -1.00000 -1.76803 -2.29022 1 0 -16 -28 -1 18 6 0; 0 32 84 516 2150 10604 48356	
16 - 821	4.69547 0.57027 0.29202 -1.00000 -1.00000 -1.00000 -2.55776 1 0 -16 -28 -5 14 4 -2; 0 32 84 532 2170 11000 49602	
16 - 822	4.70578 0.40294 0.00000 0.00000 -1.21781 -1.37947 -2.51144 1 0 -16 -28 -6 8 0 0; 0 32 84 536 2200 11120 50456	
16 - 823	4.70156 0.61803 0.61803 -1.00000 -1.61803 -1.61803 -1.70156 1 0 -16 -30 0 28 5 -8; 0 32 90 512 2260 10862 50680	
16 - 824	4.71021 0.92011 0.00000 -0.61466 -1.36223 -1.65342 -2.00000 1 0 -16 -30 -2 26 12 0; 0 32 90 520 2270 11012 51268	

Number	Eigenvalues Characteristic polynomial coefficients and spectral moments	Graph
16 - 825	4.72285 0.79840 0.16227 -1.00000 -1.00000 -1.49052 -2.19300 1 0 -16 -30 -4 22 9 -2 ; 0 32 90 528 2290 11222 52150	
16 - 826	4.74274 0.67149 0.00000 -0.50249 -1.00000 -1.68003 -2.23172 1 0 -16 -30 -7 14 6 0 ; 0 32 90 540 2330 11528 53662	
16 - 827	4.75197 1.10320 -0.38044 -1.00000 -1.00000 -1.63718 -1.83756 1 0 -16 -32 -5 34 28 6 ; 0 32 96 532 2390 11576 54614	
16 - 828	4.76644 1.00000 -0.28282 -1.00000 -1.00000 -1.48361 -2.00000 1 0 -16 -32 -7 28 22 4 ; 0 32 96 540 2420 11804 55748	
16 - 829	4.79129 0.61803 0.20871 -1.00000 -1.00000 -1.61803 -2.00000 1 0 -16 -32 -10 16 7 -2 ; 0 32 96 552 2480 12182 57806	
16 - 830	4.81730 0.53050 0.00000 -1.00000 -1.00000 -1.00000 -2.34780 1 0 -16 -32 -15 8 6 0 ; 0 32 96 572 2520 12668 59808	
16 - 831	4.81730 0.53050 0.00000 -1.00000 -1.00000 -1.00000 -2.34780 1 0 -16 -32 -15 8 6 0 ; 0 32 96 572 2520 12668 59808	
16 - 832	4.83860 0.86424 -0.34422 -1.00000 -1.00000 -1.47640 -1.88223 1 0 -16 -34 -14 20 19 4 ; 0 32 102 568 2620 12890 61992	
16 - 833	4.85952 0.57414 0.00000 -1.00000 -1.00000 -1.43367 -2.00000 1 0 -16 -34 -17 10 8 0 ; 0 32 102 580 2670 13244 63854	
16 - 834	5.03404 0.51346 -1.00000 -1.00000 -1.00000 -1.00000 -1.54751 1 0 -16 -40 -35 -4 10 4 ; 0 32 120 652 3220 16292 81900	
17 - 835	4.88090 0.36491 0.00000 0.00000 -1.00000 -2.00000 -2.24581 1 0 -17 -32 -8 8 0 0 ; 0 34 96 610 2680 13714 65576	
17 - 836	4.92272 0.61803 0.00000 -0.43759 -1.24138 -1.61803 -2.24374 1 0 -17 -34 -11 14 6 0 ; 0 34 102 622 2820 14380 69735	
17 - 837	4.93163 0.41421 0.32319 -1.00000 -1.00000 -1.25482 -2.41421 1 0 -17 -34 -13 12 5 -2 ; 0 34 102 630 2830 14590 70462	
17 - 838	4.95870 0.00000 0.00000 0.00000 -1.00000 -1.44285 -2.51585 1 0 -17 -34 -18 0 0 0 ; 0 34 102 650 2890 15130 73066	
17 - 839	4.96506 0.71801 0.00000 -1.00000 -1.00000 -1.68306 -2.00000 1 0 -17 -36 -14 18 12 0 ; 0 34 108 634 2970 15070 74215	

Number	Eigenvalues Characteristic polynomial coefficients and spectral moments	Graph
17 - 840	4.98518 0.55986 0.00000 -1.00000 -1.00000 -1.24769 -2.29735 1 0 -17 -36 -18 10 8 0 ; 0 34 108 650 3010 15502 76174	
17 - 841	5.00000 1.00000 -1.00000 -1.00000 -1.00000 -1.00000 -2.00000 1 0 -17 -38 -17 28 33 10 ; 0 34 114 646 3090 15694 77994	
17 - 842	5.01573 0.80067 -0.32787 -1.00000 -1.00000 -1.67696 -1.81156 1 0 -17 -38 -19 18 19 4 ; 0 34 114 654 3140 15982 79759	
17 - 843	5.04355 0.42259 0.00000 -1.00000 -1.00000 -1.29896 -2.16719 1 0 -17 -38 -24 4 6 0 ; 0 34 114 674 3210 16570 82782	
17 - 844	5.13555 0.65317 -1.00000 -1.00000 -1.00000 -1.00000 -1.78872 1 0 -17 -42 -33 4 17 6 ; 0 34 126 710 3550 18382 94150	
18 - 845	5.16228 0.00000 0.00000 0.00000 -1.16228 -2.00000 -2.00000 1 0 -18 -40 -24 0 0 0 ; 0 36 120 744 3600 19056 97440	
18 - 846	5.20704 0.37751 0.00000 -1.00000 -1.00000 -1.39243 -2.19211 1 0 -18 -42 -29 2 6 0 ; 0 36 126 764 3770 20052 103530	
18 - 847	5.24338 0.61803 -0.41790 -1.00000 -1.00000 -1.61803 -1.82548 1 0 -18 -44 -32 6 15 4 ; 0 36 132 776 3930 20838 108865	
18 - 848	5.27492 0.00000 0.00000 -1.00000 -1.00000 -1.00000 -2.27492 1 0 -18 -44 -39 -12 0 0 ; 0 36 132 804 4020 21684 113316	
18 - 849	5.29654 0.59795 -1.00000 -1.00000 -1.00000 -1.00000 -1.89449 1 0 -18 -46 -39 0 16 6 ; 0 36 138 804 4140 22128 116844	
19 - 850	5.46410 0.00000 0.00000 -1.00000 -1.00000 -1.46410 -2.00000 1 0 -19 -50 -48 -16 0 0 ; 0 38 150 914 4830 26690 145278	
19 - 851	5.50331 0.38493 -1.00000 -1.00000 -1.00000 -1.00000 -1.88824 1 0 -19 -52 -53 -16 7 4 ; 0 38 156 934 5020 27830 152797	
20 - 852	5.74166 0.00000 -1.00000 -1.00000 -1.00000 -1.00000 -1.74166 1 0 -20 -60 -75 -44 -10 0 ; 0 40 180 1100 6220 35860 205661	
21 - 853	6.00000 -1.00000 -1.00000 -1.00000 -1.00000 -1.00000 -1.00000 1 0 -21 -70 -105 -84 -35 -6 ; 0 42 210 1302 7770 46662 279930	

Bibliography

[ABR1] ABRAMJAN K. B.: On an invariant of a simple graph (Russian), Graph Theory (ed. N. P. Homenko), Kiev 1977, 210–212.

[ACH1] ACHARYA B. D.: A programme logic for listing of sigraphs, their characteristic polynomials, and their spectra, Abstract, Graph Theory Newsletter, 9(1979), No. 2, 1.

[ACH2] ACHARYA B. D.: Spectral criterion for cycle balance in networks, J. Graph Theory, 4(1980), No. 1, 1–11.

[ACH3] ACHARYA B. D.: A graph theoretical expression for the characteristic polynomial of a matrix, Proc. Nat. Acad. Sci. India, 50A(1980), No. 3, 164–175.

[AHR1] AHRENS J. H.: Paving the chessboard, J. Combin. Theory (A), 31(1981), 277–288.

[AIH1] AIHARA J.: General rules for constructing Hückel molecular orbital characteristic polynomials, J. Amer. Chem. Soc., 98(1976), 6840–6844.

[AIH2] AIHARA J.: A new definition of Dewar–type resonance energy, J. Amer. Chem. Soc., 98(1976), 2750–2758.

[AIH3] AIHARA J.: Three dimensional aromaticity of polyhedral boranes, J. Amer. Chem. Soc., 100(1978), 3339–3342.

[AIH4] AIHARA J.: Graph–theoretical formulation of London diamagnetism, J. Amer. Chem. Soc., 101(1979), 5913–5917.

[AIH5] AIHARA J.: Matrix representation of an olefinic reference structure for monocyclic conjugated compounds, Bull. Chem. Soc.
Japan, 52(1979), 1529–1530.

[AIH6] AIHARA J.: Theoretical relationship between diamagnetic susceptibility exaltation and aromatic stabilization, J. Amer. Chem.
Soc., 101(1979), 558–560.

[AIH7] AIHARA J.: Stability of multi–center bonds in inorganic cluster
compounds, Bull. Chem. Soc. Japan, 52(1979), 2202–2204.

[AIH8] AIHARA J.: Unified theory of aromaticity and London diamagnetism, J. Amer. Chem. Soc., 103(1981), 5704–5706.

[AIH9] AIHARA J.: Aromaticity and diatropicity, Pure Appl. Chem.,
54(1982), 1115–1128.

[AIHo1] AIHARA J., HORIKAWA T.: General graph–theoretical formula for the London susceptibility of a cyclic conjugated system
with highly degenerate orbitals, Chem. Phys. Letters, 95(1983),
561–563.

[AIHo2] AIHARA J., HORIKAWA T.: Graph–theoretical formula for
ring current induced in a polycyclic conjugated system, Bull.
Chem. Soc. Japan, 56(1983), 1853–1584.

[AkGL1] AKHIEZER N. I., GLAZMAN I. M.: **Theory of linear operators in Hilbert spaces** (Russian), Vol. I, Harkov 1977; Vol.
II, Harkov 1978.

[ALMI1] ALON N., MILMAN V. D.: λ_1 isoperimetric inequalities
for graphs, and superconcentrators, J. Combinatorial Theory B
38(1985), 73–88.

[AMA1] D'AMATO S. S.: Eigenvalues of graphs with twofold symmetry,
Mol. Phys., 37(1979), 1363–1369.

[AMA2] D'AMATO S. S.: Eigenvalues of graphs with threefold symmetry, Theoret. Chim. Acta (Berl.), 53(1979), 319–326.

[AMGT1] D'AMATO S. S., GIMARC B. M., TRINAJSTIĆ, N.: Isospectral and subspectral molecules, Croat. Chem. Acta, 54(1981),
No. 1, 1–52.

[ARN1] ARNOLD, L.: On the asymptotic distribution of the eigenvalues of random matrices. J. Math. Analysis Appl., 20(1967), 262–268.

[ARN2] ARNOLD, L.: On Wigner's semicircle law for the eigenvalues of random matrices, Z. Wahrscheinlichkeitstheorie verw. Geb., (1971), 191–198.

[ARN3] ARNOLD, L.: Deterministic version of Wigner's semicircle law for the distribution of matrix eigenvalues, Linear Algebra and its Appl., 13(1976), 185–199.

[AsGi1] ASPVALL B., GILBERT J. R.: Graph colouring using eigenvalue decomposition, SIAM J. Alg. Disc. Math., 5(1984), No. 4, 526–538.

[AuYu1] TANG AU–CHIN, KIANG YUAN–SUN: Graph theory of molecular orbitals, Sci. Sinica, 19(1976), No. 2, 207–226.

[AuYu2] TANG AU–CHIN, KIANG YUAN–SUN: Graph theory of molecular orbitals, II. Symmetrical analysis and calculations of MO coefficients, Sci. Sinica, 20(1977), 595–612.

[AuYSG] TANG AU–CHIN, KIANG YUAN–SUN, DI SHU–SUN, YEN GUO–SUN: **Graph Theory and Molecular Orbitals** (Chinese), Science Press, Peking, 1980.

[BAB1] BABAI L.: On the isomorphism problem. Preprint appended to Proc. 1977 FCT–Conference, Poznan–Kornik, Poland, 1977, 1–5.

[BAB2] BABAI L.: Isomorphism testing for graphs with distinct eigenvalues, preprint.

[BAB3] BABAI L.: Isomorphism testing and symmetry of graphs, Ann. Discrete Math., 8(1980), 101–110.

[BaGr1] BABAI L., GRIGORYEV D. YU.: Isomorphism testing for graphs with bounded eigenvalue multiplicities, preprint, 1–15.

[BAL1] BALASUBRAMANIAN K.: Spectra of chemical trees, Internat. J. Quantum Chem., 21(1982), 581–590.

[BAL2] BALASUBRAMANIAN K.: Computer generation of the characteristic polynomial of chemical graphs, J. Comput. Chem., 5(1984), 387–394.

[BAL3] BALASUBRAMANIAN K.: The use of Frame's method for the characteristic polynomials of chemical graphs, Theoret. Chim. Acta (Berl.), 65(1984), 49–58.

[BAPA1] BALASUBRAMANIAN K., PARTHASARATHY K. R.: In search of a complete invariant for graphs, **Combinatorics and Graph Theory** (Calcuta, 1980), Lecture Notes Math. 885, Springer, Berlin, 1981, pp. 42–59.

[BARA1] BALASUBRAMANIAN K., RANDIĆ M.: The characteristic polynomials of structures with pending bonds, Theoret. Chim. Acta (Berl.), 61(1982), 307–323.

[BAIT1] BANNAI E., ITO T.: On the spectra of certain distance-transitive graphs, J. Combin. Theory(B), 27(1979), No. 3, 274–293.

[BAIT2] BANNAI E., ITO T.: On the spectra of certain distance-regular graphs II, Quart. J. Math. Oxford Ser. (2), 32(1981), No. 128, 389–411.

[BAIT3] BANNAI E., ITO T.: Regular graphs with excess one, Discrete Math., 37(1981), 147–158.

[BAIT4] BANNAI E., ITO T.: **Algebraic Combinatorics I** Benjamin/Cummings Publishing Company, Menlo Park, California, 1984.

[BACK1] BANNAI E., CAMERON P. J., KAHN J.: Nonexistence of certain distance-transitive digraphs, J. Combin. Theory(B), 31(1981), 105–110.

[BEE1] BEEZER, R. A.: On the polynomial of a path, Linear Algebra and its Appl., 63(1984), 221–225.

[BECL1] BEHZAD M., CHARTRAND G., LESNIAK–FOSTER L.: **Graphs and Digraphs,** Wadsworth Internat. Group, Belmont, 1979.

[BeHa1] BEKER H., HAEMERS W.: 2-designs having an intersection number $k - n$, J. Combin. Theory(A), 28(1980), 64–81.

[Ber1] BERESINA L. JA.: The normal curvatures of a graph, J. Geom., 18(1982), 54–56.

[Big1] BIGGS N.: Coloring square lattice graphs, Bull. London Math. Soc., 9(1977), No. 1, 54–56.

[Big2] BIGGS N. L.: Some odd graph theory, (Second Internat. Conf. Combinatorial Math., New York, 1978), Ann. New York Acad. Sci., 319 (1979), 71–81.

[Big3] BIGGS N.: Girth, valency and excess, Linear Algebra Appl., 31 (1980), 55–59.

[Big4] BIGGS N. L.: Excess in vertex–transitive graphs, Bull. London Math. Soc., 14(1982), 52–54.

[BiIt1] BIGGS N., ITO T.: Graphs with even girth and small excess, Math. Proc. Cambridge Philos. Soc., 88(1980), No. 1, 1–10.

[BiKa1] BILEK O., KADURA P.: On the tight binding eigenvalues of finite S. C. and F. C. C. crystallites with (100) surfaces, Phys. Stat. Sol. (b), 85(1978), 225–231.

[BoSt1] BOCHVAR D. A., STANKEVICH I. V.: Approximate formulas for some characteristics of the electronic structure of molecules. 1. Total π–electron energy (Russian), Zh. Strukt. Khim., 21(1980), 61–66.

[BoSt2] BOCHVAR D. A., STANKEVICH I. V.: Approximate formulas for some characteristics of the electronic structure of molecules. 2. Bond orders and effective charges on atoms (Russian), Zh. Strukt. Khim., 21(1980), 67–72.

[BoGo1] BOGART K., GORDON J.: Hypergraphs, characteristic polynomials and permutation characters, Linear and multilinear algebra, 7(1979), 213–236.

[Bol1] BOLLOBÁS B.: **Graph Theory, An Introductory Course**, Springer Verlag, New York–Heidelberg–Berlin, 1979.

[BoMe1] BONCHEV D., MEKENYAN O.: A topological approach to the calculation of the π-electron energy and energy gap of infinite conjugated polymers, Z. Naturforsch., 35a(1980), 739–747.

[BoHe1] BONDY J. A., HEMMINGER R. L.: Graph reconstruction–A survey, J. Graph Theory, 1(1977), No. 3, 227–268.

[BoJó1] BOROWIECKI M., JÓŹWIAK T.: Computing the characteristic polynomial of a multigraph, Inst. Mat. i Fiz. Wyzsey Skoly Inz., Zielona Gora, Poland, Raport No. IF–1–79(1979), 1–11.

[BoJó2] BOROWIECKI M., JÓŹWIAK T.: Computing the permanental polynomial of a multigraph, Inst. Mat. i Fiz. Wyzsey Skoly Inz., Zielona Gora, Poland, Raport No. IF–6–80(1980), 1–13; Discussiones Math. 5(1982), .

[BoJó3] BOROWIECKI M., JÓŹWIAK T.: On characteristic and permanental polynomials of multigraphs, Inst. Mat. i Fiz. Wyzsey Skoly Inz., Zielona Gora, Poland, Raport No. IF–3–80(1980), 1–11.

[BoJó4] BOROWIECKI M., JÓŹWIAK T.: Computing the characteristic polynomial of a multidigraph, Demonstr. Math., 14(1981), No. 2, 361–370.

[BoJó5] BOROWIECKI M., JÓŹWIAK T.: A note on characteristic and permanental polynomials of multigraphs, In: Graph Theory, Proc. Conf. held in Lagow, Poland 1981, Springer-Verlag, Berlin 1983, pp. 75–78.

[Bor1] BORZACCHINI L.: Reconstruction theorems for graph enumerating polynomials, Publ. Inst. applic. calcolo Mauro Picone, 3(171)(1979), 3–8.

[Bos1] BOSAK J.: Partially directed Moore graphs, Math. Slovaca, 29 (1979), 181–196.

[BoGu1] BOSANAC S., GUTMAN I.: Effect of a ring on the stability of polycyclic conjugated molecules, Z. Naturforsch., 32a(1977), 10–12.

[BoSh1] BOSE R. C., SHRIKHANDE S. S.: Graphs in which each pair of vertices is adjacent to the same number of other vertices, Studia Sci. Math. Hungar., 5(1970), 181–195.

[BrGe1] BRAUER A., GENTRY I. C.: On the characteristic roots of tournament matrices, Bull. Amer. Math. Soc., 74(1968), 1133–1135.

[BrGe2] BRAUER A., GENTRY I. C.: Some remarks on tournament matrices, Linear Algebra and Its Appl., 5(1972), 311–318.

[BrMe1] BRIDGES W. G., MENA R. A.: Rational circulants with rational spectra and cyclic strongly regular graphs, Ars Combinatoria, 8 (1979), 143–161.

[BrMe2] BRIDGES W. G., MENA R. A.: Multiplicative cones—a family of three eigenvalue graphs, Aequationes Math., 22(1981), No. 2–3, 208–214.

[BrMe3] BRIDGES W. G., MENA R. A.: X^k–digraphs, J. Combin. Theory(B), 30(1981), 136–143.

[BrMe4] BRIDGES W. G., MENA R. A.: On the rational spectra of graphs with Abelian Singer groups, Linear Algebra Appl., 46(1982), 51–60.

[BrMe5] BRIDGES W. G., MENA R. A.: Rational G–matrices with rational eigenvalues, J. Combin. Theory (A), 32(1982), No. 2, 264–280.

[BrSh1] BRIDGES W. G., SHRIKHANDE M. S.: Special partially balanced incomplete block designs and associated graphs, Discrete Math., 9(1974), 1–18.

[BrDu1] BRIGHAM R. C., DUTTON R. D.: Bounds on graph spectra, J. Combin. Theory B, 37(1984), 228–234.

[BrLi1] BROWER A. E., LINT J. H. VAN : Strongly regular graphs and partial geometries, **Enumeration and designs**, Waterloo Ontario 1982, Academic Press, Toronto Ontario, 1984, 85–122.

[Bru1] BRUALDI R. A.: Matrices, eigenvalues, and directed graphs, Linear and Multilinear Algebra, 11(1982), No. 2, 143–165.

[BrLi1] BRUALDI R. A., LI Q.: Problem 31, Discrete Math., 43(1983), 329–330.

[BrGo1] BRUALDI R. A., GOLDWASSER J. L.: Permanent of the Laplacian matrix of trees and bipartite graphs, Discrete Math., 48 (1984), 1–21.

[BruLi1] BRUALDI R. A., LI Q.: Small diameter interchange graphs of matrices of zeros and ones, Linear Algebra Appl., 46(1982), 177–194.

[BuHo1] BUCKLEY F., HOFFMAN A. J.: On the mixed achromatic number and other functions of graphs, **Graph Theory and related Topics**, (Proc. Conf. Univ. Waterloo, 1977), Academic Press, New York 1979, pp. 105–119.

[Bum1] BUMILLER C.: On rank three graphs with a large eigenvalue, Discrete Math., 23(1978), 183–187.

[BuMS1] BUSSEMAKER F. C., MATHON R. A., SEIDEL J. J.: Tables of two–graphs, Technological Univ. Eindhoven, TH–Report 79–WSK–05(1979), 1–99; **Combinatorics and Graph Theory** (Calcutta 1980), Lecture Notes Math. 885, Springer, Berlin, 1981, pp. 70–112.

[Cam1] CAMERON P. J.: Strongly regular graphs, **Selected topics in graph theory**, (Ed. L. W. Beineke, R. J. Wilson), Academic Press, London–New York–San Francisco, 1978, pp. 337–360.

[CaGoS] CAMERON P. J., GOETHALS J. M., SEIDEL J. J.: Strongly regular graphs having strongly regular subconstituents, J. of Algebra, 55(1978), 257–280.

[CaMo1] CAPOBIANCO M., MOLLUZZO J. C.: **Examples and counterexamples in graph theory**, North–Holland, New York, 1978.

[ChLe1] CHEN H. H., LEE F.: Numerical representation and identification of graphs, J. Math. Phys., 22(1981), 2727–2731.

[Che1] CHENG C. S.: Maximizing the total number of spanning trees in a graph; two related problems in graph theory and optimum design theory, J. Combin. Theory (B), 31(1981), No. 2, 240–248.

[CJLF1] CHOU KUO-CHEN, JIANG SHOU-PING, LIU WEI-MIN, FEE CHIH-HAO: Graph theory of enzyme kinetics I, Steady-state reaction systems, Sci. Sinica, 22(1979), 341–358.

[ChLi1] CHOU KUO-CHEN, LIU WEI MIN: Graphical rules for non-steady state enzyme kinetics, J. Theor. Biol., 91(1981), 637–654.

[CLHa1] CLIFF A. D., HAGGETT P.: Graph theory and geography; some combinatorial and other applications, In: **Applications of Combinatorics**, (Ed. Wilson R. J.), Shiva Publ. Co., Nantwich, Cheshire, 1982, pp. 51–66.

[CLHO1] CLIFF A., HAGGETT P., ORD K.: Graph theory and geography, In: **Applications of Graph Theory**, (Ed. R. J. Wilson, L. W. Beineke) Academic Press, London–New York–San Francisco, 1979, 293–326.

[COH1] COHEN A. M.: Geometries originating from certain distance-regular graphs, London Math. Soc. Lecture Note Ser., 49(1981), 81–87.

[COL1] COLLATZ L.: Spektren periodischer Graphen, Resultate Math., 1(1978), 42–53.

[CON1] CONSTANTINE G. M.: Schur convex functions on the spectra of graphs, Discrete Math., 45(1983), No. 2–3, 181–188.

[CoLo1] COULSON C. A., LONGUET-HIGGINS H. C.: The electronic structure of conjugated systems, I. General theory, Proc. Roy. Soc. (London) Ser. A, 191(1947), 39–60.

[CoLM1] COULSON C. A., O'LEARY B., MALLION R. B.: **Hückel Theory for Organic Chemists**, Academic Press, London, 1978.

[Cox1] COXETER H. S. M.: The product of the generators of a finite group generated by reflections, Duke Math. J., 18(1951), 765–782.

[CKMT1] CRISCUOLO G., KWOK CH. M., MOWSHOWITZ A., TORTORA R.: The group and the minimal polynomial of a graph, J. Combin. Theory (B), 29(1980), 293–302.

[CVE1] CVETKOVIĆ D.: Some topics from the theory of graph spectra, Ber. Math. –Statist. Sekt. Forsch. Graz No. 100–105, No. 101(1978), 1–5.

[CVE2] CVETKOVIĆ D.: A note on construction of graphs by means of their spectra, Publ. Inst. Math. (Beograd), 27(41)(1980), 27–30.

[CVE3] CVETKOVIĆ D.: **Combinatorial Matrix Theory with Applications to Electrical Engineering, Chemistry and Physics**(Serbo-Croatian), Naučna knjiga, Beograd, 1980.

[CVE4] CVETKOVIĆ D.: Some possible directions in further investigations of graph spectra, In: **Algebraic Methods in Graph Theory**, Vol. I; Szeged, 1978, Colloq. Math. Soc. Janos Bolyai (1981), North–Holland, Amsterdam, 1981, pp. 47–67.

[CVE5] CVETKOVIĆ D.: On graphs whose second largest eigenvalue does not exceed 1, Publ. Inst. Math. (Beograd), 31(45)(1982), 15–20.

[CVE6] CVETKOVIĆ D.: Discussing graph theory with a computer II, Theorems suggested by the computer, Publ. Inst. Math. (Beograd), 33(47)(1983), 29–33.

[CVE7] CVETKOVIĆ D.: Spectral characterizations of line graphs. Variations on the theme, Publ. Inst. Math. (Beograd), 34(48), (1984), 31–35.

[CVE8] CVETKOVIĆ D.: A project for using computers in further development of graph theory, in: **Theory and Applications of Graphs**, Proc. 4th International Conf. on Theory and Applications of Graphs, Kalamazoo 1980, ed. G. Chartrand, Y. Alavi, D. L. Goldsmith, L. Lesniak-Foster, D. R. Lick John Wiley & Sons, New York 1981, 285–296.

[CVE9] CVETKOVIĆ D.: Discussing graph theory with a computer IV, Knowledge organization and examples of theorem proving, (**Graph Theory**, Proc. Fourth Yugoslav Seminar Graph Theory, Novi Sad, April 15–16, 1983, ed. D. Cvetković, I. Gutman, T. Pisanski, R. Tošić University of Novi Sad, Novi Sad 1984), 43–68.

[CvDo1] CVETKOVIĆ D., DOOB M.: Root systems, forbidden sub-graphs and spectral characterizations of line graphs, Graph Theory, Proc. Fourth Yugoslav Sem. **Graph Theory**, Novi Sad, April 15–16, 1983, (Ed. D. Cvetković, I. Gutman, T. Pisanski, R. Tošić, Inst. Math., Novi Sad 1984), pp. 69–99.

[CvDo2] CVETKOVIĆ D., DOOB M.: Developments in the theory of graph spectra, Linear and Multilinear Algebra, 18(1985) 153–181.

[CvDG1] CVETKOVIĆ D., DOOB M., GUTMAN I.: On graphs whose spectral radius does not exceed $\sqrt{2+\sqrt{5}}$, Ars Combinatoria, 14(1982), 225–239.

[CvDSa1] CVETKOVIĆ D., DOOB M., SACHS H.: **Spectra of graphs–Theory and application**, Deutscher Verlag der Wissenschaften., Berlin, 1980; Academic Press, New York, 1980; second edition: 1982; Russian translation: Naukova Dumka, Kiev, 1984.

[CvDS1] CVETKOVIĆ D., DOOB M., SIMIĆ S.: Some results on generalized line graphs, C. R. Math. Rept. Acad. Sci. Canada, 2(1980), No. 3, 147–151.

[CvDS2] CVETKOVIĆ D., DOOB M., SIMIĆ S.: Generalized line graphs, J.Graph Theory, 5(1981), No. 4, 385–399.

[CvG1] CVETKOVIĆ D., GUTMAN I.: A new spectral method for determining the number of spanning trees, Publ. Inst. Math. (Beograd), 29(43)(1981), 49–52.

[CvGS] CVETKOVIĆ D., GUTMAN I., SIMIĆ S.: On self pseudo–inverse graphs, Univ. Beograd Publ. El. tehn. fak., Ser. Mat. Fiz., No. 602–633, (1978), 111–117.

[CvKS1] CVETKOVIĆ D., KRAUS L., SIMIĆ S.: Discussing graph theory with a computer I, Implementation of graph theoretic algorithms, Univ. Beograd Publ. Elektrotehn. Fak. Ser. Mat. Fiz., No. 716–No. 734(1981), 100–104.

[CvPe1] CVETKOVIĆ D., PETRIĆ M.: A table of connected graphs on six vertices, Discrete Math., 50(1984), 37–49.

[CvPe2] CVETKOVIĆ D., PETRIĆ M.: Connectedness of the non-complete exptended p-sum of graphs, Rev. Res. Fac. Sci. Univ. Novi Sad, 13(1983), 345–352.

[CveP1] CVETKOVIĆ D., PEVAC I.: Discussing graph theory with a computer III, Man-machine theorem proving, Publ. Inst. Math. (Beograd), 34(48)(1984), 37–47.

[CvRa1] CVETKOVIĆ D., RADOSAVLJEVIĆ Z.: A construction of the 68 connected regular graphs, non-isomorphic but cospectral to line graphs, **Graph Theory**, Proc. Fourth Yugoslav Sem. Graph Theory, Novi Sad, April 15–16, 1983, (Ed. D. Cvetković, I. Gutman, T. Pisanski, R. Tošić, Inst. Math., Novi Sad 1984), pp. 101–123.

[CvSi1] CVETKOVIĆ D., SIMIĆ S.: Graph equations, In: **Beiträge zur Graphentheorie und deren Anwendungen,** vorgetragen auf dem Internat. Koll. Oberhof(DDR), 1977, 40–56.

[CvSi2] CVETKOVIĆ D., SIMIĆ S.: A bibliography of graph equations, J. Graph Theory, 3(1979). No. 4, 311–324.

[Cyv1] CYVIN S. J.: Two-dimensional Hückel molecular orbital theory, Tetrahedron Letters, 22(1981), 2709–2712.

[Cyv2] CYVIN S. J.: Two-dimensional Hückel molecular orbital theory, Theochem, 4(1982), 315–324.

[Dam1] DAMERELL R. M.: Distance-transitive and distance-regular graphs, J. Combin. Theory(B), 31(1981), 46–53.

[DaGo1] DAMERELL R. M., GEORGIACODIS M. A.: On the maximum diameter of a class of distance-regular graphs. Bull. London Math. Soc. 13(1981) 316–322.

[DaGo2] DAMERELL R. M., GEORGIACODIS M. A.: Moore geometries I. J. London Math. Soc. 23(1981) 1–9.

[Dav1] DAVIDSON R. A.: Spectral analysis of graphs by cyclic automorphism subgroups, Theoret. Chim. Acta (Berl.), 58(1981), 193–231.

[DED1] DEDO E.: La ricostruibilità del polinomio caratteristico del commutato di un graffo, Boll. Unione Mat. Ital., 18A(1981), No. 5, 423–429.

[DEPo1] DEDO E., PORCU L.: Some properties of a miltidigraph by semi-incidence matrices, Discrete Math., 55(1985), 1–11.

[DETo1] DEIFT P. A., TOMEI C.: On the determinant of the adjacency matrix for a planar sublattice, J. Combin. Theory (B), 35(1983), 278–289.

[DEL1] DELSARTE P.: Regular schemes over a finite abelian group, Geom. Dedic., 8(1979), 477–490.

[DESo1] DERFLINGER G., SOFER H.: Die HMO–Koeffizienten der linearen Polyacene in geschlossener Form, Monatsh. Chem., 99(1968), 1866–1875.

[DIA1] DIAS J. R.: Properties and derivation of the fourth and sixth coefficients of the characteristic polynomial of molecular graphs. New graphical invariants, Theoret. Chim. Acta (Berl.), 68(1985), 107-123.

[DIA2] DIAS J. R.: A periodic table for polycyclic aromatic hydrocarbons. Part VII. 1–Factors and 2–Factors of benzenoid hydrocarbons, Nouv. J. Chim., 9(1985), 125–134.

[DiKZ1] DINIC E. A., KELMANS A. K., ZAITSEV M. A.: Nonisomorphic trees with the same T–polynomial, Inform. Process. Lett., 6(1977), No. 3, 3–8.

[DIx1] DIXON W. T.: A useful theorem in simple molecular orbital theory, J. C. S. Faraday II, 72(1976), 282–287.

[DIx2] DIXON W. T.: Construction of molecular orbital energy level diagrams by the method of progressive interactions of atomic orbitals, J. C. S. Faraday II, 73(1977), 67–75.

[DIx3] DIXON W. T.: Some new theorems and methods for establishing the relationship between the symmetry of a molecular orbital and its energy, J. C. S. Faraday II, 74(1978), 511–520.

[DMI1] DMITRIEV I. S.: **Molecules without Chemical Bonds**(Russian), Khimiya, Leningrad, 1980.

[DMI2] DMITRIEV I. S.: **Molecules without Chemical Bonds** Mir, Moscow, 1981.

[DMI3] DMITRIEV I. S.: **Molekule ohne chemische Bindungen**, Verlag für Grundstoffindustrie, Leipzig, 1982.

[DOB1] DOBRYNIN V. YU.: On Cvetković's estimate of the independencé Redkol number of a graph (Russian), Vestnik LGU, Math., mech., astr., Leningrad 1983, (manuscript No. 6217–83 DEP, VINITI 23. nov. 1983).

[DOO1] DOOB M.: Graphs with a small number of distinct eigenvalues, II, manuscript.

[DOO2] DOOB M.: Seidel switching and cospectral graphs with four distinct eigenvalues, Ann. New York Acad. Sci., 319(1979), 164–168.

[DOO3] DOOB M.: A surprising property of the least eigenvalue of a graph, Linear Algebra Appl., 46(1982), 1–7.

[DOO4] DOOB M.: Applications of grpah theory to linear algebra Math. Mag. 57(1984), 67–76.

[DOO5] DOOB M.: Pseudocyclic Graphs, Proceedings of the Sixth Yugoslav Seminar on Graph Theory, Dubrovnik, 1985, 107–114.

[DOCV1] DOOB M., CVETKOVIĆ D.: On spectral characterizations and embeddings of graphs, Linear Algebra Appl., 27(1979), 17–26.

[DRFR1] DROESBEKE F., FRENNE A. DE: Étude des bournes du nombre chromatique d'un graphe (deuxième partie), **Problèmes Combinatoires et Théorie des Graphes**, Colloque Internat. C. N. R. S, Orsay, 1976, C. N. R. S., No. 260, Publ. 1978, (ed. Bermond J. C., Fournier J. C., M. Las Vergnas, Sotteau D.), 113–116.

[DYKA1] DYADYUSHA G. G., KACHKOVSKII A. D.: Application of graphs to the theory of the basicity of the terminal groups of polymethine dyes (Russian), Teoret. Eksp. Khim., 15(1979), 152–161.

[DYKA2] DYADYUSHA G. G., KACHKOVSKII A. D.: Application of graphs to the investigation of electron-donor properties of annelated terminal groups of polymethine dyes (Russian), Teoret. Eksp. Khim., 16(1980), 615–619.

[DYKA3] DYADYUSHA G. G., KACHKOVSKII A. D.: Application of graph theory to the theory of the colour of polymethine dyes (Russian), Teoret. Eksp. Khim., 17(1981), 393–398.

[EDEL1] EDWARDS C. S., ELPHICK C. H.: Lower bounds for the clique and the chromatic number of a graph, Discrete Appl. Math. 5(1983), 51–64

[EGA1] EGAWA Y.: Characterization of $H(n,q)$ by the parameters, J. Combin. Theory(A), 31(1981), 108–125.

[EGMA1] EGOROV V. N., MARKOV A. I.: Ádáms conjecture for graphs with circulant adjacency matrix (Russian), Dokl. Akad. Nauk SSSR, 249(1979), No. 3, 529–532.

[EIC1] EICHINGER B. E.: Elasticity theory, I. Distribution functions for perfect phantom networks, Macromolecules, 5(1972), No. 4, 496–503.

[EIC2] EICHINGER B. E.: Configuration statistics of Gaussian molecules, Macromolecules, 13(1980), No. 1, 1–11.

[EIC3] EICHINGER B. E.: Random elastic networks, I. Computer simulation of linked stars, J. Chem. Phys., to appear.

[EIMA1] EICHINGER B. E., MARTIN J. E.: Distribution functions for Gaussian molecules, II. Reduction of the Kirchoff matrix for large molecules, J. Chem. Phys., 69(10)(1978), 4595–4599.

[EIN1] EINBU J. M.: The enumeration of bit–sequences that satisfy local criteria, Publ. Inst. Math. (Beograd), 27(41)(1980), 51–56.

[ELB1] EL–BASIL S.: Matrix inverse of chemical graphs. II. Singular graphs, Match, 16(1984), 153–162.

[ELB2] EL–BASIL S.: Characteristic polynomials of large graphs. On alternate form of characteristic polynomial, Theoret. Chim. Acta (Berl.), 65(1984), 191–197.

[ELB3] EL–BASIL S.: Fibonacci relations. On the computation of some counting polynomials of very large graphs, Theoret. Chim. Acta (Berl.), 65(1984), 199–213.

[ELBH1] EL–BASIL S., HUSSEAN A.: Matrix inverse of chemical graphs. I. Ordering, comparability and non–alternant hydrocarbons, Match, 16(1984), 133–151.

[ERFH1] ERDOS P., FAJTLOWICZ S., HOFFMAN A. J.: Maximum degree in graphs of diameter 2, Networks, 10(1980), 87–90.

[ESHA1] ESSER F., HARARY F.: On the spectrum of a complete multipartite graph, European J. Combin., 1(1980), 211–218.

[ESHA2] ESSER F., HARARY F.: Digraphs with real and Gaussian spectra, Discrete Appl. Math., 2(1980), 113–124.

[ESHA3] ESSER F., HARARY F.: The pairing theorem for digraph spectra, Bull. Malaysian Math. Soc. (2), 4(1981), No. 1, 17–19.

[ESHA4] ESSER F., HARARY F.: Complete multipartite graphs have unimodal characteristic sequences.

[FAR1] FARRELL E. J.: Matchings in ladders, Ars Combinatoria, 6(1978), 153–161.

[FAR2] FARRELL E. J.: The tree polynomial of a graph, J. Combin., Inf. Syst. Sci., 4(1979), No. 3, 211–218.

[FAR3] FARRELL E. J.: Introduction to matching polynomials, J. Combin. Theory (B), 27(1979), No. 1, 75–86.

[FAR4] FARRELL E. J.: On a class of polynomials obtained from the circuits in a graph and its applicaton to the characteristic polynomials of graphs, Discrete Math., 25(1979), 121–133.

[FAR5] FARRELL E. J.: On a general class of graph polynomials, J. Combin. theory (B), 26(1979), 111–122.

[FAR6] FARRELL E. J.: The matching polynomial and its relation to the acyclic polynomial of a graph, Ars Combinnatoria, 9(1980), 221–228.

[FAR7] FARRELL E. J.: Matchings in complete graphs and complete bipartite graphs, J. Combin. Inf. System Sci., 5(1980), 47–51.

[FAR8] FARRELL E. J.: Defect–d matchings in graphs with cyclomatic numbers 0, 1 and 2, Discrete Math., 33(1981), 259–266.

[FAR9] FARRELL E. J.: Matchings in triangular animals, J. Combin. Inf. System Sci., 7(1982), 143–154.

[FAR10] FARRELL E. J.: A note on the circuit polynomials and characteristic polynomials of wheels and ladders, Discrete Math., 39(1982), No. 1, 31–36.

[FAR11] FARRELL E. J.: On a class of polynomials associated with the paths in a graph and its application to minimum nodes disjoint path coverings of graphs, Internat. J. Math. Sci., 6(1983), 715–726.

[FAR12] FARRELL E. J.: The subgraph polynomial and its relation to some other graph polynomials, Caribbean J. Math., 2(1983). 39–53.

[FAR13] FARRELL E. J.: A survey of the unifying effects of F-polynomials in combinatorics and graph theory, **Graph Theory**, Proc. Fourth Yugoslav Sem. Graph Theory, Novi Sad, April 15–16, 1983, (Ed. D. Cvetković, I. Gutman, T. Pisanski, R. Tošić, Inst. Math., Novi Sad 1984), pp. 137–150.

[FAR14] FARRELL E. J.: On coefficients of path polynomials, Internat. J. Math. Sci., 7(1984), 65–73.

[FAGU1] FARRELL E. J., GUTMAN, I.: A note on the circuit polynomial and its relation to the μ–polynomial, Match, 18(1985), 55–62.

[FAGR1] FARRELL E. J., GRELL J. C.: On reconstructing the circuit polynomial of a graph, Caribbean J. Math., 1(1982), No. 3, 109–119.

[FaGr2] FARRELL E. J., GRELL J. C.: Some analytical properties of the circuit polynomial of a graph, Caribbean J. Math., 2(1983), 69–76.

[FaWa1] FARRELL E. J., WAHID S. A.: Matchings in benzene chains, Discrete Appl. Math., 7(1984), 45–54.

[FaWa2] FARRELL E. J., WAHID S. A.: Matchings in pentagonal chains, Discrete Appl. Math., 8(1984), 31–40.

[FaWa3] FARRELL E. J., WAHID S. A.: On the reconstruction of the matching polynomial and the reconstruction conjecture, manuscript.

[Fin1] FINCK H. J.: Das Spektrum eines Graphen und seine Anwendung in der Chemie, Rostock. Math. Kolloq., 11(1979), 47–58.

[For1] FORSMAN W. C.: Graph theory and the statistics and dynamics of polymer chains, J. Chem. Phys., 65(1976), 4111–4115.

[FüKo1] FÜREDI Z., KOMLÓS J.: The eigenvalues of random symmetric matrices. Combinatorica, 1(1981), No. 3, 233–241.

[GaRa1] GANGOPADHYAY T., RAO H. P.: Multipartite self-complementary graphs, Ars Combinatoria, 13(1982), 87–114.

[Gar1] GARDINER A. D.: The classification of symmetric graphs, Berichte der Math.–Statistischen Sektion in Forschungszentrum Graz, No. 100–No. 105(1978), No. 102, 1–33.

[Gar2] GARDINER A. D.: When is an array realised by a distance-regular graph?, **Algebraic Methods in Graph Theory**, Vol. I, Szeged, 1978, Colloq. Math. Soc. Janos Bolyai, 25, North–Holland, Amsterdam, 1981, 209–219.

[GaRo1] GASIOROWSKI B., ROZMUS M.: Application of the characteristic polynomial of the transition matrix to the investigation of the properties of a finite graph, Demonstratio Math., 15(1982), No. 3, 625–634.

[Gil1] GILL M. K.: A note concerning Acharya's conjecture on a spectral measure of structural balance in a social system, **Combinatorics and Graph Theory**, (Calcutta, 1980), Lecture Notes Math. 885, Springer, Berlin, 1981, pp. 266–271.

[GIL2] GILL M. K.: A graph theoretical recurrence formula for computing the characteristic polynomial of a matrix, **Combinatorics
and Graph Theory**, (Calcutta, 1980), Lecture Notes Math.
885, Springer, Berlin 1981, pp. 261–265.

[GIAC1] GILL M. K., ACHARYA B. D.: A spectral characterization of
local cycle balance in networks, Technical Report, Mehta Res.
Inst., Allahabad (India), 1979.

[GIAC2] GILL M. K., ACHARYA B. D.: A recurrence formula for computing the characteristic polynomial of a sigraph, J. Combin.,
Inform. Sys. Sci., 5(1980), No. 1, 68–72.

[GIN1] GINSBURG B. D.: On the number of interior stability of graphs
(Russian), Reports ANSSSR, 85(1977), 289–292.

[GIR1] GIRKO, V. L.: **Random Matrices**, Kiev, 1975.

[GOD1] GODSIL C. D.: Matchings and walks in graphs, J. Graph Theory, 5(1981), No. 3, 285–297.

[GOD2] GODSIL C. D.: Matching behaviour is asymptotically normal,
Combinatorica, 1(1981), No. 4, 369–376.

[GOD3] GODSIL C. D.: Hermite polynomials and the duality relation
for the matching polynomial, Combinatorica, 1(1981), No. 3,
257–262.

[GOD4] GODSIL C. D.: Equiarboreal graphs, Combinatorica, 1(1981),
No. 2, 163–167.

[GOD5] GODSIL C. D.: Some graphs with characteristic polynomials
which are not solvable by radicals, J. Graph Theory, 6(1982),
No. 2, 211–214.

[GOD6] GODSIL C. D.: Eigenvalues of graphs and digraphs, Linear
Algebra Appl., 46(1982), 43–50.

[GOD7] GODSIL C. D.: Real graph polynomials, **Progress in Graph
Theory**, Proc. conf. Combinatorics, Waterloo, Ontario, 1982,
(1984) 281–293

[GOD8] GODSIL C. D.: Inverses of trees, Combinatorica, 5(1985) 33–39.

[GOD9] GODSIL C. D.: Spectra of trees, In: Proc. Conf. on Convexity and Graph Theory, Israel, 1981, Annals Discrete Math., 20(1984) 151-159.

[GOGU1] GODSIL C. D., GUTMAN I.: Topological resonance energy is real, Z. Naturforsch., 34a(1979), 776–777.

[GOGU2] GODSIL C. D., GUTMAN I.: On the matching polynomial of a graph, **Algebraic methods in Graph Theory**Vol. I; Szeged 1978, Colloq. Math. Soc. Janos Bolyai, 25, North–Holland, Amsterdam, 1981, pp. 241–249; Also published in: Math. Research Report Univ. Melbourne, 35(1978), 1–10.

[GOGU3] GODSIL C. D., GUTMAN I.: On the theory of matching polynomial, J. Graph Theory, 5(1981), No. 2, 137–144.

[GOGU4] GODSIL C. D., GUTMAN I.: Some remarks on the matching polynomial and its zeros, Croat. Chem. Acta, 54(1981), 53–59.

[GOGU5] GODSIL C. D., GUTMAN I.: Contributions to the theory of topological resoance energy, Acta Chim. Acad. Sci. Hung., 110(1982), 415–424.

[GHMK1] GODSIL C. D., HOLTON D. A., MCKAY B. D.: The spectrum of a graph, **Combinatorial Mathematics**V (Ed. C. H. C. Little), Springer Verlag, Berlin–Heidelberg–New York, 1977, pp. 91–117.

[GOMK1] GODSIL C. D., MCKAY B. D.: Constructing cospectral graphs, Aequat. Math, 25(1982), 257–268.

[GOMK2] GODSIL C. D., MCKAY B. D.: A new graph product and its spectrum, Bull. Austral. Math. Soc., 18(1978), 21–28.

[GOMK3] GODSIL C. D., MCKAY B. D.: Graphs with regular neighbourhoods, **Combinatorial Mathematics VII**, (Proc. Seventh Austral. Conf., Univ. Newcastle, Newcastle, 1979), Lecture Notes Math. 829, Springer, Berlin, 1980, pp. 127–140.

[GOMK4] GODSIL C. D., MCKAY B. D.: Feasibility conditions for the existence of walk–regular graphs, Linear Algebra Appl., 30(1980), 51–61.

[GoMK5] GODSIL C. D., MCKAY B. D.: Spectral conditions for the reconstructibility of a graph, J. Combin. Theory(B), 30(1981), No. 3, 285–289.

[GoKr1] GOHBERG I. C., KREIN M. G.: **Introduction to the theory of linear non-selfadjoint operators in Hilbert spaces**(Russian), Nauka, Moscow 1965.

[GoJW1] GOLDMAN J. R., JOICHI J. T., WHITE D. E.: Rook polynomial, Möbius inversion and the umbral calculus, J. Combin. Theory (A), 21(1976), 230–239.

[GoMi1] GONDRAN M., MINOUX M.: Valeurs propres et vecteurs propres en théorie des graphes, **Problèmes Combinatoires et Théorie des Graphes**, Colloque Internat. C. N. R. S, Orsay, 1976, C. N. R. S., No. 260, Publ. 1978, (ed. Bermond J. C., Fournier J. C., M. Las Vergnas, Sotteau D.), 181–183.

[GoMi2] GONDRAN M., MINOUX M.: Valeurs propres et vecteurs propres dans les dioïdes et leur interprétation en théorie des graphes, Bull. Dir. Etud. Rech. (C), Math. Informat., 2(1977), 25–41.

[Gou1] GOULD P. R.: On the geographical interpretation of eigenvalues, Inst. Brit. Geogr. Publ., 42(1967), 63–92.

[GrLo1] GRAHAM R. L., LOVÀSZ L.: Distance matrix polynomials of trees, Proc. Internat. Conf., Western Mich. Univ., Kalamazoo, Mich. 1976, Lecture Notes in Math., No. 642, Springer, Berlin 1978, 186–190.

[GrLo2] GRAHAM R. L., LOVÀSZ L.: Distance matrix polynomials of trees, **Problèmes Combinatoires et Théorie des Graphes**, (Colloq. Internat. C. N. R. S, Univ. Orsay, 1976), Colloq. Internat. C. N. R. S., 260(1978), 189–190.

[GrLo3] GRAHAM R. L., LOVÀSZ L.: Distance matrix polynomials of trees, Adv. in Math., 29(1978), No. 1, 60–88.

[GrA1] GRAOVAC A.: On the construction of a Hermitian matrix associated with the acyclic polynomial of a conjugated hydrocarbon, Chem. Phys. Letters, 82(1981), 248–251.

[GrBa1] GRAOVAC A., BABIĆ D.: On the matching spectrum of ro-
tagraphs, Z. Naturforsch., 40a(1985) 66–72.

[GrGu1] GRAOVAC A., GUTMAN I.: The determinant of the adjacency
matrix of the graph of a conjugated molecule, Croat. Chem.
Acta, 51 (1978), 133–140.

[GrGu2] GRAOVAC A., GUTMAN I.: The determinant of the adjacency
matrix of a molecular graph, Match, 6(1979), 49–73.

[GrGu3] GRAOVAC A., GUTMAN I.: Estimation of the HOMO–LUMO
separation, Croat. Chem. Acta, 53(1980), 45–50.

[GrGP1] GRAOVAC A., GUTMAN I., POLANSKY O. E.: Topological
effects on MO energies. IV. The total π-electron energy of S- and
T-isomers, Monatsh. Chem., 115(1984), 1–13.

[GrGP2] GRAOVAC A., GUTMAN I., POLANSKY O. E.: An interlacing
theorem in simple molecular orbital theory, J. C. S. Faraday II,
to appear.

[GrGT1] GRAOVAC A., GUTMAN I., TRINAJSTIĆ N.: **Topological
Approach to the Chemistry of Conjugated Molecules**,
Springer Verlag, Berlin–Heidelberg–New York, 1977.

[GrKT1] GRAOVAC A., KASUM D., TRINAJSTIĆ N.: On acyclic polynomials of (N)-heteroannulenes, Croat. Chem. Acta, 54(1981),
91–95.

[GrPT1] GRAOVAC A., POLANSKY O. E., TRINAJSTIĆ N., TYU-
TYULKOV N.: Graph theory in chemistry, II. Graph-theoretical
description of heteroconjugated molecules, Z. Naturforsch.,
30a(1975), 1696–1699.

[GPTy1] GRAOVAC A., POLANSKY O. E., TYUTYULKOV N. N.:
Acyclic and characteristic polynomial of regular conjugated polymers and their derivatives, Croat. Chem. Acta, 56(1983), No. 3,
325–356.

[GrTr1] GRAOVAC A., TRINAJSTIĆ N.: Möbius molecules and graphs,
Croat. Chem. Acta, 47(1975), 95–104.

[GrTr2] GRAOVAC A., TRINAJSTIĆ N.: Graphical description of Möbius molecules, J. Mol. Struct., 30(1976), 416–420.

[GrTR1] GRAOVAC A., TRINAJSTIĆ N., RANDIĆ M.: Graph-theoretical search for benzenoid polymers with zero energy gap, Croat. Chem. Acta, 53(1980), No. 4, 571–579.

[GrHK1] GREENWELL D. L., HEMMINGER R. L., KLERLEIN J.: Forbidden subgraphs, In: Proc. Fourth South–Eastern Conf. on Combinatorics, Graph Theory, and Computing, Florida, Atlantic University, 1973, pp. 389–394.

[Gut1] GUTMAN I.: The acyclic polynomial of a graph, Publ. Inst. Math. (Beograd), 22(36)(1977), 63–69.

[Gut2] GUTMAN I.: Bounds for total π-electron energy of polymethines, Chem. Phys. Letters, 50(1977), 488–490.

[Gut3] GUTMAN I.: Topological properties of benzenoid systems. An identity for the sextet polynomial, Theoret. Chim. Acta (Berl.), 45 (1977), 309–315.

[Gut4] GUTMAN I.: Topological properties of bond order, Z. Naturforsch., 32a(1977), 765–767.

[Gut5] GUTMAN I.: A topological formula for total π–electron energy, Z. Naturforsch., 32a(1977), 1072–1073.

[Gut6] GUTMAN I.: A comparative study of bond orders, Match, 3(1977), 121–132.

[Gut7] GUTMAN I.: A rule for charge distribution in non-alternant hydrocarbons, Chem. Phys. Letters, 58(1978), 137–139.

[Gut8] GUTMAN I.: Relations between bond orders, Bull. Soc. Chim. Beograd, 43(1978), 379–384.

[Gut9] GUTMAN I.: Bounds for Hückel total π–electron energy, Croat. Chem. Acta, 51(1978), 299–306.

[Gut10] GUTMAN I.: Topology and stability of conjugated hydrocarbons. The dependence of total π–electron energy on molecular topology (Serbo–Croatian), Bull. Soc. Chim. Beograd, 43(1978), 761–774.

[GUT11] GUTMAN I.: Contribution to the problem of the spectra of compound graphs, Publ. Inst. Math. (Beograd), 24(38)(1978), 53–60.

[GUT12] GUTMAN I.: The energy of a graph, **10. Steiermarkisches Math. Symposium**; Stift Rein, Graz, 1978, Ber. Math.–Statist. Sekt. Forsch. Graz, No. 100–No. 105(1978), Bericht No. 103, 1–22.

[GUT13] GUTMAN I.: Relation between the Coulson and Pauling bond orders, J. Chem. Phys., 68(1978), 1321–1322.

[GUT14] GUTMAN I.: Graph–theoretical formulation of Forsman's equations, J. Chem. Phys., 68(1978), 2523–2524.

[GUT15] GUTMAN I.: Bounds for reactivity indices, Theoret. Chim. Acta (Berl.), 47(1978), 217–222.

[GUT16] GUTMAN I.: Topological formulas for free-valence index, Croat. Chem Acta, 51(1978), 29–33.

[GUT17] GUTMAN I.: Electronic properties of Möbius systems, Z. Naturforsch., 33a(1978), 214–216.

[GUT18] GUTMAN I.: On polymethine graphs, Match, 5(1979), 161–176.

[GUT19] GUTMAN I.: The matching polynomial, Match, 6(1979), 75–91.

[GUT20] GUTMAN I.: Two theorems on topological resonance energy, Bull. Soc. Chim. Beograd, 44(1979), 173–178.

[GUT21] GUTMAN I.: Effect of cycles on total π-electron energy of alternant conjugated hydrocarbons, J. C. S. Faraday II, 75(1979), 799–805.

[GUT22] GUTMAN I.: Difficulties with topological resonance energy, Chem. Phys. Letters, 66(1979), 595–597.

[GUT23] GUTMAN I.: On the dependence of π-electron charge distribution on molecular topology, Bull. Soc. Chim. Beograd, 44(1979), 627–630.

[GUT24] GUTMAN I.: Topological studies on heteroconjugated molecules. Alternant systems with one heteroatom, Theoret. Chim. Acta (Berl.), 50(1979), 287–297.

[GUT25] GUTMAN I.: A graph theoretical study on linear polyenes and their derivatives, Acta Chim. Acad. Sci. Hung., 99(1979), 145–153.

[GUT26] GUTMAN I.: Effect of cycles on topological resonance energy, Croat. Chem. Acta, 53(1980), 581–586.

[GUT27] GUTMAN I.: Spectral properties of some graphs derived from bipartite graphs, Match, 8(1980), 291–314.

[GUT28] GUTMAN I.: Characteristic and matching polynomials of some compound graphs, Publ. Inst. Math. (Beograd), 27(41)(1980), 61–66.

[GUT29] GUTMAN I.: Note on a topological property of the HOMO-LUMO separation, Z. Naturforsch., 35a(1980), 458–460.

[GUT30] GUTMAN I.: Polynomial matrix method for the estimation of π–electron energies of some linear conjugated molecules, J. C. S. Faraday II, 76(1980), 1161–1169.

[GUT31] GUTMAN I.: Total π–electron energy of a class of conjugated polymers, Bull. Soc. Chim. Beograd, 45(1980), 67–68.

[GUT32] GUTMAN I.: Contribution to the spectral theory of trees (Serbo-Croatian), Doct. thesis, Univ. Beograd, El. tehn. fak., 1981.

[GUT33] GUTMAN I.: Contribution to the quantum theory of alternant molecules (Serbo-Croatian), Univ. Kragujevac, Collection Sci. Papers Fac. Sci., 2(1981), 23–32.

[GUT34] GUTMAN I.: On the topological resonance energy of hetero-conjugated molecules, Croat. Chem. Acta, 54(1981), 75–80.

[GUT35] GUTMAN I.: Bounds for the smallest positive eigenvalue of a bipartite molecular graph, Match, 11(1981), 75–86.

[GUT36] GUTMAN I.: On perturbation graph theory, Chem. Phys. Letters, 81(1981), 303–305.

[GUT37] GUTMAN I.: Topological studies on heteroconjugated molecules. III. On the law of alternating polarity, Z. Natur-forsch., 36a (1981), 1112–1114.

[GUT38] GUTMAN I.: Topological properties of benzenoid systems. VI. On Kekulé structure count, Bull. Soc, Chim. Beograd, 46(1981), 411–415.

[GUT39] GUTMAN I.: A note on analogies between the characteristic and the matching polynomial of a graph, Publ. Inst. Math. (Beograd), 31(45)(1982), 27–31.

[GUT40] GUTMAN I.: Topological properties of benzenoid molecules, Bull. Soc. Chim. Beograd, 47(1982), 453–471.

[GUT41] GUTMAN I.: Characteristic and matching polynomials of benzenoid hydrocarbons, J. C. S. Faraday II, 79(1983), 337–345.

[GUT42] GUTMAN I.: Topological properties of benzenoid systems. XXI. Theorems, conjectures, unsolved problems, Croat. Chem. Acta, 56 (1983), 365–374.

[GUT43] GUTMAN I.: New approach to the McClelland approximation, Match, 14(1983), 71–81.

[GUT44] GUTMAN I.: General formulation of the first order perturbation graph theory, Chem. Phys. Letters, 103(1984), 475–478.

[GUT45] GUTMAN I.: Topological studies on heteroconjugated molecules. A new pairing theorem, Z. Naturforsch, 39a(1984), 152–154.

[GUT46] GUTMAN I.: Note on algebraic structure count, Z. Natur-forsch., 39a (1984), 794–796.

[GUT47] GUTMAN I.: Topological studies on heteroconjugated molecules. IV. Effect of a heteroatom on molecular orbital energies, Bull. Soc. Chim. Beograd, 49(1984), 157–163.

[GUT48] GUTMAN I.: Polynomials in graph theory, In: **Graph theory and its chemical applications**, (Ed. Bonchev D., Tyutyulkov N. N.), Nauka i Izkustvo, Sofia, to appear.

[GUT49] GUTMAN I.: Bounds for total π-electron energy of conjugated hydrocarbons, Z. Phys. Chem. (Leipzig), 226(1985), 59–64.

[GUCV1] GUTMAN I., CVETKOVIĆ D.: Relations between graphs and special functions, Univ. Kragujevac, Collection Sci. Papers Fac. Sci., 1(1980), 101–119.

[GUFW1] GUTMAN I., FARRELL E. J., WAHID S. A.: On the matching polynomials of graphs containing benzenoid chains, J. Combin. Inf. System Sci., 8(1983), 159–168.

[GUGR1] GUTMAN I., GRAOVAC A.: On structural factors causing stability differences between conjugated isomers, Croat. Chem. Acta, 49 (1977), 453–459.

[GUGM1] GUTMAN I., GRAOVAC A., MOHAR B.: On the existence of a Hermitian matrix whose characteristic polynomial is the matching polynomial of a molecular graph, Match, 13(1982), 129–150.

[GUGP1] GUTMAN I., GRAOVAC A., POLANSKY O. E.: On the theory of S- and T-isomers, Chem. Phys. Letters, 116(1985), 206–209.

[GUHA1] GUTMAN I., HARARY F.: Generalizations of the matching polynomial, Utilitas Math., 24(1983), 97–106.

[GUHE1] GUTMAN I., HERNDON W. C.: Note on the contribution of a ring to the stability of conjugated molecules, Chem. Phys. Letters, 105(1984), 281–284.

[GUHO1] GUTMAN I., HOSOYA H.: On the calculation of the acyclic polynomial Theoret. Chim. Acta (Berl.), 48(1978), 279–286.

[GUKR1] GUTMAN I., KRUSZEWSKI J.: On the occurence of eigenvalue one in the graph spectrum of benzenoid systems. Nouv. J. Chim., 9(1985) to appear.

[GUME1] GUTMAN I., MALLION R. B., ESSAM J. W.: Counting the spanning trees of a labelled molecular graph, Mol. Phys., 50(1983), 859–877.

[GUMT1] GUTMAN I., MILUN M., TRINAJSTIĆ N.: On topological resonance energy, Croat. Chem. Acta, 48(1976), 87–95.

[GuMT2] GUTMAN I., MILUN M., TRINAJSTIĆ N.: Non-parametric resonance energies of arbitrary conjugated molecules, J. Amer. Chem. Soc., 99 (1977), 1692–1704.

[GuMo1] GUTMAN I., MOHAR B.: Artifact in the topological resonance energy method, Chem. Phys. Letters, 69(1980), 375–377.

[GuMo2] GUTMAN I., MOHAR B.: More difficulties with topological resonance energy, Chem. Phys. Letters, 77(1981), 567–570.

[GuMo3] GUTMAN I., MOHAR B.: Some observations on the topological resonance energy of benzenoid hydrocarbons, Croat. Chem. Acta, 55 (1982), 375–382.

[GuPe1] GUTMAN I., PETROVIĆ S.: On total π-electron energy of benzenoid hydrocarbons, Chem. Phys. Letters, 97(1983), 292–294.

[GuPM1] GUTMAN I., PETROVIĆ S., MOHAR B.: Topological properties of benzenoid systems. XX. Matching polynomials and topological resonance energies of cata-condensed benzenoid hydrocarbons, Univ. Kragujevac, Collection Sci. Papers Fac. Sci., 3(1982), 43–90.

[GuPM2] GUTMAN I., PETROVIĆ S., MOHAR B.: Topological properties of benzenoid systems. XXa. Matching polynomials and topological resonance energies of peri-condensed benzenoid hydrocarbons, Univ. Kragujevac, Collection Sci. Papers Fac. Sci., 4(1983), 189–225.

[GuPo1] GUTMAN I., POLANSKY O. E.: On the matching polynomial of the graph $G(R_1, R_2, \ldots, R_n)$, Match, 8(1980), 315–322.

[GuPo2] GUTMAN I., POLANSKY O. E.: Cyclic conjugation and the Hückel molecular orbital model, Theoret. Chim. Acta (Berl.), 60(1981), 203–226.

[GuRo1] GUTMAN I., ROUVRAY D. H.: An approximate topological formula for the HOMO–LUMO separation in alternant hydrocarbons, Chem. Phys. Letters, 62(1979), 384–388.

[GuSh1] GUTMAN I., SHALABI A.: Edge erasure technique for computing the characteristic polynomials of molecular graphs, Match,
18(1985) 3–16.

[GuTN1] GUTMAN I., TEODOROVIĆ A. V., NEDELJKOVIĆ LJ.: Topological properties of benzenoid systems. Bounds on approximate formulae for total π-electron energy, Theoret. Chim. Acta
(Berl.), 65(1984), 23–31.

[GuTR1] GUTMAN I., TRINAJSTIĆ N.: Application of the Cauchy inequalities in simple molecular orbital theory, J. C. S. Faraday II,
73 (1977), 435–437.

[HAE1] HAEMERS W.: Eigenvalues methods, In: **Packing and Covering in Combinatorics**, Math. Centre tracts, 106(1979) (Ed. A.
Schrijver), Amsterdam, 1979, pp. 15–38;also published in: Stape-
len en overdekken, Syllabus, Stichting Math. Centrum, Amsterdam, 1978, pp. 15–33.

[HAE2] HAEMERS W.: Some remarks on subdesigns of symmetric designs, J. Statistical Planning and Inference, 3(1979), 361–366.

[HAE3] HAEMERS W.: An upper bound for the Shannon capacity of
graph **Algebraic Methods in Graph Theory**, Szeged, 1978,
Vol I, Colloq. Math. Soc. Janos Bolyai, 25, North–Holland, Amsterdam, 1981, 267–272.

[HAE4] HAEMERS W.: On the problem of Lovàsz concerning the Shannon capacity of a graph, IEEE Trans. Inform. Theory, 25(1979),
231–232.

[HAE5] HAEMERS W.: Eigenvalue techniques in design and graph
theory, Thesis, Tech. Univ. Eindhoven, The Netherlands, 1979;
Mathematical Centre Tracts 121, Mathematisch Centrum, Amsterdam 1980.

[HaLi1] HAEMERS W., LINT J. H. VAN: A partial geometry $pg(9, 8, 4)$.
Algebraic and Geometric Combinatorics(ed. E. Mendelsohn) Annals of Discrete Math. 15(1982), 205–212.

[HaRo1] HAEMERS W., ROOS C.: An inequality for generalized
hexagons, Geom. Dedic., 10(1981), 219–222.

[HAL1] HALL G. G.: A graphical model of a class of molecules, Internat. J. Math. Educ. Sci. Technology, 4(1973), 233–240.

[HAL2] HALL G. G.: On the eigenvalues of molecular graphs, Mol. Phys., 33(1977), No. 2, 551–557.

[HAL3] HALL G. G.: Eigenvalues of molecular graphs, Inst. Math. Appl., 17(1981), 70–72.

[HATA1] HANDA H., TAKENAKA Y.: On the spectrum of a graph, Keio Engrg. Rep., 31(1978), No. 9, 99–105.

[HAY1] HAY A.: On the choice of methods in the factor analysis of connectivity matrices: a comment, Inst. Brit. Geogr. Publ., 66(1975), 163–167.

[HEI1] HEILBRONNER E.: A simple equivalent bond orbital model for the rationalization of the C_2S spectra of the higher n-alkanes, in particular of polyethylene, Helv. Chim. Acta, 60 (1977), 2248–2257.

[HEI2] HEILBRONNER E.: Some comments on cospectral graphs, Match, 5 (1979), 105–133.

[HELI1] HEILMANN O. J., LIEB E. H.: Monomers and dimers, Phys. Rev. Letters, 24(1970), 1412–1414.

[HELI2] HEILMANN O. J., LIEB E. H.: Theory of monomer-dimer systems, Commun. Math. Phys., 25(1972), 190–232.

[HEIN1] HEINRICH P.: Zur Charakterisierung der Jordanschen Normalform der Adjazenzmatrix eines gerichteten zyklenfreien Graphen, Rostocker Math. Kolloq., 10(1978), 49–52.

[HEIN2] HEINRICH P.: Über die Jordansche Normalform der Adjazenzmatrix fur spezielle Klassen gerichteter Graphen, Preprint, Bergakademie, Freiberg, 1978.

[HEIN3] HEINRICH P.: Eine Beziehung zwischen den Weg-Kreis-Systemen in einem gerichteten Graphen und den Unterdeterminanten seiner charakteristischen Matrix, Math. Nachr., 93(1979), 7–20.

[HEIN4] HEINRICH P.: Zur Struktur gerichteter Graphen, Koll. Geometrie und Kombinatorik, TH Karl-Marx-Stadt, April 1979.

[HER1] HERNDON W. C.: On the concept of graph–theoretical individual ring resonance energies, J. Amer. Chem. Soc., 104(1982), 3541–3542.

[HEEL1] HERNDON W. C., ELLZEY M. L.: Closed-shell biradical structures, Tetrahedron Letters, 15(1974), 1399–1402.

[HEEL2] HERNDON W. C., ELLZEY M. L.: Procedures for obtaining graph-theoretical resonance energies, J. Chem. Inf. Comput. Sci., 19(1979), No. 4, 260–264.

[HEER1] HERNDON W. C., ELLZEY M. L., RAGHUVEER K. S.: Topological orbitals, graph theory and ionization potentials of saturated hydrocarbons, J. Amer. Chem. Soc., 100(1978), 2645–2654.

[HEPA1] HERNDON W. C., PÁRKÁNYI C.: Perturbation–graph theory, I. Resonance energies of heteroannulene π-systems, Tetrahedron, 34 (1978), 3419–3425.

[HEY1] HEYDEMANN M. C.: Charactérisation spectrale du joint d'un cycle par un stable, **Problèmes Combinatoires et Théorie des Graphes**, Colloq. Internat. C. N. R. S., Orsay 1976, C. N. R. S. No. 260(1978), (Ed. J. C. Bermond, J. C. Fournier, M. Las Vergnas, D. Sotteau), 225–227.

[HOF1] HOFFMAN A. J.: On signed graphs and gramians, Geom. Dedic., 6(1977), No. 4, 455–470.

[HOJA1] HOFFMAN A. J., JAMIL B. A.: On the line graph of the complete tripartite graph, Linear and Multilinear Algebra, 5(1977/78), 19–25.

[HOJO1] HOFFMAN A. J., JOFFE P.: Nearest S-matrices of given rank and the Ramsey problem for eigenvalues of bipartite S-graphs, (**Problèmes combinatoires et Théorie des Graphs**; Colloq. CNRS, Univ. Orsay, Orsay, 1976), Colloques Internat. CNRS, Paris, 260(1978), 237–240.

[HOSE1] HOMENKO N. P., SHEVCHENKO K. M.: Calculation of the number of subgraphs of an arbitrary graph that are isomorphic to another graph (Russian), Dokl. Akad. Nauk Ukrain. SSR (Ser. A), 1976, No. 9, 785–790, 864.

[HoSe2] HOMENKO N. P., SHEVCHENKO E. N.: On the problem of identifying and enumerating (Russian), Ukrain. Mat. Zh., 30(1978), No. 2, 201–211, 282 (English translation: Ukrainian Math. J., 30(1978), No. 2, 152–160).

[Hon1] HONG YUAN: An eigenvector condition for reconstructibility, J. Com bin. Theory(B), 32(1982), No. 3, 353–354.

[HoAi1] HORIKAWA T., AIHARA J.: Contribution of quadruply degenerate π-electron orbitals to London susceptibility, Bull. Chem. Soc. Japan, 56(1983), 1547–1548.

[Hos1] HOSOYA H.: Graphical and combinatorial aspects of some orthogonal polynomials, Nat. Sci. Rept. Ochanomizu Univ., 32(1981), No. 2, 127–138.

[HoHo1] HOSOYA H., HOSOI K.: Topological index as applied to π-electron systems. III. Mathematical relations among various bond orders, J. Chem. Phys., 64(1976), 1065–1073.

[HoMo1] HOSOYA H., MOTOYAMA A.: An effective algorithm for obtaining polynomials for dimer statistics. Application of operator technique on the topological index to two- and three-dimensional rectangular lattices. J. Math. Phys., 26(1985) 157–167.

[HoMu1] HOSOYA H., MURAKAMI M.: Topological index as applied to π-electron system. II. Topological bond order, Bull. Chem. Soc. Japan, 48(1975), No. 12, 3512–3517.

[HoOh1] HOSOYA H., OHKAMI N.: Operator technique for obtaining the recursion formulas of characteristic and matching polynomials as applied to polyhex graphs, J. Comput. Chem, 4(1983), 585–593.

[HoRa1] HOSOYA H., RANDIĆ M.: Analysis of the topological dependency of the characteristic polynomial in its Chebyshev expansion, Theoret. Chim. Acta (Berl.), 63(1983), 473–495.

[ILST1] ILIĆ P., SINKOVIĆ B., TRINAJSTIĆ N.: Topological resonance energies of conjugated structures, Israel J. Chem., 20(1980), 258–269.

[ITO1] ITO T.: Bipartite distance-regular graphs of valency three, Linear Algebra Appl., 46(1982), 195–213.

[JAHA1] JAN H. S., HALL A. S.: Linkage characteristic polynomials: assembly theorems, uniqueness, Trans. ASME: J. Mech. Des., 104(1982), No. 1, 11–20.

[JIA1] JIANG YUAN-SHENG: Graph theory of molecular orbitals. Calculation of a_N — the determinant of adjacency matrix and the stability of molecules, Sci. Sinica, 23(1980), No. 7, 847–861

[JIA2] JIANG YUAN-SHENG: Graphical-deleting method for computing characteristic polynomial of a graph, Sci. Sinica(B), 25(1982), No. 7, 681–689.

[JIA3] JIANG YUAN-SHENG: Problem on isospectral molecules, Sci. Sinica (B), 27(1984), 236–248.

[JITH1] JIANG YUAN-SHENG, TANG A., HOFFMAN, R.: Evaluation of moments and and their application to Hückel molecular orbital theory, Theoret. Chem. Acta. (Berl.), 65(1984) 255–265.

[JIIF1] JIDO Y., INAGAKI T., FUKUTOME H.: Application of the transfer matrix method to the Hückel molecular orbital analysis of complex conjugated molecules, Progress Theoret. Phys., 48(1972), No. 3, 808–825.

[JOE1] JOELA H.: The transformation of aromatic molecules into the subspace of their double bonds, Theoret. Chim. Acta (Berl.), 39 (1975), 241–246.

[JOKS1] JOHNSON C. R., KELLOG R. B., STEPHENS A. B.: Complex eigenvalues of a non-negative matrix with a specified graph, II, Linear and Multilinear Algebra, 7(1979), No. 2, 129–143; Erratum: ibid. 8(1979/ 80), No. 2, 171.

[JOLE1] JOHNSON C. R., LEIGHTON F. T.: An efficient linear algebraic algorithm for the determination of isomorphism in pairs of undirected graphs, J. Research NBS, A. Math. Sci., 80b(1976), No. 4, 447–483.

[JoNe1] JOHNSON C. R., NEWMAN M.: A note on cospectral graphs, J. Combin. Theory(B), 28(1980), No. 1, 96–103.

[Jon1] JONES O.: Contentment in graph theory: covering graphs with cliques, Proc: Kon. Ned. Akad. Wetensch. A, 80(1977), 406–424.

[Jóź1] JÓŹWIAK T.: Charakteristisches und per–charakteristisches Polynom und ihre Anwendung in der Theorie der Graphen, 24. Int. Wiss. Kolloq., TH Ilmenau, Ilmenau DDR 1979, Heft 5, 15–17.

[Juh1] JUHÁSZ F.: On the asymptotic behaviour of the spectra of non-symmetric random (0–1)-matrices (Hungarian), Alkalmaz. Mat. Lapok, 6(1980), No. 3–4, 345–349.

[Juh2] JUHÁSZ F.: On the spectrum of a random graph, (**Algebraic methods in Graph Theory**, Vol. I, II; Szeged, 1978; North–Holland, Amsterdam), Colloq. Math. Soc. Janos Bolyai, 25(1981), 313–316.

[Juh3] JUHÁSZ F.: On the asymptotic behaviour of the spectra of non-symmetric random (0–1)-matrices, Discrete Math., 41(1982), No. 2, 161–165.

[Juh4] JUHÁSZ F.: The asymptotic behaviour of Lovàsz theta function for random graphs, Computer and Autom. Inst., Hungar. Acad. Sci., preprint, pp. 1–4; Combinatorica, 2(1982), No. 2, 153–155.

[JuMa1] JUHÁSZ F., MALYUSZ K.: Problems of cluster analysis from the viewpoint of numerical analysis, (Proc. Conf. Numerical methods, Kesztely, 1977), 405–415.

[Kac1] KAC A. O.: A class of polynomial invariants of a graph (Russian), Latv. Mat. Ezhegodnik, 23(1979), 137–146, 274.

[Kad1] KADURA P.: Zur Bandstruktur endlicher F. C. C.—Kristallite unterschiedlicher Gestalt, Wiss. Z. Friedrich-Schiller-Univ. Jena, Math. Naturwiss. R., 29(1980), 695–704.

[KAS1] KASSMAN A. J.: Generation of the eigenvectors of the topological matrix from graph theory, Theoret. Chim. Acta (Berl.), 67(1985) 255-262.

[KATG1] KASUM D., TRINAJSTIĆ N., GUTMAN I.: Chemical graph theory. III. On the permanental polynomial, Croat. Chem. Acta, 54(1981), 321–328.

[KACH1] KAULGUD M. V., CHITGOPKAR V. H.: Polynomial matrix method for the calculation of π-electron energies for linear conjugated polymers, J. C. S. Faraday II, 73(1977), 1385–1395.

[KACH2] KAULGUD M. V., CHITGOPKAR V. H.: Polynomial matrix method for the calculation of charge densities and bond orders in linear conjugated π-electron systems, J. C. S. Faraday II, 74(1978), 951–957.

[KESK1] KEMENY J. G., SNELL J. L., KNOPP A. W.: **Denumerable Markov chains**, Second edition, Springer-Verlag, Berlin 1976.

[KEQ1] KEQIN F.: A problem on algebraic graph theory, Discrete Math. 54(1985), 101–107.

[KIA1] KIANG YUAN-SUN: Determinant of adjacency matrix and Kekulé structures, Internat. J. Quantum Chem. Symposium, 14(1980), 541–547.

[KIA2] KIANG YUAN-SUN: Calculation of the determinant of the adjacency matrix and the stability of conjugated molecules, Internat. J. Quantum Chem., 18(1980), 331–338.

[KIA3] KIANG YUAN-SUN: Partition technique and molecular graph theory, Internat. J. Quantum Chem. Symposium, 15(1981), 293–304.

[KICH1] KIANG YUAN-SUN, CHEN ER-TING: Evaluation of HOMO–LUMO separation and homologous linearity of conjugated molecules, Pure Appl. Chem., 55(1983), No. 2, 283–288.

[KIN1] KING R. B.: Symmetry factoring of the characteristic equations of graphs corresponding to polyhedra, Theoret. Chim. Acta (Berl.), 44(1977), 223–243.

[KIN2] KING R. B.: Conjugates of regular complete multipartite graphs
and hyperpolygonal graphs, Houston J. Math., 5(1979), No. 2,
209–217.

[KIN3] KING R. B. (EDITOR): **Chemical Applications of Topology
and Graph Theory**, (Proc. Conference on Chemical Applications of Topology and Graph Theory, Athens, Ga., USA, 1983)
Elsevier, Amsterdam, 1983.

[KIRO1] KING R. B., ROUVRAY D. H.: Chemical applications of group
theory and topology, 7. A graph theoretical interpretation of the
bonding topology in polyhedral boranes, carboranes, and metal
clusters, J. Amer. Chem. Soc., 99(1977), 7834–7840.

[KIRO2] KING R. B., ROUVRAY D. H.: A graph-theoretical interpretation of the bonding topology in polyhedral boranes, metal
clusters and organic cations, Match, 7(1979), 273–287.

[KIR1] KIRBY E. C.: The characteristic polynomial; Computer-aided
decomposition of the secular determinant as a method of evaluation for homo-conjugated systems, J. Chem. Research (S) (1984),
4–5; (M) (1984), 123–146.

[KLI1] KLIMOVICKII S. M.: Investigation of a characteristic polynomial of a graph for certain cases (Russian), Moskov. Inst. Inz.
Zhelezno- dorog. Transporta Trudy, No. 640(1979), 132–137,
158–159.

[KNMS1] KNOP J. V., MÜLLER W. R., SZYMANSKI K., RANDIĆ
M., TRINAJSTIĆ N.: Note on acyclic structures and their self-
returning walks, Croat. Chem. Acta, 56(1983), No. 3, 405–409.

[KPRT1] KNOP J. V., PLAVŠIĆ D., RANDIĆ M., TRINAJSTIĆ N.:
Chemical graph theory. V. On the classification of topological
biradicals, Croat. Chem. Acta, 56(1983), No. 3, 411–441.

[KNTR1] KNOP J. V., TRINAJSTIĆ N.: Chemical graph theory. II. On
graph theoretical polynomials of conjugated structures, Internat.
J. Quantum Chem. Symposium, 14(1980), 503–520.

[KOC1] KOCH H. VON: Sur quelques points de la théorie des
déterminants, Acta Math., 24(1900), 89–122.

[KOH1] KOHOV V. A.: Stability numbers and stability spectrum of a graph (Russian), Trudy Moskov. Otdelen. Lenin. Energet. Inst., No. 415(1979) 111–115, 135.

[KOL1] KOLMYKOV V. A.: A graph product and its spectrum (Russian), Voronezh University, Voronezh, 1983, 14pp, (manuscript No. 6710–83 DEP, VINITI 12. dec. 1983).

[KMSS1] KOLMYKOV V. A., MEN'SHIH V. V., SUBOTIN V. F., SUMIN M. V.: Investigations of spectral properties of trees (Russian), Voronezh University, Voronezh, 1983, 30pp, (manuscript No. 6709–83 DEP, VINITI 12. dec. 1983).

[KoSu1] KOLMYKOV V. A., SUBOTIN V. F.: Spectra of graphs and cospectrality (Russian), Voronezh University, Voronezh, 1983, 23pp, (manuscript No. 6708–83 DEP, VINITI 12. dec. 1983).

[Kov1] KOVÁCS P.: The non–existence of certain regular graphs of girth 5, J. Combin. Theory(B), 30(1981), No. 3, 282–284.

[KrRu1] KREIN M. G., RUTMAN M. A.: Linear operators in a Banach space leaving a connus invariant (Russian), Uspekhi Mat. Nauk, 3(23)(1948), No. 1, 3–95.

[Kri1] KRISHNAMURTHY E. V.: A form invariant multivariable polynomial representation of graphs, In: **Combinatorics and Graph Theory**(Calcutta 1980), Lecture Notes Math. 885, Springer-Verlag, Berlin, 1981, pp. 18–32.

[KrTr1] KRIVKA P., TRINAJSTIĆ N.: On the distance polynomial of a graph, Apl. Mat., 28(1983), No. 5, 357–363.

[KuRS1] KUMAR V., RAO S. B., SINGHI N. M.: Graphs with eigenvalues at least −2, Linear Algebra Appl., 46(1982), 27–42.

[Kun1] KUNZ H.: Location of the zeros of the partition function for some classical lattice systems, Phys. Letters (A), 32(1970), 311–312.

[Lab1] LABARTHE J. J.: Graphical method for computing the determinant and inverse of a matrix. Generating functions for harmonic oscillator integrals, J. Phys. (A), 11(1978), 1009–1015.

[Lam1] Lam C. W. H.: On some solutions of $A^k = dI + \lambda J$, J. Combin.Theory (A), 23(1977), 140–147.

[Lam2] Lam C. W. H.: Distance transitive digraphs, Discrete Math., 29 (1980), No. 3, 265–274.

[LaLi1] Lam C. W. H., Lint J. H. van: Directed graphs with unique pathsof fixed length, J. Combin. Theory(B), 24(1978), 331–337.

[Lan1] Lane K. D.: Computing the index of a graph, Proc. Fourteenth Southeastern Conf. on Combinatorics, graph Theory, and Computing (Boca Raton, Fla., 1983), Congr. Numer. 40(1983), 143–154.

[LeHP1] Leahey W. J., Herndon W. C., Phan V. T.: A note on permanents, Linear and Multilinear Algebra, 3(1975), 193–196.

[LiFe1] Li Qiao, Feng Ke Qin: On the largest eigenvalue of graphs, Acta Math. Appl. Sinica (Chinese), 2(1979), 167–175.

[Lie1] Liebeck M. W.: Lie algebras, 2-graphs and permutation groups, Bull. London Math. Soc., 12(1980), 29–33.

[Lieb1] Liebler R. A.: On the uniquenes of the tetrahedral association scheme, J. Combin. Theory(B), 22(1977), 246–262.

[LiWZ1] Lin Liang-tang, Wang Nan-qin, Zhang Qian-er: Isospectral molecule, Acta Univ. Amoiensis Scientiarium Naturalium (Chinese), 2(1979), 65–75.

[Lin1] Lint J. H. van: $\{0, 1, *\}$ problems in combinatorics, In: **Surveys in combinatorics 1985**, (Ed. Ian Anderson) London Mathematical Society Lecture Note series 103, Cambridge University Press, Cambridge, 1985, pp. 113–135.

[Liu1] Liu Wei-min: Application of graph theory of molecular orbitals to conjugated molecule possessing a proper axis of rotation, Sci. Sinica, 22(1979), No. 5, 539–554.

[Lor1] Lorimer P.: Vertex-transitive graphs of valency 3, European J. Combin., 4(1983), No. 1, 37–44.

[Lov1] Lovász L.: Chromatic number of hypergraphs and linear algebra, Studia Sci. Math. Hungar., 11(1976), No. 1–2, 113–114.

[Lov2] LOVÁSZ L.: On the Shannon capacity of a graph, IEEE Trans. Inform. Theory, 25(1979), No. 1, 1–7.

[Lov3] LOVÁSZ L.: **Combinatorial problems and excerises.**North Holland, Amsterdam, 1979.

[MAE1] MAEHARA H.: Regular embeddings of a graph, Pacific J. Math., 107(1983), No. 2, 393–402.

[MAL1] MALLION R. B.: Théorie des graphes: Sur les conditions pour l'existence des valeurs propres nulles dans les spectres des graphes cycliques (C_n) representant les hydrocarbures annulaires, Bull. Soc. Chim. Fr., (1974), 2799–2800.

[MAL2] MALLION R. B.: On the number of spanning trees in a molecular graph, Chem. Phys. Letters, 36(1975), 170–174.

[MAL3] MALLION R. B.: Some chemical applications of the eigenvalues and eigenvectors of certain finite, planar graphs, In: **Applications of Combinatorics**, (Ed. R. J. Wilson), Shiva Publ. Co., Nantwich, Cheshire, 1982, pp. 87–114.

[MaRo1] MALLION R. B., ROUVRAY D. H.: Molecular topology and the Aufbau principle, Mol. Phys., 36(1978), 125–128.

[MaRo2] MALLION R. B., ROUVRAY D. H.: On a new index for characterising the vertices of certain non-bipartite graphs, Studia Sci. Math. Hung., 13(1978), 229–243.

[MaEi1] MARTIN J. E., EICHINGER B. E.: Distribution functions for Gaussian molecules, I. Stars and random regular nets, J. Chem. Phys., 69(10) (1978), 4588–4594.

[MAT1] MATHON R.: 3-class association schemes, **Proc. Conf. on Algebraic Aspects of Combinatorics**, 1975, (Ed. D. Corneil, E. Mendelsohn), Utilitas Math., Winnipeg, 1975, 123–155.

[McCl1] McCLELLAND B. J.: On the graphical factorization of Hückel characteristic equations, Mol. Phys., 45(1982), No. 1, 189–190.

[McCl2] McCLELLAND B. J.: Eigenvalues of the topological matrix. Splitting of graphs with symmetrical components and alternant graphs, J. C. S. Faraday II, 78(1982), 911–916.

[McRR] McEliece R. J., Rodemich E. R., Runsey H. C. –Jr.: The Lovász bound and some generalizations, J. Combin. Inform. Sys. Sci., 3 (1978), 134–152.

[McKa1] McKay B. D.: Transitive graphs with fewer than twenty vertices, Math. Comp., 33(1979), 1101–1121.

[McKa2] McKay B. D.: The expected eigenvalue distribution of a random labelled regular graph, Univ. Melbourne, Math. Res. Report, No. 9, (1979), 1–15.

[McKa3] McKay B. D.: The expected eigenvalue distribution of a large regular graph, Linear Algebra Appl., 40(1981), 203–216.

[McKa4] McKay B. D.: Spanning trees in regular graphs, Vanderbilt Univ., Techn. Rep. CS–81–01(1981), 1–24; European J. Combin., 4 (1983), No. 2, 149–160.

[McKa5] McKay B. D.: Spanning trees in random regular graphs, Vanderbilt Univ., Computer Sci. Technical Report, CS–81–05(1981), 1–6.

[McKa6] McKay B. D.: Subgraphs of random graphs with specified degrees, to appear.

[Men1] Mena R. A.: The characteristic polynomials of the plane of order 2, J. Combin. Inform. System Sci., 7(1982), No. 1, 91–93.

[MeSi1] Merrifield R. E., Simmons H. E.: Generalized molecular duplexes, Chem. Phys. Letters, 74(1980), 348–353.

[Mer1] Merris R.: On the pseudocharacteristic polynomial of K. Bogart and J. Gordon, Linear and Multilinear Algebra, 7(1979), 237–239.

[Mer2] Merris R.: The Laplacian permanental polynomial for trees, Czech. Math. J., 32(1982), No. 3, 397–403. .

[MeRW1] Merris R., Rebmann K. R., Watkins W.: Permanental polynomials of graphs, Linear Algebra Appl., 38(1981), 273–288.

[Mic1] Michalski M.: Spectra and direct derivations of graphs, J. Combin. Inform. System Sci., 8(1983), 116–122.

[MIN1] MINGOS D. M. P.: A topological Hückel model for organometallic complexes, Part I. Bond lengths in complexes of conjugated olefins, J. C. S. Dalton, (1977), 20–25.

[MIN2] MINGOS D. M. P.: A topological Hückel model for organometallic complexes, Part II. Alternant bonding networks in η3- and η-4-olefin complexes, J. C. S. Dalton, (1977), 25–30.

[MIZ1] MIZOGUCHI N.: The circuit rule for the magnetic susceptibility of a polycyclic conjugated system, Chem. Phys. Letters, 106(1984) 451–454.

[MKVP1] MLADENOV I. M., KOTAROV M. D., VASSILEVA–POPOVA J. G.: Method for computing the characteristic polynomial, Internat. J. Quantum Chem., 18(1980), 339–341.

[MNU1] MNUHIN V. B.: Spectra of graphs under certain unary operations (Russian), In: **Nekotorye Topologicheskie i Kombinator. Svoistva Grafov**, Akad. Nauk Ukrain. SSR Inst. Mat. Preprint, No. 8 (1980), 38–44.

[MNU2] MNUHIN V. B.: On the reconstruction of graph polynomials (Russian), 27. Internat. Colloq., TH Ilmenau, Ilmenau 1982, Heft 6, 87–90.

[MOH1] MOHAR B.: Generalization of the Sachs theorem and the acyclic polynomials of graphs (Slovenian), Univ. Ljubljana, Ljubljana, 1979, pp. 1–24.

[MOH2] MOHAR B.: Spectra of graphs (Slovenian), Univ. Ljubljana, Ljubljana, 1980, pp. 1–26.

[MOH3] MOHAR B.: The spectrum of an infinite graph, Linear Algebra Appl., 48(1982), 245–256.

[MOOM1] MOHAR B., OMLADIČ M.: The spectrum of infinite graphs with bounded vertex degrees, In: **Proc. Eyba Conference, Eyba**, 1–6. oct. 1984, DDR, to appear.

[MOOM2] MOHAR B., OMLADIČ M.: Divisors and the spectrum of infinite graphs, manuscript.

[MoTr1] MOHAR B., TRINAJSTIĆ N.: On computation of the topological resonance energy, J. Comput. Chem, 3(1982), No. 1, 28–36.

[Moo1] MOON A.: Characterization of odd graphs $O(k)$ by parameters. Discrete Math. 42(1982) 91–97.

[Moo2] MOON, A.: Characterization of the graphs of the Johnson scheme $G(3k,3)$ and $G(3k+1,3)$. J. Combin. Theory B 33(1982) 213-222.

[Neu1] NEUMAIER A.: Strongly regular graphs with smallest eigenvalue $-m$, Arch. Math. (Basel), 33(1979/80), No. 4, 392–400.

[Neu2] NEUMAIER A.: Cliques and claws in edge-transitive strongly regular graphs, Math. Z., 174(1980), 197–202.

[Neu3] NEUMAIER A.: Distance matrices and n-dimensional designs, European J. Combin., 2(1981), 165–172.

[Neu4] NEUMAIER A.: Distance matrices, dimension, and conference matrices, Nederl. Akad. Wetensch. Indag. Math., 43(1981), 385–391.

[Neu5] NEUMAIER A.: Classification of graphs by regularity, J. Combin. Theory(B), 30(1981), 318–331.

[Neu6] NEUMAIER A.: The second largest eigenvalue of a tree, Linear Algebra Appl., 46(1982), 9–25.

[NeSe1] NEUMAIER A., SEIDEL J. J.: Discrete hyperbolic geometry, Combinatorica, 3(1983), 219–237.

[Nij1] NIJENHUIS A.: On permanents and the zeros of rook polynomials, J. Combin. Theory (A), 21(1976), 240–244.

[NoWW1] NORINDER U., WENNERSTRÖM O., WENNERSTRÖM H.: Hückel theory applied to large linear and cyclic conjugated π-systems, Tetrahedron, 41(1985) 713–726.

[Nuf1] NUFFELEN C. VAN: On the rank of the adjacency matrix, **Problèmes Combinatoires et Théorie des Graphes**, Colloque Internat. C. N. R. S., Orsay, 1976, C. N. R. S. No. 260(1978), (Ed. J. C. Bermond, J. C. Fournier, M. Las Vergnas, D. Sotteau), pp. 321–322.

[NUF2] NUFFELEN C. VAN: Remarks on the adjacency matrix of a graph, **Régards sur la théorie des graphes,** Proc. colloq. held at Cerisy, (1980), 275–277.

[NUF3] NUFFELEN C. VAN: Some bounds for fundamental numbers in graph theory, Bull. Soc. Math. Belg., Ser. B, 32(1980), No. 2, 251–257.

[NUF4] NUFFELEN C. VAN: Rank and diameter of a graph, Bull. Soc. Math. Belg., Ser. B, 34(1982), No. 1, 105–111.

[NUF5] NUFFELEN C. VAN: Rank and domination number, In: **Graph and other combinatorial topics,** Proc. Third Czech. Symp., Prague 1982, Teubner–Texte Math. 59, 1983, pp. 209–211.

[NUF6] NUFFELEN C. VAN: The rank of the adjacency matrix of a graph, Bull. Soc. Math. Belg., Ser. B, 35(1983), 219–225.

[OLI1] OLIVEIRA G. N. DE: Note on the characteristic roots on tournament matrices, Linear Algebra and Its Appl., 8(1974), 271–272.

[PAWI1] PATTERSON H. D., WILLIAMS R. E.: Some theoretical results on general block designs, In: **Proc. Fifth British Comb. Conf.,** 1975, (Ed. C. St. J. A. Nash–Williams, J. Sheehan), Utilitas Math., Winnipeg, 1976, pp. 489–496.

[PE1] PETRIĆ M.: Spectral method and the problem of determining the number of walks in a graph (Serbo-Croatian), Master Thesis, Univ. Beograd, Fac. Sci., 1980.

[PE2] PETRIĆ M.: A note on the number of walks in a graph, Univ. Beograd, Publ. Elektrotehn. Fak., Ser. Mat. Fiz., No. 716–No. 734(1981), 83–86.

[PETR1] PETROV K.: On some strongly regular graphs, C. R. Acad. Bulgare Sci., 36(1983), No. 4, 421–423.

[PET1] PETROVIĆ M. M.: The spectrum of an infinite labelled graph (Serbo-Croatian), Master thesis, Univ. Beograd, Fac. Sci., 1981.

[PET2] PETROVIĆ M. M.: The spectrum of infinite complete multipartite graphs, Publ. Inst. Math. (Beograd), 31(45)(1982), 169–176.

[PET3] PETROVIĆ M. M.: Spectra of some operations on infinite graphs, Glasnik Mat., 18(38)(1983), 27–33.

[PET4] PETROVIĆ M. M.: Contribution to the spectral theory of graphs (Serbo–Croatian), Doct. thesis, Univ. Beograd, Fac. Sci., 1984.

[PET5] PETROVIĆ M. M.: Finite type graphs and some graph operations, I, Univ. Beograd, Publ. Elektrotehn. Fak., Ser. Mat. Fiz., No. 735–No. 762 (1982), No. 758, 142–147.

[PET6] PETROVIĆ M. M.: On the spectrum and characteristic function of an infinite graph, Univ. Kragujevac, Collection Sci. Papers Fac. Sci., 4(1983), 43–58.

[PET7] PETROVIĆ M. M.: On graphs whose spectral spread does not exceed 4, Publ. Inst. Math. (Beograd), 34(48)(1983), 169–174.

[PET8] PETROVIĆ M. M.: On graphs whose second spread does not exceed 3/2, **Graph Theory**, Proc. Fourth Yugoslav Sem. Graph Theory, Novi Sad, 15–16. April 1983, (Ed. D. Cvetković, I. Gutman, T. Pisanski, R. Tošić, Inst. Math., Novi Sad 1984), pp. 245–257.

[PET9] PETROVIĆ M. M.: Some classes of graphs with two, three or four positive eigenvalues, Univ. Kragujevac, Collection Sci. Papers Fac. Sci., 5(1984), 31–39.

[PET10] PETROVIĆ M. M.: Graphs with bounded reduced positive energy, In: **Proc. Fourth Yugoslav Sem. Appl. Math.**, Split, May 28–30, 1984, 147-153.

[PiWo1] PICARDELLO M. A., WOESS W.: Martin boundaries of random walks. Ends of trees and groups, to appear.

[PLE1] PLESNIK J.: On homogenious tournaments, Acta Fac. Rer. Nat. Univ. Comenianae Math., 21(1968), 26–34.

[POL1] POLANSKY O. E.: On the band structure of Möbius polymers, Z. Naturforsch, 38a(1983), 909–915.

[POL2] POLANSKY O. E.: Relations between the roots of the acyclic and the characteristic polynomials of Hückel and Möbius cycles

— Eigenvalues of certain Hückel and Möbius forms of polyacenes, Match, 14(1983), 47–70.

[Pol3] POLANSKY O. E.: Topological effects displayed in absorption and photoelectron spectra, J. Molec. Struct., 113(1984), 281–298.

[Pol4] POLANSKY O. E.: On the characteristic polynomials of topologically related isomers formed within three different models, Match, 18(1985), 111–165.

[Pol5] POLANSKY O. E.: Some structural conditions for the central moiety in a certain topological model, Match, 18(1985), 16.7–215.

[Pol6] POLANSKY O. E.: Einige neue Beziehungen zwischen den mit bestimmten Graphen assozierten charakteristischen Polynomen, Match, 18(1985), 217–247.

[PoGu1] POLANSKY O. E., GUTMAN I.: On the calculation of the largest eigenvalue of molecular graph, Match, 5(1979), 145–159.

[PoMa1] POLANSKY O. E., MARK G.: A topological model for some bis-metallo-complexes, Match, 18(1985), 249–259.

[PoZa1] POLANSKY O. E., ZANDER M.: Topological effects on MO energies, J. Mol. Struct., 84(1982), 361–385.

[Poly1] POLY C. LE CONTE DE: Graphes d'amitié et plans en blocs symmetriques, Math. Sci. Humaines, 51(1975), 25–33, 87.

[Por1] PORCU L.: Sul raddopio di un grafo, Instituto Lombardo (Rend. Sci.) A, 110(1976), 453–480.

[Pöt1] PÖTSCHKE D.: Maximaler Eigenwert und R-Isomorphieklassen von endlichen Graphen, ZKI Inf., Akad. Wiss. DDR, Zentralinst. Kybern. und Informationsprozesse, No. 3(1982), 14–50.

[Pöt2] PÖTSCHKE D.: Maximal eigenvalue characterizes random isomorphism classes of finite graphs, **Graphs and other combinatorial topics**, Proc. Third Czech. Symp., Prague, 1982, Teubner–Texte Math. 59 (1983), 223–235.

[Pöt3] PÖTSCHKE D.: Die topologische Entropie als Isomorphieinvariante endlicher Graphen. Rostock Math. Kolloq. 21(1982), 47–57.

[Pou1] POUZET M.: Quelques remarques sur les résultats de Tutte concernant le problème de Ulam, Publ. Dept. Math. Lyon, 14(1977), No. 2, 1–8.

[PoSu1] POWERS D. L., SULAIMAN M. M.: The walk partition and colorations of a graph, Linear Algebra Appl., 48(1982)145–159.

[PrDe1] PRABHU G. M., DEO N.: On the power of a perturbation for testing nonisomorphism of graphs. BIT, 24(1984), 302–307.

[Qua1] QUANTUM CHEMISTRY GROUP, KIRIN UNIVERSITY: Graph theory of molecular orbitals (Chinese), Huaxue Tongbao, 1(1977), 29–38.

[Rad1] RADOSAVLJEVIĆ Z.: Graph factorization with application to the theory of graph spectra (Serbo-Croatian), Master thesis, Univ. Beograd, El. tehn. fak., 1980.

[Rad2] RADOSAVLJEVIĆ Z. S.: Inequivalent regular factors of regular graphs on 8 vertices, Publ. Inst. Math. (Beograd), 29(1981), 171–190.

[RaSi1] RADOSAVLJEVIĆ Z., SIMIĆ S.: Nonregular nonbipartite integral graphs having maximal vertex degree four, I, II. to appear.

[RSST1] RADOSAVLJEVIĆ Z., SIMIĆ S., SYSLO M., TOPP J.: A note on generalized line graphs, Pul. Inst. Math. (Beograd), 34(48)(1983), 193–198.

[RaBa1] RAMARAJ R., BALASUBRAMANIAN K.: Computer generation of matching polynomials of chemical graphs and lattices, J. Comput. Chem. 6(1985), 122-141.

[Ran1] RANDIĆ M.: Random walks and their diagnostic value for characterization of atomic environment, J. Comput. Chem., 1(1980), No. 4, 386–399.

[Ran2] RANDIĆ M.: On evaluation of the characteristic polynomial for large molecules, J. Comput. Chem., 3(1982), No. 3, 421–435.

[Ran3] RANDIĆ M.: Characteristic equations and graph isomorphism, 1982, preprint, 19pp.

[RAN4] RANDIĆ M.: An alternative form of the characteristic polynomial and the problem of graph recognition, Theoret. Chim. Acta (Berl.), 62(1983), 485–498.

[RAN5] RANDIĆ M.: On computing the characteristic polynomial, J. Comput. Chem., 1983, manuscript.

[RAN6] RANDIĆ M.: On the characteristic equations of the characteristic polynomial. SIAM J. Alg. Disc. Meth., 6(1985), 145–162.

[RAWG1] RANDIĆ M., WOODWORTH W. L., GRAOVAC A.: Unusual random walks, Internat. J. Quantum Chem., 24(1983), 435–452.

[RASV1] RAO S. B., SINGHI N. M., VIJAYAN K. S.: The minimal forbidden subgraphs for generalized line-graphs, **Combinatorics and Graph Theory** (Calcutta 1980), Lecture Notes Math. 885, Springer, Berlin 1981, pp. 459–472.

[RASV2] RAO S. B., SINGHI N. M., VIJAYAN K. S.: Spectral characterization of the line graph of $K(l, n)$, **Combinatorics and Graph theory**(Calcutta 1980), Lecture Notes Math. 885, Springer, Berlin, 1981, pp. 473–480.

[RAU1] RAUT G.: Spectra of complete k-ary trees (Romanian), Stud. Cerc. Mat., 35(1983), No. 3, 183–188.

[RAY1] RAY-CHAUDHURY D. K.: Some characterization theorems for graphs and incidence structures, **Combinatorics**, Vol. II, Amsterdam, 1978, pp. 821–842.

[RECo1] READ R. C., CORNEIL D. G.: The graph isomorphism disease, J. Graph Theory, 1(1977). No. 4, 339–363.

[REBR1] REID K. B., BROWN E.: Double regular tournaments are equivalent to skew-Hadamard matrices, J. Combinatorial Theory A, 12(1972), 332–338.

[RIMA1] RIGBY M. J., MALLION R. B.: On the eigenvalues and eigenvectors of certain finite, vertex-weighted, bipartite graphs, J. Combin. Theory (B), 27(1979), No. 2, 122–129.

[RIMD1] RIGBY M. J., MALLION R. B., DAY A. C.: Comment on a graph—theoretical description of heteroconjugated molecules, Chem. Phys. Letters, 51(1977). No. 1, 178–182.

[RiMW1] RIGBY M. J., MALLION R. B., WALLER D. A.: On the quest for an isomorphism invariant which characterises finite chemical graphs, Chem. Phys. Letters, 59(1978), No. 2, 316–320.

[Rio1] RIORDAN J.: **An Introduction to Combinatorial Analysis**, Wiley, London 1958.

[Rob1] ROBERTS F. S.: **Graph Theory and Its Applications to Problems of Society**, Soc. Ind. Appl. Math., Philadelphia 1978, pp. 94–99.

[RoBr1] ROBERTS F. S., BROWN T. A.: Signed digraphs and the energy crisis, Amer. Math. Monthly, 82(1975), 577–594.

[Rol1] ROLLAND P. T.: On the uniqueness of the tetrahedral association scheme, Ph. D. Thesis, City University of New York, New York, 1976.

[Roi1] ROITMAN M.: An infinite family of integral graphs. Discrete Math 52(1984) 313–315.

[Roo1] ROOS C.: On the feasibility of certain intersection matrices, Delft Progr. Rep., 5(1980), 215–226.

[RoZa1] ROOS C., ZANTEN A. J. VAN: On the existence of certain distance-regular graphs, J. Combin. Thoery (B), 33(1982), No. 3, 197–212.

[RoZa2] ROOS C., ZANTEN A. J. VAN: On a class of distance-regular graphs whose members have small diameters, part II, Delft Progress Report, 8(1983), No. 1, 3–16.

[RoZa3] ROOS C., ZANTEN A. J. VAN: On the existence of certain generalized Moore geometries, part I, Discrete Math., 51(1984), 179–190.

[Rot1] ROTH J. P.: Spectre du laplacien sur un graph, C. R. Acad. Sci. Paris, Ser. I Math., 296(1983), No. 19, 793–795.

[Row1] ROWLINSON P.: On the number of simple eigenvalues of a graph, Proc. Roy. Soc. Edinbourgh, 94A(1982), 247–250.

[Row2] ROWLINSON P.: Simple eigenvalues of intransitive graphs, Bull. London Math. Soc., 16(1984), 122–126.

[Row3] ROWLINSON P.: Certain 3-decompositions of complete graphs, with an application to finite fields, Proc. Roy. Soc. Edinburgh, 99A(1985), 277–281.

[Row4] ROWLINSON P.: On 4-cycles and 5-cycles in regular tournaments, Bull. London Math. Soc., to appear.

[Sac1] SACHS H.: Automorphismengruppe und Spektrum von endlichen Graphen, 24. Internat. Wiss. Kolloq., TH Ilmenau, Ilmenau 1979, Heft 5, 11–12.

[SaSt1] SACHS H., STIEBITZ M.: Automorphism group and spectrum of a graph, **Algebraic Methods in Graph Theory**, Szeged, 1978, Vol. II, Colloq. Math. Sc. Janos Bolyai, 25, North–Holland, Amsterdam, 1981, pp. 657–670.

[SaSt2] SACHS H., STIEBITZ M.: Konstruktion schlichter transitiver Graphen mit maximaler Anzahl einfacher Eigenwerte, Math. Nachr., 100(1981), 145–150.

[SaLe1] SALEM L., LEFORESTIER C.: Exact Hückel molecular orbitals of the finite square-cut FCC crystal, J. Amer. Chem. Soc 107(1985) 2526-2528.

[Sam1] SAMUEL I.: On a computational method of pairs of eigenvectors associated with a double degenerate eigenvalue of a diagonalizable square matrix. Applications to Hückel's matrix of alternant hydrocarbons, Match, 18(1985), 63-70.

[SHNT1] SCHAAD L. J., HESS B. A., NATION J. B., TRINAJSTIĆ N., GUTMAN I.: On the reference structure for the resonance energy of aromatic hydrocarbons, Croat. Chem. Acta, 52(1979), 233–248.

[Sch1] SCHAAR G.: Algebraische Aspekte der Graphentheorie, 24. Int. Wiss. Kolloq., TH Ilmenau, Ilmenau, 1979, Heft 5, 19–22.

[Schn1] SCHNEIDER R.: On A. D. Aleksandrov's inequalities for mixed discriminants, J. Math. Mech., 15(1966), 285–290.

[Schr1] SCHRIJVER A.: Association schemes and Shannon capacity; Eberlein polynomials and Erdos–Ko–Rado theorem, **Algebraic Methods in Graph Theory**, Szeged, 1978, Vol. II, Colloq.

Math. Soc. Janos Bolyai, 25, North–Holland, Amsterdam, 1981, pp. 671–688.

[SCHR2] SCHRIJVER A.: A comparison of the Delsarte and Lovász bounds, IEEE Trans. Inform. Theory, 25 (1979), 425–429.

[SCHU1] SCHULZ M.: Automorphismen und Eigenwerte von Graphen, Diplomarbeit, Humboldt–Univ., Berlin, Sektion Mathematik, 1979.

[SCHU2] SCHULZ M.: An example of a cubic graph with two orbits and simple eigenvalue $\sqrt{3}$, Discrete Math., 31(1980), No. 2, 221–222.

[SCHW1] SCHWENK A. J.: Spectral reconstruction problems, Ann. N. Y. Acad. Sci., 328(1979), 183–189.

[SCHW2] SCHWENK A. J.: Removal-cospectral sets of vertices in a graph, Proc. of the Tenth Southeastern Conference on Combinatorics, Graph Theory and Computing, 1979, Utilitas Math.; Winnipeg, Manitoba, (1979), 849–860.

[SCHW3] SCHWENK A. J.: On unimodal sequences of graphical invariants, J. Combin. Theory(B), 30(1981), 247–250.

[SCHE1] SCHWENK A. J., HERNDON W. C., ELLZEY M. L. –JR.: The construction of cospectral composite graphs, Ann. N. Y. Acad. Sci., 319 (1979), 490–496.

[SCWI1] SCHWENK A. J., WILSON R.: On the eigenvalues of a graph. In: **Selected topics in graph theory**(Ed. L. Beineke, R. Wilson), Academic Press, London 1979, 307–**336**.

[SED1] SEDLAČEK J.: Some comments about skeletons of connected graphs, (Czech) Časopis Pěst. Mat., 104(1979), 75–85.

[SEI1] SEIDEL J. J.: Strongly regular graphs, an introduction, (Proc. Seventh British Combin. Conf., Cambridge, 1979), London Math. Soc. Lecture Note Ser., 38(1979), 157–180.

[SEI2] SEIDEL J. J.: Delsarte's theory of association schemes, (Proc. Third Czech. Symposium on Graph Theory, ed. M. Fiedler) Tuebner, Leipzig, 1983, 249–258.

[SEBW1] SEIDEL J. J., BLOKHUIS A., WILBRINK H. A.: Graphs and association schemes, algebra and geometry, Eindhoven Univ. of Technology, EUT–Report 83–WSK–02(1983), 1–117.

[SETA1] SEIDEL J. J., TAYLOR D. E.: Two-graphs, a second survey, **Algebraic Methods in Graph Theory**, Szeged, 1978, Vol. II, Colloq. Math. Soc. Janos Bolyai, 25, North-Holland, Amsterdam 1981, pp. 689-712.

[SEFK1] SENO A., FUKUNAGA K., KASAI T.: An algorithm for graph isomorphism, Bull. Univ. Osaka Prefecture Ser. A, 26(1977), No. 2, 61–74.

[SHA1] SHANNON, C. E.: The zero-error capacity of a noisy channel, IRE Trans. Information Theory, 3(1956), 3–15.

[SHI1] SHINODA S.: A characteristic polynomial of a subdivision graph, Chuo Bull. Fac. Sci. Eng., Chuo Univ., 19(1976), 125–130.

[SHI2] SHINODA S.: On the characteristic polynomial of the adjacency matrix of the subdivision graph of a graph, Discrete Appl. Math., 2(1980), No. 4, 349–351.

[SHRI1] SHRIKHANDE M. S.: Strongly regular graphs and group divisible designs, Pacific J. Math., 54(1974), No. 2, 199–208.

[SIE1] SIEMONS J.: Automorphism group of graphs, Arch. Math., 41(1982), 379–384.

[SIME1] SIMMONS H. E., MERRIFIELD R. E.: Paraspectral molecular pairs, Chem. Phys. Letters, 62(1979), 235–237.

[SIN1] SINANOGLU O.: A theorem for qualitative deductions in organic and inorganic chemistry regarding the relative stabilities, distortions and reactions of molecules, Chem. Phys. Letters, 103(1984) 315–322.

[SIN2] SINANOGLU O.: Structural covariance of graphs, Theoret. Chim. Acta (Berl.), 65(1984), 255–265.

[SISR1] SINHA B. P., SRIMANI P. K.: Some studies on the characteristic polynomial of a graph, Int. J. Electr., 54(1983), No. 3, 377–400.

[SMI1] SMITH J. H.: Symmetry and multiple eigenvalues of graphs, Glas. Mat. Ser. III, 12(1977), No. 1, 3–8.

[STA1] STANTON D.: A partially ordered set and q-Krawchouk polynomials, J. Combin. Theory(A), 30(1981), 276–284.

[STE1] STETSENKO V. YU.: On a spectral property of an irreducible operator (Russian), Uspekhi Mat. Nauk, 22(135)(1967), No. 3, 242–244.

[STI1] STIEBITZ M.: Automorphismengruppe und Spektrum eines Graphen, 24. Internat. Wiss. Kolloq., TH Ilmenau, Ilmenau, 1979, Heft 5, p. 13.

[STI2] STIEBITZ M.: Eihfache Eigenwerte und Automorphismengruppe von Graphen, Dissertation, TH Ilmenau, Ilmenau, 1980.

[STO1] STOCKMEYER P. K.: Which reconstruction results are significant. In: **The Theory and Application of Graphs**, Proc. Fourth Internat. Conf. on Theory and Appl. of Graphs, (Ed. G. Chartrand et al.), John Wiley, New York 1981, pp. 543–555.

[STON1] STONE M. H.: **Linear Transformations in Hilbert spaces**, AMS Coll. Publ., vol. XV, New York 1932.

[SUST1] SUBOTIN V. F., STEKOL'SCIK R. B.: Spectrum of the Coxeter transformation and regularity of graph representations (Russian), Trudy Mat. fak. Voronezh. Univ., 1975, vol. 16, 62–65.

[SUST2] SUBOTIN V. F., STEKOL'SCIK R. B.: The Jordan form, Coxeter transformation and applications to the representations of finite graphs (Russian), Funkcional. Analiz Prilozheniya, 12(1978), 84–85.

[SUSU1] SUBOTIN V. F., SUMIN M. V.: Investigation of polynomials connected with representations of graphs (Russian), Voronezh Univ., Voronezh, 1982, 37pp.

[SYS1] SYSLO M. M.: Adjacency matrix equations and related problems: research notes, Comment. Math. Univ. Carolin., 24(1983), No. 2, 211–222.

[TAN1] TANNER R. M.: Explicit concentrators from generalized N-gons, SIAM J. Discrete Methods, 5(1984), 287–293.

[TAY1] TAYLOR A. E.: **Introduction to functional analysis**, John Wiley, New York 1958.

[TaLe1] TAYLOR D. E., LEVINSTON R.: Distance regular graphs, In: Proc. Internat. Conf. Combinat. Theory, Austral. Nat. Univ., Lecture Notes Math. 686, Springer, Berlin 1978, pp. 313–323.

[TaHP1] TAUSCH M. W., HASS E. C., PLATH P. J.: Beschreibung von Valenzisomerisierung durch Strukturvariation auf vollständigen Graphen, Match, 7(1979), 289–298.

[TeCh1] TENG TSUNG-HAO, CHU KUAN-CHICH: The energy levels and molecular orbitals of conjugated molecules with periodical structures (Chinese), Acta Chimica Sinica, 36(1978), No. 3, 159–169.

[TER1] TERWILLIGER P.: Eigenvalue multiplicities of highly symmetric graphs, Discrete Math., 41(1982), No. 3, 295–302.

[TER2] TERWILLIGER P.: The Johnson graph $J(d,r)$ is unique if $(d,r) \neq (2,8)$, Discrete Math. 58(1986), 175–189.

[THO1] THOMPSON D. M.: Eigengraphs; constructing strongly regular graphs with block designs, Utilitas Math., 20(1981), 83–115.

[TIN1] TINKLER K. J.: The physical interpretation of eigenvalues of dichotomous matrices, Inst. Brit. Geogr. Publ., 55(1972), 17–46.

[TIN2] TINKLER K. J.: On the choice of methods in the factor analysis of connectivity matrices: a reply, Inst. Brit. Geogr. Publ., 66(1975), 168–171.

[TOI1] TOIDA S.: A note on Ádám's conjecture, J. Combin. Theory(B), 23 (1977), 239–246.

[TOR1] TORGAŠEV A.: On spectra of infinite graphs, Publ. Inst. Math. (Beograd), 29(43)(1981), 269–282.

[TOR2] TORGAŠEV A.: On infinite graphs with three and four non-zero eigenvalues, Bull. Acad. Serbe Sci. et Arts (Sci. Math.) (76), 11 (1981), 39–48.

[TOR3] TORGAŠEV A.: On infinite graphs with five non–zero eigenvalues, Bull. Acad. Serbe Sci. et Arts(Sci. Math.)(79), 12(1982),
31–38.

[TOR4] TORGAŠEV A.: The spectrum of line graphs of some infinite
graphs, Publ. Inst. Math. (Beograd), 31(45)(1982), 209–222.

[TOR5] TORGAŠEV A.: Finiteness of spectra of graphs obtained by
some operations on infinite graphs, Publ. Inst. Math. (Beograd),
33(47) (1983), 227–234.

[TOR6] TORGAŠEV A.: On the automorphism group of an infinite
graph, Publ. Inst. Math. (Beograd), 34(48)(1983), 233–237.

[TOR7] TORGAŠEV A.: On infinite graphs whose spectrum is greater
than −2, Bull. Acad. Serbe Sci. et Arts (Sci. Math.), 13(1984),
21–35.

[TOR8] TORGAŠEV A.: On infinite graphs whose spectrum is uniformly
bounded by $\sqrt{2 + \sqrt{5}}$, In: **Graph Theory**, Proc. Fourth Yugoslav Sem. Graph Theory, Novi Sad, 15–16. April 1983, (Ed.
D. Cvetković, I. Gutman, T. Pisanski, R. Tošić, Inst. Math.,
Novi Sad 1984), pp. 299–309.

[TOR9] TORGAŠEV A.: Graphs whose second least negative eigenvalue
is greater than −1, Univ. Beograd Publ. Elektrotehn. Fak., Ser.
Mat. Fiz., No. 735–No. 762(1982), No. 759, 148–154.

[TOR10] TORGAŠEV A.: A note on infinite generalized line graphs, In:
Graph Theory, Proc. Fourth Yugoslav Sem. Graph Theory,
Novi Sad, 15–16. April 1983, (Ed. D. Cvetković, I. Gutman, T.
Pisanski, R. Tošić, Inst. Math., Novi Sad 1984), pp. 291–297.

[TOR11] TORGAŠEV A.: Graphs whose energy does not exceed 3,
Univ. Novi Sad, Zbornik radova Prir. Mat. Fak. (Ser. Math.),
13(1983), 353–360.

[TOR12] TORGAŠEV A.: Graphs with the reduced spectrum in the unit
interval, Publ. Inst. Math. (Beograd), 36(50), (1984), 15–26.

[TOR13] TORGAŠEV A.: Infinite graphs with the least limiting eigenvalue greater than −2, Linear Algebra and its Appl., 82(1986),
133–141.

[Tri1] TRINAJSTIĆ N.: New developments in Hückel theory, Internat. J. Quantum Chem. Symposium, 11(1977), 469–477.

[Tri2] TRINAJSTIĆ N.: Computing the characteristic polynomial of a conjugated system using the Sachs theorem, Croat. Chem. Acta, 49(1977), No. 4, 593–633.

[Tri3] TRINAJSTIĆ N.: Hückel theory and topology. In: **Semiempirical methods of electronic structure calculations**, Part A: Techniques, Modern Theoretical Chemistry, Vol. 7, Plenum Press, New York, 1977 (Ed. G. J. Segal), 1–27.

[Tri4] TRINAJSTIĆ N.: **Chemical Graph Theory**, Vol. I, II. CRC Press, Boca Raton, 1983.

[Tür1] TÜRKER, L.: Graph theoretical approach to HMO coefficient polynomials of certain molecules, METU J. Pure Appl. Sci., 13(1980) 209–224.

[Tür2] TÜRKER, L.: The topological estimation of p-bands of certain cata-condensed aromatic hydrocarbons, METU J. Pure Appl. Sci., 13(1980) 249–264.

[Tür3] TÜRKER L.: An upper bound for total π-electron energy of alternant hydrocarbons, Match, 16(1984), 83–94.

[Vij1] VIJAYAN K. S.: On a class of distance transitive graphs, J. Combin. Theory(B), 25(1978), 125–129.

[Wah1] WAHID S. A.: On the matching polynomials of graphs, M. Sc. thesis, Univ. West Indies, St. Augustine (Trinidad), 1983.

[Wal1] WALTER C. D.: Intersection numbers for coherent configurations and the spectrum of a graph, J. Combin. Theory (B), 35(1983), 201–204.

[WaYa1] WANG J. F., YANG D. S.: On the number of spanning trees of circulant graphs, International J. Comput. Math. 16(1984), 229–241.

[Wan1] WANG K.: On the matrix equation $A^m = \lambda J$, J. Combin. Theory (A), 29(1980), 134–141.

[WAN2] WANG K.: On the polynomials of graphs, SIAM J. Algebraic Discrete Methods, 4(1983), 522–528.

[WAN3] WANG K.: Characteristic polynomials of symmetric graphs, Linear Algebra Appl., 51(1983), 121–125.

[WAN4] WANG K.: Spectra of some graphs, SIAM J. Alg. Discrete Meth. 5(1984), No. 1, 57–60.

[WAN5] WANG K.: On the g-circulant solutions to the matrix equation $A^n = \lambda J$, J. Combinatorial Th. B 33(1982), 187–296.

[WaLZ1] WANG NAN-QIN, LIN LIAN–TANG, ZHANG QIAN-ER: Graphical method of π-electron energy spectrum with conjugation in straight chain molecule of polymer (Chinese), Acta Univ. Amoiensis Scientiarium Naturalium, 2(1979), 76–85.

[WAT1] WATANABE M.: On the characteristic polynomial of a multigraph, Math. Rep. Toyama Univ., 2(1979), 87–94.

[WAT2] WATANABE M.: Note on integral trees, Math. Rep. Toyama Univ., 2(1979), 95–100.

[WaSc1] WATANABE M., SCHWENK A. J.: Integral starlike trees, J. Austr. Math. Soc. Ser. A, 28(1979), 120–128.

[WeHP1] WEIGEL H. W., HASS E. C., PLATH P. J.: Zur Diagonalisierung von Adjazenzmatrizen zur Gewinnung von Strukturinformation, Match, 6(1979), 137–155.

[WIL1] WILD U. P.: Degenerate eigenvalues of the Hückel matrix, Theoret. Chim. Acta (Berl.), 54(1980), 245–249.

[WILS1] WILSON R. J.: Graph theory and chemistry, **Combinatorics**Colloquia Mathematica Soc. Janos Bolyai, 18; (Ed. by A. Hajnal, V. Sós), North Holland, Amsterdam–Oxford–New York, 1978, Vol. II, 1147–1164.

[WIN1] WINKLER P. M.: Proof of the squashed cube conjecture, Combinatorica 3(1983) 135–139.

[YaSH1] YAMAGUCHI T., SUZUKI M., HOSOYA H.: Tables of the topological characteristics of polyhexes (condensed aromatic hydrocarbons), Natural Sci. Rept. Ochanomizu Univ., 26(1975), No.
1, 39–60.

[YaLY1] YAN DEYUE, LI GOUYING, YIANG YUANSHENG: Graphical
method in kinetics of polymerization, Sci Sinica, 24(1981), No.
1, 46–60.

[YAN1] YAN GUO-SEN: The graph theoretical formulas for determinant
expansions, Internat. J. Quantum Chem. Symposium, 14(1980),
549–555.

[YaWa1] YAN JI-MIN, WANG ZOU-XIN: The characteristics of LCAO-
MO's secular matrices for symmetric molecules and their applications (Chinese), Acta Chimica Sinica, 39(1981), No. 1, 17–22.

[YAP1] YAP H. P.: The characteristic polynomial of the adjacency
matrix of a multi–digraph, Nanta Math., 8(1975), No. 1, 41–46.

[YUAN1] YUAN H.: The Kth largest eigenvalue of a tree, Linear Algebra and its app., 73(1986) 151–155.

[ZAG1] ZAGAGLIA-SALVI N.: Sulle matrici torneo associate a matrici di
permutazione, (Italian), Note di Matematica, 2(1982), 177–188.

[ZAG2] ZAGAGLIA-SALVI N.: Alcune propreità delle matrici torneo re-
golari (Italian), Atti del Convegno "Geometria combinatoria e di
incidenza: fondamenti e applicazioni," La Mendola, 4–11 Juglio
1982, Vita e pensiero, 1984, 635–643.

[ZAS1] ZASLAVSKY T.: The geometry of root systems and signed
graphs, Amer. Math. Monthly, 88(1981), No. 1, 88–105.

[ZAS2] ZASLAVSKY T.: Complementary matching vectors and the
uniform matching extension property, European J. Combin.,
2(1981), 91–103, 305.

[ZAS3] ZASLAVSKY T.: Matchings and Chebyshev polynomials,
preprint (1981), pp. 1–27.

[ZAS4] ZASLAVSKY T.: Generalized matchings and generalized Hermite
polynomials, preprint (1981), pp. 1–21.

[ZAS5] ZASLAVSKY T.: Signed graphs, Discrete Appl. Math., 4(1982), 47–74, 248.

[ZAS6] ZASLAVSKY T.: Uniform distribution of a subgraph in a graph, Ann. Discrete Math., 17(1983), 657–664.

[ZEI1] ZEID A.: Bounds on the eigenvalues for certain classes of dynamic systems, Doct. Dissertation, Mich. State Univ., Michigan, 1981.

[ZEI2] ZEID M. A.: The spectrum of a skew symmetric adjacency matrix of an oriented graph, Congressus Numerantium, 35(1982), 415–425.

[ZHA1] ZHANG FUJI: Eigenvalue network and molecular orbitals, Sci. Sinica, 22(1979), No. 10, 1160–1168.

[ZHA2] ZHANG FUJI: Two theorems of elimination of nodes and edges for a Coates graph (Chinese), Sci. Sinica, 21(1979), 966–968.

[ZHA3] ZHANG FUJI: Two theorems of comparision of bipartite graphs by their energy, Kexue Tongbao, 28(1983), 726–730.

[ZHLA1] ZHANG FUJI, LAI ZAIKANG: The totally ordered subset in a sort of Hasse graphs of trees (Chinese), J. Xinjiang Univ., (1983), 12–20.

[ZHLA2] ZHANG FUJI, LAI ZAIKANG: Three theorems of comparisions of trees by their energy, Science Exploration, 3(1983), 35–36.

[ZHLW1] ZHANG QIAN-ER, LIN LIAN-TANG, WANG NAN-QIN: Graphical method of Hückel matrix, Sci. Sinica, 22(1979), No. 10, 1169–1184.

[ZHLW2] ZHANG QIAN-ER, LIN LIAN-TANG, WANG NAN-QIN: Graphical methods of the MO theory of AH (Chinese), Acta Univ. Amoiensis Scientiarium Naturalium, 2(1979), 55–64.

Bibliographic Index

Index